P9-DUN-496

WITHDRAWN
UTSA Libraries

SYNTHETIC PRODUCTION AND UTILIZATION OF AMINO ACIDS

SYNTHETIC PRODUCTION AND UTILIZATION OF AMINO ACIDS

Edited by

Takeo Kaneko
Shiseido Co., Ltd.

Yoshiharu Izumi
Osaka University

Ichiro Chibata
Tanabe Seiyaku Co., Ltd.

Tamio Itoh
Ajinomoto Co., Inc.

A HALSTED PRESS BOOK

KODANSHA LTD.
Tokyo

JOHN WILEY & SONS
New York-London-Sydney-Toronto

Copyright © 1974 by Kodansha Ltd.

All rights reserved
No part of this book may be reproduced in any form, by photostat, microfilm, or any other means, without the written permission of Kodansha Ltd. (except in the case of brief quotation for criticism or review)

This book is a translated version of the Japanese book, *Amino-san Kogyo* (ed. T. Kaneko *et al.*), Kodansha, 1973, revised and modified in part in conjunction with the authors.

Library of Congress Cataloging in Public Data
Main entry under title:
Synthetic production and utilization of amino acids.
Translation of Aminosan kogyo.
"A Halsted Press book."
1. Amino acids. 2. Chemistry, Organic—Synthesis.
I. Kaneko, Takeo, 1907- ed.
TP248.A6A4613 1974 661'.894 74-9924
ISBN 0-470-45590-X
KODANSHA EDP NO. 3043-298289-2253 (0)

Published in Japan by
KODANSHA LTD.
12-21 Otowa 2-chome, Bunkyo-ku, Tokyo 112, Japan
Published by
HALSTED PRESS
a Division of John Wiley & Sons, Inc.
605 Third Avenue, New York, N.Y. 10016, U.S.A.

PRINTED IN JAPAN

Contributors

LIBRARY
University of Texas
At San Antonio

Ichiro CHIBATA, Research Laboratory of Applied Biochemistry, Tanabe Seiyaku Co., Ltd., *Higashi-Yodogawa-ku, Osaka 532, Japan*

Tamio ITOH, Euro-Americas Department, Ajinomoto Co., Inc., *Chuo-ku, Tokyo 104, Japan*

Yoshiharu IZUMI, Institute for Protein Research, Osaka University, *Suita-shi, Osaka 565, Japan*

Takeo KANEKO, Shiseido Laboratories, Shiseido Co., Ltd., *Kohoku-ku, Yokohama-shi, Kanagawa-ken 222, Japan*

Tadashi SHIRAI, Pharmaceutical Division, Research Laboratories, Nippon Kayaku Co., Ltd., *Kita-ku, Tokyo 115, Japan*

Koji TOI, Central Research Laboratories, Ajinomoto Co., Inc., *Kawasaki-shi, Kanagawa-ken 210, Japan*

Katsuyuki TOKI, Sumitomo Chemical Co., Ltd., *Takarazuka-shi, Hyogo-ken 665, Japan*

Ryonosuke YOSHIDA, Product Development Laboratories, Ajinomoto Co., Inc., *Kawasaki-shi, Kanagawa-ken 210, Japan*

Preface

Proteins are universally present where life exists, and can be correlated directly with the vital processes of living organisms. They are indispensable components of the human body, being both synthesized internally from their component amino acids and ingested whole as food.

When considered as one of the three important types of nutrient materials, proteins may be assessed for their quality or nutrient value based on the kinds and amounts of their indispensable amino acids, as demonstrated by the nutritional studies of Rose. Chicken egg protein is thought to approximate closely to an "ideal" protein, and its ratio of indispensable amino acids is thus commonly regarded as the one best balanced for the human diet. Indispensable amino acids, which are more or less lacking in proteins of low nutrient value, may also be supplemented to such proteins to attain a proper amino acid balance and to enhance the nutrient value of the proteins.

Amino acids are thus essential substances for all vital phenomena and cannot be replaced nutritionally by alternative food or feed materials. Accordingly, it is natural that extensive studies have been carried out in the fields of human and animal nutrition and medicine in order to determine the precise role of amino acids, and to devise means for their artificial production.

The amino acid industry of Japan has developed in a unique way, quite different from that of the chemical industry in general. This fact derives largely from the special food customs of the Japanese people, which have arisen under the influence of Buddhism and have led to the development of a refined taste sense.

Japan's three most important traditional seasoning sources are processed dried fish, processed soybean products, and *konbu* (sea tangle). Components responsible for the taste of processed dried fish have been identified, and development of means for their production has led to the establishment in the past 10–15 years of several new industries, such as the inosinic acid industry. The relevant developments involved both chemical synthesis and fermentation techniques. In the case of seasonings and foods based on soybeans, both soy sauce (which affords many Japanese dishes with their characteristic taste) and *miso* (soybean paste) are the products of fermentative processes which have been perfected over the long period since their first introduction from China. The flavor-giving components of soy sauce and *miso* are largely, but not entirely, amino acids which form as a result of protein hydrolysis. It also appears that the hydration and buffering action of such amino acids assists in the moderation of salty tastes. *Konbu* is widely used in the preparation of Japanese foods, and its characteristic flavor derives almost entirely from L-glutamic acid, as discovered by K. Ikeda in 1908.

The first real step towards the development of an amino acid industry in Japan may be attributed to S. Suzuki, who undertook the production of L-glutamic acid. At that time, however, L-glutamic acid was generally obtained by the hydrolysis of gluten or soybean protein. More recently, new methods have been established, such as the fermentation techniques developed by S. Kinoshita *et al.* and the large-scale production of L-glutamic acid from acrylonitrile developed by Ajinomoto Co., Inc., following the establishment of an industrial method for optical resolution of the racemic amino acid. Further expansion and modernization of the industry are also expected as a result of new advances in petroleum chemistry and fermentation technology.

During World War II there was a food crisis in Japan, and protein deficiency diseases were especially widespread in the early post-war period. Based on this experience, and due to the other factors mentioned above, Japanese scientists became extremely interested in protein nutrition and, in turn, in methods for the artificial production of indispensable amino acids. The successful results have already led to considerable expansion and diversification of the amino acid industry. The products obtained have been developed as nutrients (including food and feed additives, transfusions, etc.), as physiological and therapeutic agents (including medicines, cosmetics, etc.), and as raw materials for the chemical industry (used in the production of surface-active agents, chelating agents, polyamino acids, peptides, etc.). Further technological improvements are expected to contribute in reducing the production cost of the amino acids and in facilitating large-scale operation, and new uses and techniques will thus no doubt be found.

The manufacturing methods of amino acids at present may be classified into three broad types: (1) extraction of natural L-amino acids from protein hydrolyzates, (2) L-amino acid production by fermentation, and (3) chemical synthesis. The present book has been designed specifically to deal with the third category. Part 1 outlines known procedures for the chemical synthesis of amino acids, while part 2 covers the broad field of amino acid utilization. The extraction method (1), although important in the past for the production of optically active amino acids (and still so in the case of cystine, histidine and leucine), is not dealt with here. Also, the fermentation method (2), which belongs strictly to the field of microbiology and is so essentially different from general chemical procedures, is omitted. For a full treatment of this aspect of amino acid production, the reader is referred to the recent book, *The Microbial Production of Amino Acids* (ed. K. Yamada *et al.*), pp. 568, Kodansha and Wiley, 1972.

As supervising editor for the present project, I wish to express my appreciation to the authors for their contributions, and would also like to thank the translator, Dr. F. Nagayama, and foreign staff of Kodansha for their invaluable assistance in the preparation of the English manuscripts comprising this book.

May, 1974 T. KANEKO

Contents

PART 1

Synthetic Production of Amino Acids

CHAPTER 1

Introduction

Y. Izumi*

There are three general methods available for the preparation of amino acids: these are (1) separation from hydrolyzed protein, (2) fermentation and (3) chemical synthesis. The most economical approach for the industrial production of a particular amino acid depends on both the volume of demand and the physical and chemical properties of the amino acid concerned. At present almost all commercially available amino acids are produced by fermentation or chemical synthesis, with only a few exceptions.

The chemical synthesis of amino acids (R–CH(NH_2)COOH) can be divided into the synthesis of the R-group, which is of course different for each amino acid, and that of the –CH(NH_2)COOH group, which is required for all amino acids. A brief survey of synthetic methods for the latter group is presented in this chapter, including historical aspects where these are of interest. The synthesis of particular R-groups and the choice of the appropriate synthetic method for the –CH(NH_2)COOH group in each case are dealt with in Chapter 4 in relation to industrial production.

Since almost all industrially produced amino acids are intended ultimately for human or animal consumption, the economic value of methods of chemical synthesis will be limited unless efficient methods of optical resolution are available. This remained a problem for many years, and it was not until such methods were developed that chemical synthesis of amino acids superseded the hydrolysis of protein. However, this problem can be

* Yoshiharu Izumi, Institute for Protein Research, Osaka University, Suita-shi, Osaka 565, Japan

avoided by utilizing fermentation processes for the production of amino acids, since the bacteria themselves produce only the desired L-form (except in the case of alanine, where the DL-form is produced in most cases). However, although the raw materials for fermentation processes are cheap, problems of scaling-up such processes mean that chemical synthesis is still competitive for medium- or large-scale plants in the production of L-amino acids. For instance, the synthetic method for L-glutamic acid, involving preferential crystallization as a procedure for optical resolution, seems preferable to fermentation methods for large-scale production since economies of scale are more easily realized with this process.

Some amino acids, such as L-histidine and L-cystine, are still obtained by the hydrolysis of protein, due to the low level of demand or to the difficulty of their synthesis. The DL- or D-forms of these amino acids are usually prepared from the L-form obtained in this way.

For the production of DL-forms or of amino acids which do not occur naturally, the synthetic method is obviously superior. Since D-amino acids are derived from the DL-forms by optical resolution, these can also be best obtained synthetically. Thus, the methods of chemical synthesis and fermentation are essentially complementary for industrial purposes, the choice depending on factors such as scale of operations.

There are two important factors to be considered in developing a synthetic method for use on an industrial scale; first, the degree of change involved in the synthetic process should be kept to a minimum by suitable selection of the starting material, and second, the starting material should not suffer the loss of too many functional groups. A method which is convenient on a laboratory scale may not always be suitable industrially. Synthetic methods involving acetamidomalonate and related compounds provide a good illustration of the above problem, since they provide convenient routes to many amino acids in good yield. However, they are not economically practicable for industrial use since too many functional groups are lost during the synthetic process, i.e. one acyl group, two alkoxy groups and one carboxyl group.

However, methods which are not economically viable at present due to poor availability or high price of the starting materials, or low process yield, may subsequently be reappraised as a result of technical developments. The re-evaluation of the Strecker and Bucherer methods provides an example. Since the development of the oxo reactions and the Reppe reaction greatly improved the industrial synthesis of ketones and aldehydes, the Strecker and Bucherer reactions have become more important in industrial amino acid synthesis.

In order to avoid optical resolution steps, some L-amino acids are prepared directly from other L-amino acids. The syntheses of L-arginine from L-ornithine, and L-cysteine from L-cystine are typical examples. Other methods involve the synthesis of an L-amino acid from another amino acid and the corresponding keto acid in the presence of a transaminase. These may be classified as chemical methods. The asymmetric synthesis of L-amino acids on an industrial scale has not yet been achieved, though there have been intensive research efforts on a laboratory scale. Only catalytic asymmetric synthesis is likely to represent an industrially important development.

Knowles *et al.* recently succeeded in the synthesis of L-phenylalanine and its derivatives from α-acetamidocinnamate by asymmetric catalytic hydrogenation using a chiral Rh-phosphine complex as the catalyst. This method has already been applied by Monsanto Co. to the synthesis of L-Dopa, but the utility of the method is limited to aromatic amino acids. A combination of racemization of the unwanted enantiomer and optical resolution is still more effective for the production of optically active amino acids.

(1) The Strecker method

The method bearing his name was first used by Strecker in 1850 to prepare alanine.[1)] The first step in the reaction involves the formation of an α-amino nitrile by the addition of hydrogen cyanide and ammonia to an aldehyde or ketone, and the second is the hydrolysis of the amino nitrile to yield the amino acid. Since no major fragments are lost during the reaction, and the molecular weight of the product is higher than that of the starting material, the Strecker method is extremely useful for industrial purposes. It produces an extremely good yield of amino acid under mild reaction conditions when compared with the Bucherer method (see below), and is used widely for the industrial synthesis of glutamic acid among others.

$$\mathrm{RCHO} + \mathrm{NH_3} + \mathrm{HCN} \longrightarrow \underset{\displaystyle \mathrm{NH_2}}{\mathrm{RCHCN}} \longrightarrow \underset{\displaystyle \mathrm{NH_2}}{\mathrm{RCHCOOH}}$$

The Strecker reaction can be conveniently carried out on a laboratory scale using a mixture of ammonium chloride and sodium or potassium cyanide.

(2) The Bucherer method

This was developed by Bucherer[2)] as an improvement of the Strecker method, in 1934. In this process, stable hydantoin compounds are used as intermediates in places of the unstable amino nitriles. Hydantoin derivatives are obtained by the reaction of aldehydes with hydrogen cyanide and ammonium carbonate. Since they are easily crystallized, they are often separated from the reaction mixture by crystallization and subsequently hydrolyzed with alkali under high pressure and temperature to yield the amino acid.

$$\mathrm{RCHO} + \mathrm{HCN} + (\mathrm{NH_4})_2\mathrm{CO_3} \longrightarrow \underset{\substack{|\quad\quad |\\ \mathrm{NH}\quad \mathrm{NH}\\ \diagdown\ \diagup \\ \mathrm{CO}}}{\mathrm{R\text{-}CH{-}CO}} \xrightarrow{\mathrm{H_2O}} \underset{\substack{|\\ \mathrm{NH_2}}}{\mathrm{RCHCOOH}}$$

Since higher yields can generally be obtained than with the Strecker method, the Bucherer method is of considerable industrial importance, for instance in the production of methionine.

(3) Amination of α-halogeno acids

This approach was first used in 1858 for the synthesis of glycine from monobromo- and monochloroacetic acids by Perkin and Duppa[3] and Cahours,[4] respectively, and is still used today for the industrial production of glycine. Many variations, such as the use of a large excess of ammonia, or the addition of excess ammonium carbonate, have been adopted in attempts to avoid or reduce the formation of secondary and tertiary amines as by-products.

$$\underset{\substack{|\\ \mathrm{X}}}{\mathrm{RCHCOOH}} \xrightarrow{\mathrm{NH_3}} \underset{\substack{|\\ \mathrm{NH_2}}}{\mathrm{RCHCOOH}}$$

The amination of α-halo-β-hydroxy acids is an important synthetic route to β-hydroxyamino acids on a laboratory scale. Thus, α-bromo-β-hydroxy acids can be obtained from α,β-unsaturated acids by reaction with hydrobromide, and α-bromo-β-methoxy acids from α,β-dibromo acids by reaction with sodium alcoholate or from the mercury acetate addition product by bromination with potassium bromide in methanol.

$$\mathrm{RCH{=}CHCOOH} \xrightarrow{\mathrm{R'OX}} \underset{\substack{|\quad |\\ \mathrm{R'O}\ \ \mathrm{X}}}{\mathrm{RCHCHCOOH}}$$

The development of ion-exchange resins for desalting solutions of readily soluble amino acids has made this process more convenient, but it has become less important since the industrial development of the Reppe and oxo reactions for aldehyde synthesis permitted more extensive use of the Strecker and Bucherer reactions for amino acid synthesis.

The amination of halogeno acids can be achieved with hexamethylenetetramine in dioxane,[5] or with succinimide[6] or phthalimide[7] in the presence of sodium alcoholate, followed by acid hydrolysis. However, these methods are not industrially useful.

$$\underset{X}{\mathrm{RCHCOOR''}} \xrightarrow{\mathrm{R'(CO)_2NH}} \underset{\mathrm{N(CO)_2R'}}{\mathrm{RCHCOOR''}} \xrightarrow[\mathrm{H^+}]{\mathrm{H_2O}} \underset{\mathrm{NH_2}}{\mathrm{RCHCOOH}}$$

R′ = o-phenylene (benzene ring) or CH_2–CH_2

(4) Erlenmeyer,[8] Wheeler,[9] Sasaki[10] and thiohydantoin methods

These methods are based on the reactions of aldehydes with activated methylene groups in azlactones (I), hydantoins (II), diketopiperazines (III) and thiohydantoins (IV), respectively, in acetic anhydride and in the presence of a basic catalyst such as sodium acetate. The condensation product (V) is catalytically hydrogenated, then hydrolyzed with acid to yield the amino acid. These methods, especially that of Erlenmeyer, are convenient for the synthesis of aromatic amino acids.

I: CH_2–CO, N, O, C(R) (azlactone ring)

II: CH_2–CO, NH, NH, CO (hydantoin ring)

III: CO-NH, CH_2, CH_2, NH-CO (diketopiperazine ring)

IV: CH_2–CO, NH, NH, CS (thiohydantoin ring)

$$\mathrm{R'CHO} + \mathrm{CH_2{-}CO(N{-}R'')} \xrightarrow[\mathrm{(CH_3CO)_2O}]{\mathrm{CH_3COONa}} \underset{\mathrm{V}}{\mathrm{R'CH{=}C{-}CO(N{-}R'')}} \xrightarrow{\mathrm{H_2}} \mathrm{R'CH_2\text{-}CH\text{-}CO(N{-}R'')}$$

$$\xrightarrow[\mathrm{H^+\ or\ OH^-}]{\mathrm{H_2O}} \underset{\mathrm{NH_2}}{\mathrm{R'CH_2CHCOOH}} \qquad \mathrm{R''} = \mathrm{I,\ II,\ III,\ IV}$$

Acetyl glycine is commonly used as a starting material in the Erlenmeyer reaction. Aldehydes can usually be reacted with acetyl glycine in acetic anhydride in the presence of sodium acetate, in place of the azlactone reaction.[11] Hydrogenation and hydrolysis with hydriodic acid are often carried out as a one-step process on a small scale, but are generally separate processes. Acetyl phenylalanine, however, can be obtained in quantitative yield by hydrogenation of benzilidene azlactone with a Raney nickel catalyst in aqueous sodium hydroxide.[12]

$$\text{RCH=C(N=C(CH}_3\text{)O)CO (azlactone)} \xrightarrow{H_2} \text{RCH}_2\text{-CH(N=C(CH}_3\text{)O)CO} \xrightarrow{H_2O,\ H^+} \text{RCH}_2\text{CH(NH}_2\text{)COOH}$$

$$\text{azlactone} \xrightarrow{NaOH} \text{RCH=C(NHCOCH}_3\text{)COONa} \xrightarrow[\text{Raney nickel}]{H_2} \text{RCH}_2\text{CH(NHCOCH}_3\text{)COONa} \xrightarrow{H_2O,\ H^+} \text{RCH}_2\text{CH(NH}_2\text{)COOH}$$

(5) Use of α-keto acid derivatives

Reductive amination of α-keto acids in aqueous ammonia with a hydrogenation catalyst such as Raney nickel, is one of the most convenient and widely applicable methods of amino acid synthesis.[13] With the development of economic methods for the synthesis of α-keto acids, this method will be of considerable importance. The α-imino acid (VI) initially produced from the ketone and ammonia is usually immediately hydrogenated *in situ*. The reaction conditions and economic factors for the synthesis of glutamic acid by this process have been widely investigated, and the highest yield was obtained by reductive amination of sodium α-ketoglutarate.

$$\text{RCOCOOH} + \text{NH}_3 \longrightarrow \underset{\text{VI}}{\text{RC(=NH)COOH}} \xrightarrow[\text{catalyst}]{H_2} \text{RCH(NH}_2\text{)COOH}$$

Amino acids can also be produced by the reduction of α-oximino acids (VII)[14] and α-keto acid hydrazones (VIII),[15] but these methods are less economically attractive.

$$\underset{\text{VII}}{\begin{array}{l}\text{RCCOOH}\\ \;\;\| \\ \text{NOH}\end{array}} \xrightarrow{H_2} \begin{array}{l}\text{RCHCOOH}\\ \;\;| \\ \text{NH}_2\end{array}$$

$$\underset{\text{VIII}}{\begin{array}{l}\text{RCCOOH}\\ \;\;\| \\ \text{NNHC}_6\text{H}_5\end{array}} \xrightarrow{H_2} \begin{array}{l}\text{RCHCOOH}\\ \;\;| \\ \text{NH}_2\end{array}$$

α-Nitroso (IX)[16] and α-phenylazo-β-keto ester (X)[17] compounds, obtained by the reaction of nitric acid or alkyl nitrates and diazonium compounds, respectively, with β-keto esters or malonate, are tautomers of α-oximino esters and phenylhydrazone derivatives, respectively, and so may be included in this category.

$$\begin{array}{l}\;\;\text{R}'\\ \;\;| \\ \text{RCOCHCOOC}_2\text{H}_5\end{array} \xrightarrow{\text{Bu-O-NO}_2} \underset{\text{IX}}{\begin{array}{l}\;\;\;\;\;\text{R}'\\ \;\;\;\;\;| \\ \text{RCOCCOOC}_2\text{H}_5\\ \;\;\;\;\;| \\ \;\;\;\;\;\text{NO}\end{array}}$$

$$\longrightarrow \begin{array}{l}\text{R}'\\ | \\ \text{CCOOC}_2\text{H}_5\\ \| \\ \text{NOH}\end{array} \xrightarrow{H_2} \begin{array}{l}\text{R}'\text{CHCOOH}\\ \;\;\;| \\ \;\;\;\text{NH}_2\end{array}$$

$$\text{RCOCH}_2\text{COOC}_2\text{H}_5 \xrightarrow{\text{C}_6\text{H}_5\text{N=NCl}} \underset{\text{X}}{\begin{array}{l}\text{RCOCHCOOC}_2\text{H}_5\\ \;\;\;\;\;| \\ \;\;\;\;\;\text{N=NC}_6\text{H}_5\end{array}}$$

$$\longrightarrow \begin{array}{l}\text{RCOCCOOC}_2\text{H}_5\\ \;\;\;\;\;\| \\ \;\;\;\;\;\text{NNHC}_6\text{H}_5\end{array} \xrightarrow{H_2} \begin{array}{l}\text{RCHCHCOOH}\\ \;\;\;\;| \;\;\; | \\ \;\;\text{HO}\;\;\text{NH}_2\end{array}$$

Amino acids can also be obtained from α-keto acids by reaction with urea to yield the hydantoins (XI) in special cases.[18]

L-Amino acids can be obtained enzymatically from α-keto acids by reaction with transaminase,[19] e.g. α-ketoglutarate is converted to L-glutamic acid by transaminase in the presence of L-aspartic acid.

$$C_2H_5OOCCH_2CH_2COCOOC_2H_5 + NH_2CONH_2 \longrightarrow C_2H_5OOCCH_2CH{=}\underset{\substack{|\\NH}}{C}-\underset{\substack{|\\NH}}{CO} \quad (\text{NH, NH bridged by CO})$$

$$\xrightarrow{H_2} \underset{\text{XI}}{\text{R-CH-CO (NH NH bridged by CO)}} \xrightarrow[OH^-]{H_2O} \underset{\substack{|\\NH_2}}{RCHCOOH}$$

$$HOOCCH_2CH_2COCOOH + HOOCCH_2\underset{\substack{|\\NH_2}}{C}HCOOH$$

$$\xrightarrow{\text{transaminase}} HOOCCH_2CH_2\underset{\substack{|\\NH_2}}{C}HCOOH + HOOCCH_2COCOOH$$

(6) Use of acetoamidomalonate and related compounds

The use of acylaminomalonate is based on investigations of amino acid synthesis from aminomalonate (XII) by Putochin[20] in 1923 and from acetoamidomalonate (XIII) by Snyder *et al.*[21] The use of acetamidomalonates is most favorable in this group due to ease of crystallization and synthesis. The synthesis of most types of amino acids by this method is simple, and good results are generally obtained.

$$\begin{array}{l} COOC_2H_5 \\ | \\ HCNHR \\ | \\ COOC_2H_5 \end{array} \qquad \begin{array}{ll} R = H & \text{XII} \\ R = CH_3CO & \text{XIII} \end{array}$$

The method is economically disadvantageous, however, in that a major part of the acetamidomalonate molecule is lost during the synthetic process, and the residue is not industrially useful.

Reactions involving acylamidomalonate and related compounds can be divided into two groups: one is the reaction of alkyl halides with the activated methylene group, and the second is the addition of a double bond to the active methylene group. These reactions, which are shown in the following schemes, will be discussed in more detail in later sections.

(a) Reaction with alkyl halides

$$R''X + HC(R)(COOC_2H_5)NHCOR' \xrightarrow{C_2H_5ONa} R''C(R)(COOC_2H_5)NHCOR' \xrightarrow[H^+]{H_2O} R''CH(NH_2)COOH$$

$R = COOC_2H_5$ or CN

$$\text{(indol-3-yl)}CH_2N(CH_3)_2 + HC(COOC_2H_5)_2NHCOCH_3 \xrightarrow{OH^-} \text{(indol-3-yl)}CH_2C(COOC_2H_5)_2NHCOCH_3$$

$$\longrightarrow \text{(indol-3-yl)}CH_2CH(NH_2)COOH$$

(b) Reaction with unsaturated compounds

$$R''COCH{=}CH_2 + HC(R)(COOC_2H_5)NHCOR' \longrightarrow R''COCH_2CH_2C(R)(COOC_2H_5)NHCOR'$$

$$\xrightarrow{HCl} R''COCH_2CH_2CH(NH_2)COOH$$

$R = COOC_2H_5$ or CN

$$HCHO + HC(COOC_2H_5)_2NHCOCH_3 \xrightarrow{OH^-} HOCH_2C(COOC_2H_5)_2NHCOCH_3$$

$$\longrightarrow HOCH_2CH(NH_2)COOH$$

(7) Use of ethyl acetoacetate

This method was extensively investigated by Feofilaktov.[22] The reaction of an α-alkyl acetoacetate and benzene diazonium chloride, with spontaneous loss of the acetyl group during the reaction, yields α-keto ester phenylhydrazone, which is readily hydrogenated to an amino acid ester, as mentioned in (5). Since large parts of the acetoacetate and phenylhydrazone molecules are lost during the reaction, this method has drawbacks.

$$CH_3COCH(R)COOC_2H_5 \xrightarrow{C_6H_5N=NCl} RCH(N{=}NC_6H_5)COOC_2H_5 \longrightarrow RC(=NNHC_6H_5)COOC_2H_5$$

(8) Use of nitromalonate (XIV)[23] and nitroacetate (XV)[24]

Since these compounds have active methylene groups, they can be used in the same types of reactions as acetoamidomalonate, and the alkylated derivatives hydrogenated to yield amino acids. These compounds are not convenient for use on a laboratory scale because it is difficult to crystallize the derivatives.

$$\text{(indol-3-yl)}CH_2N(CH_3)_2 + HC(R')(NO_2)COOC_2H_5 \longrightarrow \text{(indol-3-yl)}CH_2C(R')(NO_2)COOC_2H_5 \xrightarrow{H_2,\ H_2O} \text{(indol-3-yl)}CH_2CH(NH_2)COOH$$

$R' = COOC_2H_5$ XIV

$R' = H$ XV

(9) Addition of ammonia to unsaturated acids

The addition of ammonia to α,β-unsaturated acids occurs on the β-carbon, so this approach is limited to the synthesis of β-amino acids such as aspartic acid and β-alanine.

(10) Hydrogenation of α-acetamidoacrylate derivatives

The hydrogenation of α-acetamidoacrylate derivatives has become a useful method[12,25] since the development of asymmetric hydrogenation of chiral Wilkinson's complexes by Knowles *et al.*[26] For example, L-Dopa can be prepared from α-acetamido-3,4-dihydroxycinnamic acid in nearly quantitative yield.

$$\underset{\displaystyle NHCOCH_3}{RCH{=}\underset{|}{C}COOH} \xrightarrow[\text{catalyst}]{H_2} \underset{\displaystyle NHCOCH_3}{RCH_2\underset{|}{C}HCOOH}$$

(11) Use of a glycine–copper complex

The alkali-catalyzed condensation of glycine–copper complex with acetaldehyle, discovered by Akabori *et al.*,[27] is the main industrial process for the synthesis of threonine. Pure threonine is obtained, avoiding the by-production of the allo isomer.

$$RCHO + \begin{matrix} CH_2{-}CO \\ | \quad\quad \searrow O \\ NH_2 \rightarrow Cu/2 \end{matrix} \longrightarrow \begin{matrix} RCH{-}CH{-}CO \\ | \quad\quad | \quad\quad \searrow O \\ OH \quad NH_2 \rightarrow Cu/2 \end{matrix}$$

(12) Use of glycine

Glycine[28] and its esters[29] condense with benzaldehyde in the presence of strong alkali to yield phenylserine.

$$2\ C_6H_5CHO + \underset{\displaystyle NH_2}{\underset{|}{C}H_2COOH} \xrightarrow[NaOH]{} \begin{matrix} C_6H_5CHCHCOOH \\ | \quad | \\ HO \quad N{=}CHC_6H_5 \end{matrix}$$

$$\longrightarrow \begin{matrix} C_6H_5CHCHCOOH \\ | \quad | \\ HO \quad NH_2 \end{matrix}$$

(13) Use of methylenaminoacetonitrile

Methylenaminoacetonitrile can be condensed with the aldehyde which is the intermediate in glycine synthesis by the Strecker reaction. Hydrolysis of the condensation product yields β-hydroxyamino acid.[30]

$$CH_3CHO + CH_2{=}NCH_2CN \longrightarrow \begin{matrix} CH_3CHCHCN \\ | \quad | \\ HO \quad N{=}CH_2 \end{matrix} \longrightarrow \begin{matrix} CH_3CHCHCOOH \\ | \quad | \\ HO \quad NH_2 \end{matrix}$$

(14) The oxo reaction

This method is potentially of wide applicability for the synthesis of amino acids, and gives excellent yields. Wakamatsu *et al.*[31] have obtained acylamino acids from aldehyde, acylamide and carbon monoxide by means of the oxo reaction using a cobalt carbonyl catalyst. The development of an asymmetric cobalt carbonyl catalyst may well result in industrially useful processes for the asymmetric synthesis of many amino acids.

$$RCHO + R'CONH_2 + CO \xrightarrow{Co_2(CO)_8} \underset{\displaystyle |\atop \displaystyle NHCOR'}{RCHCOOH}$$

(15) Oxidation of amino alcohols

Acylamino alcohols can be oxidized to acylamino acids by oxidation with potassium permanganate.[32]

$$\underset{\displaystyle |\atop \displaystyle NHCOR'}{RCHCH_2OH} \longrightarrow \underset{\displaystyle |\atop \displaystyle NHCOR'}{RCHCOOH}$$

(16) Hofmann degradation

Alkyl cyanoacetates can be converted to alkyl malonylamidates, and the resulting amide is readily converted to the amino acid by treatment with hypobromite, i.e. the Hofmann reaction.[33]

$$\underset{\displaystyle |\atop \displaystyle CN}{RCHCOOC_2H_5} \xrightarrow[H_2SO_4]{H_2O} \underset{\displaystyle |\atop \displaystyle CONH_2}{RCHCOOC_2H_5} \xrightarrow{\text{i) KOBr; ii) KOH}}$$

$$\underset{\displaystyle |\atop \displaystyle NCO}{RCHCOOK} \xrightarrow{H_2O} \underset{\displaystyle |\atop \displaystyle NH_2}{RCHCOOH}$$

(17) The Schmidt reaction

Amino acids can be obtained from alkyl acetoacetates through the corresponding hydroazoic acid by means of the Schmidt reaction.[34]

$$\underset{\displaystyle |\atop \displaystyle COCH_3}{RCHCOOC_2H_5} + HN_3 \xrightarrow[H_2SO_4]{} \underset{\displaystyle |\atop \displaystyle NHCOCH_3}{RCHCOOC_2H_5} \xrightarrow{H_2O} \underset{\displaystyle |\atop \displaystyle NH_2}{RCHCOOH}$$

$$RCH(COOH)_2 + HN_3 \xrightarrow[H_2SO_4]{} \underset{\displaystyle |\atop \displaystyle NH_2}{RCHCOOH}$$

(18) Curtius degradation

Amino acids can be prepared by the Curtius degradation of alkyl malonazidates.[35]

$$RCH\begin{matrix}\diagup COOC_2H_5 \\ \diagdown COOC_2H_5\end{matrix} \xrightarrow{KOH} RCH\begin{matrix}\diagup COOK \\ \diagdown COOC_2H_5\end{matrix} \xrightarrow{NH_2NH_2} RCH\begin{matrix}\diagup COOK \\ \diagdown CONHNH_2\end{matrix}$$

$$\xrightarrow{HNO_2} RCH\begin{matrix}\diagup COOK \\ \diagdown CON_3\end{matrix} \xrightarrow[\text{ii) acid hydrolysis}]{\text{i) heat}} \underset{\displaystyle NH_2}{\underset{|}{RCHCOOH}}$$

Malonate or acetoacetate can be used as the starting material.[36]

$$\underset{\displaystyle COOC_2H_5}{\underset{|}{R\text{-}CH\text{-}CN}} \xrightarrow{NH_2NH_2} \underset{\displaystyle CONHNH_2}{\underset{|}{RCHCN}} \xrightarrow{HNO_2} \underset{\displaystyle CON_3}{\underset{|}{RCHCN}} \longrightarrow \underset{\displaystyle NH_2}{\underset{|}{RCHCOOH}}$$

Methods (16) through (18) are now of purely historical interest except in a few special cases.

References

1. A. Strecker, *Ann. Chem.*, **75**, 27 (1850).
2. H. T. Bucherer and W. Steiner, *J. Prakt. Chem.*, **140**, 291 (1934).
3. W. H. Perkin and B. F. Duppa, *Ann. Chem.*, **108**, 106 (1858).
4. M. A. Cahours, *Compt. Rend.*, **46**, 1044 (1858).
5. G. Hillmann and A. Hillmann, *Z. Physiol. Chem.*, **283**, 71 (1948).
6. S. Gabriel and K. Kroseberg, *Chem. Ber.*, **22**, 426 (1889).
7. S. Akabori, Y. Izumi and T. Fujiwara, *Nippon Kagaku Kaishi* (Japanese), **75**, 993 (1954).
8. E. Erlenmeyer, *Ann. Chem.*, **275**, 1 (1893).
9. H. L. Wheeler and C. Hoffmann, *Am. Chem. J.*, **45**, 368 (1911).
10. T. Sasaki, *Chem. Ber.*, **54**, 163 (1921).
11. R. M. Herbst and D. Shemin, *Org. Syn. Coll.*, vol. 2, p. 491, 1943.
12. T. Okuda and Y. Fujii, *Bull. Chem. Soc. Japan*, **30**, 698 (1957).
13. F. Knoop and H. Oesterlin, *Z. Physiol. Chem.*, **170**, 186 (1927).
14. D. Shemin and R. M. Herbst, *J. Am. Chem. Soc.*, **60**, 1951 (1938).
15. E. Fischer and R. Groh, *Ann. Chem.*, **383**, 363 (1911).
16. R. H. Barry, A. M. Mattocks and W. H. Hartung, *J. Am. Chem. Soc.*, **70**, 693 (1948).
17. Y. Izumi and S. Konishi, *Nippon Kagaku Kaishi* (Japanese), **74**, 957 (1953).

18. C. Mentzer and D. Billet, *Compt. Rend.*, **232**, 2104 (1951).
19. S. Sakurai, *J. Biochem.*, **45**, 379 (1958).
20. N. J. Putochin, *Chem. Ber.*, **56**, 2213 (1923).
21. H. R. Snyder, E. E. Howe, G. W. Cannon and M. A. Nyman, *J. Am. Chem. Soc.*, **65**, 2211 (1943).
22. V. V. Feofilaktov and F. Blanko, *J. Gen. Chem. U.S.S.R.*, **11**, 859 (1941).
23. T. Okuda, *Bull. Chem. Soc. Japan*, **32**, 1165 (1959).
24. D. A. Lyttle and D. I. Weisblat, *J. Am. Chem. Soc.*, **69**, 2118 (1947).
25. G. Nadeau and R. Gaudry, *Can. J. Res.*, **27B**, 421 (1949).
26. W. S. Knowles, M. J. Sabacky and B. D. Vincyard, *Chem. Comm.*, **1972**, 10.
27. M. Sato, K. Okawa and S. Akabori, *Bull. Chem. Soc. Japan*, **30**, 937 (1972).
28. E. Erlenmeyer and F. Früstück, *Ann. Chem.*, **284**, 86 (1895).
29. E. D. Bergmann, H. Bendas and W. Taub, *J. Chem. Soc.*, **1951**, 2673.
30. P. B. Mahajani and J. N. Ray, *J. Indian Chem. Soc.*, **33**, 455 (1956).
31. H. Wakamatsu, J. Uda and N. Yamakami, *Chem. Comm.*, **1971**, 1540.
32. J. H. Billman and E. E. Parker, *J. Am. Chem. Soc.*, **66**, 538 (1944).
33. Y. Huang, K. Lin and L. Li, *J. Chinese Chem. Soc.*, **5**, 31, 38, 46 (1947).
34. K. F. Schmidt, *Chem. Ber.*, **57**, 704 (1924).
35. T. Curtius, *J. Prakt. Chem.*, **125**, 211 (1930).
36. P. E. Gagnon, R. Gaudry and F. E. King, *J. Chem. Soc.*, **1944**, 13.

CHAPTER **2**

Optical Resolution of DL-Amino Acids

I. Chibata*

Since industrially produced amino acids are used mainly in the nutritional and pharmaceutical fields, natural form L-isomer is generally desirable. However, most available synthetic procedures produce amino acids in the racemic form. Therefore, optical resolution of such racemic amino acids is

* Ichiro CHIBATA, Research Laboratory of Applied Biochemistry, Tanabe Seiyaku Co., Ltd., Higashi-Yodogawa-ku, Osaka 532, Japan

a highly significant part of the synthetic procedure. Recently, DL-amino acids and their derivatives have been produced inexpensively on a large scale, following the development of the chemical industry and progress in synthetic organic chemistry. D-Amino acids or unnatural amino acids have also been synthesized and utilized in the pharmaceutical and other fields. In this case, optical resolution of the synthesized DL-amino acids is necessary. The synthetic method is thus advantageous for the production of optically active amino acids only when each of the three steps of chemical synthesis, optical resolution and racemization, can be carried out efficiently and in an integrated manner. For example, even a very good synthetic process may be unsatisfactory if it necessitates a complicated procedure for optical resolution of the product. Therefore, it is most desirable that an intermediate in the chemical synthetic process be easily resolved into the optical antipodes, and the undesired antipode be easily racemized to the DL-form and then reused for the resolution step.

Optical resolution methods for amino acids may be broadly classified into four types: (1) physicochemical methods, (2) chemical methods, (3) enzymatic methods, and (4) biological methods. Most of the principles underlying these methods were discovered by Louis Pasteur in the 19th century and no entirely new principle has been discovered since, although technical application of the methods has been developed remarkably in recent years. The biological method is now seldom employed for practical purposes since usually only the D-isomer, and not the L-isomer, is obtained. Therefore, it is generally expected that the physicochemical, chemical and enzymatic methods will be improved and developed further as practical resolution processes.

This chapter describes those optical resolution methods which are of interest from the standpoint of the industrial production of optically active amino acids. Comprehensive reviews of optical resolution methods have also previously been given.[1,2)]

2.1 Physicochemical methods

Although many physicochemical methods such as crystallization procedures, chromatographic procedures, etc. have been developed, the crystallization procedure is the most important for industrial use. The chromatographic procedures have been tested for their possible industrial application; however, while it is possible to employ them for laboratory purposes, none has yet been found to meet industrial demands.

2.1.1 Crystallization procedures

Among the crystallization procedures for optical resolution, the mechanical separation procedure employed sucessfully by Pasteur with a magnifier and tweezers, is well known historically. However, the method now used for practical and industrial purposes is the "preferential crystallization procedure" or "selective crystallization procedure", which enables the desired optically active isomer to crystallize preferentially from a supersaturated solution of racemic amino acid.

Another method is the "replacing crystallization procedure", which possesses characteristics of both the chemical resolution and preferential crystallization procedures. Resolution is accomplished by seeding a solution of DL-amino acid with another optically active amino acid as a third component and then crystallizing the L- or D-isomer from the solution according to the kind of optically active amino acid added.

A. Mechanical separation procedure

As attempted for the resolution of sodium ammonium DL-tartrate by Pasteur in 1848, this procedure comprises (1) crystallization of the D-isomer and L-isomer independently as a racemic mixture from a solution of the racemic modification under appropriate conditions, (2) distinguishing the difference of both crystal forms visually, and (3) separating both isomers mechanically.[3)]

One example of an amino acid isolated by this method is asparagine.[4)] However, the technique is not so suitable in practice since it can be applied only to a limited number of amino acids, and requires considerable skill and patience on the part of the operator in order to obtain crystals with good hemihedral facets that are suitable for mechanical separation. An improved method has been reported whereby optically active crystals having greater mesh size are seeded to a supersaturated solution of the racemic amino acid and the large separated crystals of the isomer are collected by sifting.[5,6)] In addition, glutamic acid has been resolved by a method which involves giving differential electro-charges to a crystalline racemic mixture by contact electrification. The charged particles are passed through an electrostatic field, and a fraction predominantly rich in one of the isomers is separated.[7)] These methods must also be investigated in relation to the method of preferential crystallization described next.

B. Preferential crystallization procedure

This method serves to crystallize a desired optically active isomer from a supersaturated solution of DL-amino acid. The resolution is carried out in the following manner. The desired isomer is crystallized from a supersaturated solution of the racemic amino acid by seeding the same isomer, or it is crystallized from a supersaturated solution containing the desired optically active isomer in excess, i.e. by preferential crystallization of the excess isomer. In the latter case, the supersaturation state of the isomer existing in excess is broken, and the isomer which has been crystallized plays the role of seed crystals in the former procedure. Both procedures are thus the same in principle, and resolution is carried out advantageously in practice by combining the two together.

Among the various physicochemical resolution methods, preferential crystallization is the most advanced. It is also the only one actually employed for the industrial production of optically active amino acids. Since the procedure is adaptable to automatic and continuous operation, it is expected to develop further as an industrial production method on a large scale. However, it does have the disadvantage that it cannot be applied to all kinds of amino acids as a general method. Also, no guiding rule has yet been found that predicts the kind of amino acids or derivatives which can be resolved effectively by this method.

Generally it is recognized that preferential crystallization can be applied in cases where the solubility of the racemic modification is higher than that of the optically active isomers.[8] That is to say, resolution is generally difficult when the racemic modification forms a racemic compound, although some possibility still remains in this case. The most desirable situation is that of resolution carried out under conditions where the racemic modification crystallizes as a racemic mixture (conglomerate). The method is expected to be applicable in cases where the racemic amino acid is readily crystallized as a racemic mixture and the crystals have adequate solubility and suitable characteristics for easy handling.

Until now, there has been no rule to regulate whether the racemic modification crystallizes as a racemic mixture or racemic compound; neither has there been a theory to connect the chemical structure of amino acids with the possibility of their optical resolution by the preferential crystallization method. However, it is now possible in some cases to know whether a racemic mixture or a racemic compound will be formed when a racemic amino acid is crystallized under fixed conditions. Consequently, it may be possible to predict whether optical resolution is feasible or not by distinguishing the amino acids, and their salts and derivatives, crystallized under various conditions, i.e. as either racemic mixtures or racemic compounds. For the purpose of distinguishing both racemic forms, X-ray diffraction diagrams can be used; however, determination of the melting point, solubility and infrared spectrum is more convenient. Roozeboom[9] distinguished racemic modifications in the solid phase into three forms, racemic mixtures, racemic compounds and racemic solid solutions, by the phase rule. The easiest method for identifying the form is to compare the melting points after admixture of a small amount of optically active isomer to the racemic modification. If the melting point is higher than that of the racemic modification, it may be judged as a racemic mixture. If the melting point is lower than that of the racemic modification, it may be judged as a racemic compound. Both forms can be also distinguished on the basis of solubility relationships. Namely, if D- or L-isomer added to a saturated solution of the racemic modification does not dissolve, the original racemic modification is a racemic mixture, and the solution remains optically inactive. If it dissolves, the racemic modification may be judged as a racemic compound. In this case, the supernatant liquid becomes optically active according to the amount of isomer dissolved. These methods are not applicable to the case of identifying racemic solid solutions. The method of distinction on the basis of infrared spectra utilizes the fact that while the spectra of a racemic mixture and of the optically active forms are identical, those of a racemic compound and of the optically active forms are different.[10)]

The technique of optical resolution by preferential crystallization was discovered by Gernez[11,12)] in 1866. In the case of amino acids, Piutti[4)] first isolated D-asparagine from a mother liquor after separation of crude L-asparagine in 1886. As a typical example of successive optical resolution by reciprocal separation of the L- and D-isomers, one may quote the optical resolution of DL-histidine monohydrochloride by Duschinsky.[13)] Application of this type of resolution for free amino acids or simple salts has been restricted only to such amino acids as histidine hydrochloride,[13)] glutamic acid,[14,15)] threonine,[16)] valine hydrochloride,[17)] homocysteinic acid[18)] and aspartic acid.[19,20)] In these cases, the method has occasionally been referred to as "direct resolution".

When the free amino acids themselves cannot be resolved directly by this method, optical resolution becomes possible in some instances after conversion to appropriate derivatives. One such example is the resolution of ammonium salts of the acyl derivatives of certain amino acids, including tryptophan[21)] phenylalanine,[22)] valine,[22)] methionine,[22)] serine[23)] and phenylglycine.[24)] However, resolution in the form of derivatives is disadvantageous inasmuch as it requires two additional processes, except in cases where the derivative is an intermediate in the process of amino acid synthesis; that is, it requires conversion to the derivative and then return to the free amino acid. From this viewpoint, resolution in the form of simple salts or addition compounds is most desirable. As described above, however, there are so far only a few examples where the simple salts of amino acids form resolvable racemic mixtures. In order to overcome this problem, a method of resolution using aromatic sulfonic acid has recently been developed. Namely, benzenesulfonate of DL-alanine,[25)] *p*-toluenesulfonate[26)] and *m*-xylenesulfonate[27] of DL-serine, and *p*-aminobenzenesulfonate,[28] antraquinone-β-sulfonate,[29] 1-chloronaphthalene-4-sulfonate[29] and β-naphthalenesulfonate[29] of lysine have all been resolved successfully by this crystallization procedure. In fact, it becomes very easy to screen the salts crystallized as racemic mixtures, because the aromatic sulfonic acids vary greatly in their properties and readily form salts with all kinds of amino acids. Furthermore, the process is promising for industrial application since the salts are themselves readily converted to free amino acids.

Several requirements must be fulfilled before resolution by preferential crystallization can be judged successful. First, and most important in principle, it is necessary to define the conditions and compound to form the racemic mixture suitable for the crystallization procedure. Next, it is necessary to establish the conditions where the desired optical isomer crystallizes in high optical purity from a supersaturated solution, and the opposite antipode remains supersaturated in the solution. However, it is impossible to define conditions that are generally applicable since the solubility of the racemic modification and stability of its supersaturation state are dependent on the type of each racemic amino acid. So, the conditions must be investigated for each particular case, with reference to actual data on this kind of resolution.

To avoid crystallization of the opposite antipode, and to crystallize only the desired optical form, the yield of the desired optically active isomer resolved in one operating cycle is limited due to the limitations of the supersaturation state. Although the yield of an optically active isomer varies with the material used, generally speaking, the maximum yield is about 25% of the optically active isomer existing in the racemic amino acid used. If 10 parts of L-isomer are separated by seeding a small amount of crystals of the L-isomer to a supersaturated solution containing 5 parts of the L-isomer, 50 parts of the racemic modification and 100 parts of water, it can be said that the yield per unit volume and resolution rate from the DL-modification are satisfactory and the resolution process is successful. By adding 10 parts of the DL-modification into this mother liquor, the same conditions as in the previous operation can be obtained, except that the solution contains the D-isomer in excess. Therefore, D-isomer can easily be obtained by the same operation as described above. This method is therefore very convenient, the D- and L-isomers being obtained reciprocally by repetition of the same operation, if the condition is satisfied that the amount of optically active isomer initially in excess is the same as that of the isomer resolved. Such reciprocal resolution is a singular advantage of this resolution method, since it cannot be expected of other methods.

Another advantage is that the optical isomer separated by the above procedure can be purified to 100% optical purity. Generally, however, the purification of optically impure amino acids cannot be accomplished by a simple method such as the repetition of recrystallization. For example, in most cases where the racemic amino acid forms a racemic compound, the same procedure as optical resolution is required for purification of the optically impure amino acid. However, the compound resolvable by this preferential crystallization procedure and forming a racemic mixture, can be purified using the property of a saturated solution of the racemic mixture that it no longer dissolves the optically active form. Namely, optical purification is performed without loss of optically active isomer by immersing or recrystallizing the crude crystals in enough solvent to dissolve the racemic modification in the optically crude crystals.[26-28,30-34]

As mentioned above, in order to avoid crystallization of the opposite antipode, the amount of optical isomer resolved per single operation should be limited. To improve on this, attempts have been made to crystallize both isomers under conditions that allow the optical isomers to be obtained separately. For example, optical resolution of DL-glutamic acid hydrochloride was carried out in the following manner. Two seed plates made by crystallizing D- and L-glutamic acid hydrochloride individually on the surface of brick or unglazed ceramic material, were placed in a supersaturated solution of the racemic amino acid. Crystallization of the D- and L-isomers then took place on the respective seed plates.[35] Continuous resolution was also achieved by passing a supersaturated solution of the racemic amino acid reciprocally through two fluidized beds consisting of the L- or D-isomer.[36] Besides these methods, using a crystallizer separated into two blocks by a screen through which a solution can pass freely but crystals cannot, seed crystals of D- and L-isomer were added to each block, and the D- and L-isomers then crystallized on the respective blocks.[37] Also, there is a reported example of resolution where the D- and L-isomers were crystallized at different mesh sizes, dried and screened.[5,6] Although the crude products thus obtained are often optically impure and need further purification, such resolution methods should be investigated further together with equipmental problems[36-39] related to crystallization and continuous operation on an industrial scale.

In this section, the resolution methods described are considered mainly from the viewpoint of producing L-amino acids. However, when the D-form is required or when enzymatic resolution cannot be applied, for instance in the optical resolution of certain unnatural amino acids, the preferential crystallization method and chemical method are still of importance. For example, the former method can be applied to obtain D-serine,[23,27] an important starting material for the synthesis of cycloserine, or to obtain D-phenylglycine,[24] which is in great demand as a starting material for synthetic penicillin derivatives.

The methodology of industrial production of optically active amino acids by combined synthesis of DL-amino acids and optical resolution by preferential crystallization is developing fast, as seen in the production of threonine, glutamic acid, etc. As a typical example, the general flowsheet for L-glutamic acid production is shown in Fig. 2–1.[40] These processes are said to be operated automatically and continuously in a sequence control system, and the method is considered useful for the mass production of sodium L-glutamate, competitive with the fermentation method.

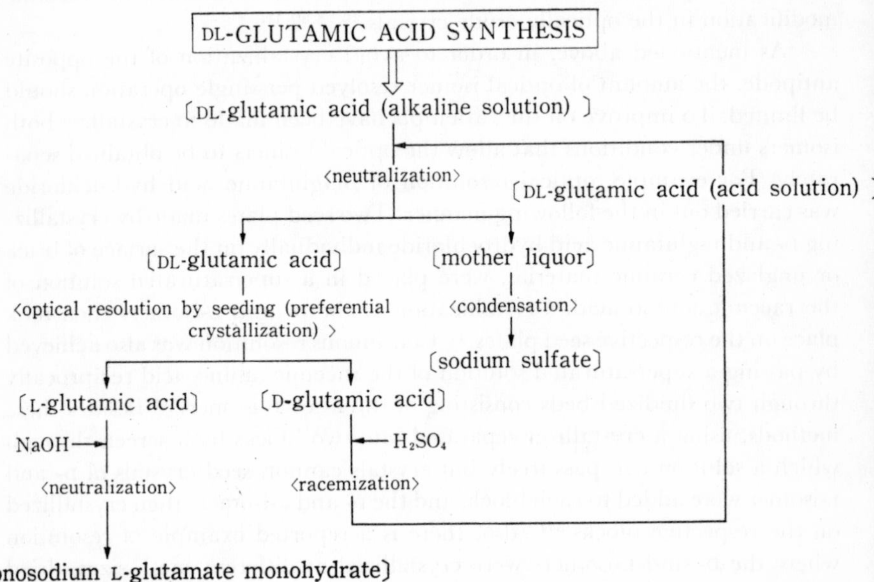

Fig. 2–1. General flowsheet for L-glutamic acid production.

C. Replacing crystallization procedure

The replacing crystallization procedure consists of preparing a solution containing a DL-amino acid and some other kind of optically active amino acid, and crystallizing only the one isomer having the opposite configuration to the optically active amino acid used initially. This technique is thus rather similar to the method of chemical resolution. It was first reported for DL-aspartic acid, DL-glutamic acid, DL-leucine and DL-tyrosine,[41] but the purity of the optically active isomers resolved was not high. Later, however, crystals of the optical antipodes in a state of high optical purity were obtained by adding optically active γ-methylglutamate to a solution of DL-glutamic acid.[42] In addition, examples have been given for the resolution of DL-glutamic acid using diester of glutamic acid.[43] Resolution of DL-alanine[44,45] and DL-aspartic acid[46] has also been performed by stereoselective ligand exchange reactions using metal chelate compounds having optically active ligands. These methods have not yet been applied to industrial production but merit further investigation.

2.1.2 Chromatographic procedures

As a procedure for physicochemical resolution without accompanying crystallization, attempts have been made to separate optical isomers by utilizing their differences in physical adsorption, chemical binding, or partition coefficent to an adsorbent. The adsorbents used for this purpose require chirality, and natural substances having chirality such as wool, starch, cellulose, saccharides, etc. are preferable. For example, by using a starch column, phenylglycine[47] and many acylamino acids[48] have been resolved. DL-Methyllysinate dihydrochloride and proline have also been resolved on a lactose column.[49] In these cases, however, the degree of resolution was unsatisfactory. As an example of the use of cellulose, it has been reported that DL-amino acid can be separated into two spots by paper chromatography and thin layer chromatography. However, this method is not effective as an optical resolution technique. Recently, optical resolution has been accomplished using polymers having chirality, such as an ion exchange resin synthesized by coupling optically active amino acids.[50] The resolution of amino acids such as lysine,[51] methionine,[52] etc. was initially attempted, and many studies on the application of these techniques have been reported. However, the methods have not yet reached the stage of practicability and are still in a state of investigation. In the future, with the development of research on asymmetric adsorbents, it may however be possible to apply this type of resolution on an industrial scale.

2.2 Chemical Methods

As shown in the equation below, the chemical method is performed as follows. DL-Amino acids or their derivatives (DL-A or DL-B; A = acidic, B = basic) are treated with optically active acids ((+)-a) or bases ((+)-b), and one of the diastereoisomeric salts formed is fractionally crystallized, mainly as a result of the difference in solubility between the two diastereoisomeric salts. Usually the less soluble salt is crystallized and purified, and optically active amino acid then separated from the salt in the usual way.

$$\text{DL-A} + (+)\text{-b} \longrightarrow \text{L-A}\cdot(+)\text{-b} + \text{D-A}\cdot(+)\text{-b}$$

$$\text{DL-B} + (+)\text{-a} \longrightarrow \text{L-B}\cdot(+)\text{-a} + \text{D-B}\cdot(+)\text{-a}$$

This method, like the physicochemical method mentioned above, was devised by Pasteur, who first employed it for the optical resolution of DL-tartaric acid in 1853.[53)] In the case of amino acids, Fischer succeeded in the optical resolution of DL-alanine and DL-glutamic acid in 1899. The *N*-benzoyl derivatives were treated with optically active alkaloids such as brucine and strychnine, the soluble and less soluble salts were prepared and the alkaloids then removed from the salts by alkali treatment. The optically active *N*-benzoylamino acids formed were hydrolyzed with hydrochloric acid and optically active amino acids so obtained.[54)] Subsequently, this method has been applied to most amino acids, and has been adopted not only in the laboratory but also in industry. Examples of the many resolutions of different amino acids have been summarized in the review mentioned above.[1)]

One problem of this method is that there is no general rule regarding the resolution procedure and results obtained, even though many examples have appeared. Also, it is sometimes difficult to obtain optically active agents cheaply and in large quantities. Furthermore, the optical purity often becomes lower at higher crystallization yields because, in many cases, there is no marked difference in solubility or in the velocity of crystal growth between the diastereoisomeric salts. Difficulty in the filtration process after crystallization depends on the form of the crystals, but it clearly influences the resolution results. On the other hand, this method has the merit that it shows a wider applicability than the above crystallization procedures which require crystallization as a racemic mixture. When the optical purity of the products obtained in a single fractionation is unsatisfactory, optically pure isomer can be obtained by repetition of the recrystallization procedure.

Considering the diastereoisomeric salts on the basis of the properties of the original amino acids, it can be said that neutral amino acids form diastereoisomeric salts with a resolving agent of a strong acid such as camphorsulfonic acid, bromo- and chlorocamphorsulfonic acids, but do not form salts with an ordinary resolving agent such as an optically active weak acid or basic reagent. Accordingly, neutral amino acids should be converted to a suitable derivative, i.e. an acidic derivative (e.g. an acyl derivative) or a basic derivative (e.g. an ester or amide) in order to prepare the diastereoisomeric salts for resolution. On the other hand, acidic or basic amino acids form salts directly with ordinary resolving agents, as described above. This procedure is therefore highly advantageous for such amino acids. Examples include the resolution of glutamic acid with L-1-hydroxy-2-aminobutane[55)] or L-(+)-1,1,3-triphenyl-2-aminopropanol-1,[56] and the resolutions of lysine[57] and histidine[58] with *d*-tartaric acid.

From the standpoint of resolving agents, it can be said that camphor derivatives, alkaloids and tartaric acid are widely employed. However, since a large amount of cheap resolving agent is required for industrial production, the use of expensive alkaloids is not so advantageous, even when the alkaloids can be recovered. For this reason, optically active amino acids or their derivatives, e.g. optically active glutamic acid[59-62] or its derivative, L-pyrrolidone carboxylic acid (L-pyroglutamic acid),[63] or lysine base,[64-66] amino acid amides,[67] and acylamino acid derivatives,[68] have been used increasingly as resolving agents. Recently, DL-glutamic acid,[69] DL-proline,[70] DL-alanine[70] and DL-isoleucine[70] were resolved using L-tyrosinehydrazide. This method was also applied to special kinds of amino acids, such as α-amino-β-hydrooxy-γ-benzyloxybutyric acid.[71] As readily available resolving agents, synthetic intermediates of industrial products have also been used instead of optically active amino acids. For example, resolution of DL-tryptophan, DL-proline, DL-alanine, DL-leucine, DL-serine, DL-threonine and DL-valine has been carried out using the *d*- and *l*-forms of threo-*p*-nitrophenyl-2-aminopropane-1,3-diol, which is an intermediate in the synthesis of chloramphenicol.[72,73] The resolution of many amino acids was also performed using fencylamine or phenylethylamine.[68,74-77]

As mentioned above, acidic and basic amino acids are functionally capable of forming salts for resolution, and their optically active forms themselves function as resolving agents. Thus, the resolution of DL-lysine by L-glutamic acid[59-61] and its derivatives,[63] or conversely of DL-glutamic acid by L-lysine[64] and of DL-ornithine by L-glutamic acid,[62] represents a typical example of the chemical method.

In the formation of salts, an equimolar amount of resolving agent is usually added to the DL-form, and the diastereoisomeric salts, L-A•(+)-b and D-A•(+)-b, are formed according to the scheme given above. However, it is also necessary to investigate the effect of the amount of resolving agent by changing its ratio to the DL-form, because a change in the molar ratio of the resolving agent, (+)-b, to DL-A sometimes brings about more satisfactory results. For example, by using 0.5 equivalent of resolving agent, only L-A•(+)-b or D-A•(+)-b is crystallized preferentially, while the residual D-A or L-A is forced to remain in solution. One successful example is the resolution of DL-α-aminocaprolactam by means of pyrrolidone carboxylic acid. That is to say, less than 0.5 mol of L-pyrrolidone carboxylic acid was added to DL-α-aminocaprolactam, and the L–L salt was selectively crystallized, while the D-α-aminocaprolactam remained in solution in toluene.[78] This resolution technique is suitable for industrial production.

Generally speaking, the chemical method can be said to be comprised of crystallization and separation of the less soluble salt based on the difference in solubility between the diastereoisomeric salts. In practical crystallization for resolution, however, the crystallizability or crystallization time represents a serious problem in addition to solubility relationships. In an extreme case, crystallization of the diastereoisomeric salt having the higher solubility may proceed when it has good crystallizability and is fast as regards crystal growth. Even diastereoisomeric salts show significant differences in solubility, and resolution may not be effected successfully when the salt has bad crystallizability and causes difficulty at the crystallization and separation steps. In such cases, instead of crystallizing one of them, it may be an effective procedure to dissolve the more soluble salt in a suitable solvent after the two diastereoisomeric salts have been formed into a solid mixture by spray drying. As a example, the resolution of DL-lysine has been achieved by using optically active α-phenoxypropionic acid.[79] That is to say, an equivalent amount of α-phenoxypropionic acid was added to DL-lysine, the D–D-salt and L–D-salt were spray-dried to a solid mixture, and resolution was carried out by dissolving the soluble D–D-salt in methanol. This dissolution method is important as a technique for large scale production, since the control of diffusion, a limiting step of dissolution, is much easier to effect than control of the crystallization velocity.

This resolution method is, as described above, usually achieved by forming diastereoisomeric salts between an acid and a base. In some cases, however, resolution has been undertaken by forming a molecular compound or a complex salt. As an example, resolution has been achieved by the formation of a complex of DL-glutamic acid with pyrrolidone carboxylic acid,[80] and also by the formation of a complex of DL-tryptophan with inosine.[81] Although not practical, another resolution procedure has been reported that utilizes the difference in solubility between diastereoisomers formed by the treatment of acylamino acid with an optically active metal–salt complex.[82]

Besides the formation of salts or complexes with optically active resolving agents, diastereoisomeric derivatives have been synthesized by reacting DL-amino acids with optically active reagents, and resolution was performed by utilizing differences in properties between the derivatives. For example, 1-menthoxyacetyl amino acids,[83] L- and D-mandelyl amino acids,[84] α-alkylamino acid 1-menthylester[85] and D-pantoyl amino acids[86] have been synthesized and resolved. However, this method is considered to be less advantageous, due to the requirement of an hydrolysis or other conversion step to recover the optically active amino acid.

The above-mentioned procedures are concerned with separation in the form of solid and liquid phases. Resolution methods utilizing differences in distribution between liquid phases[87] and gas chromatography[88] have on occasions been reported; however, these methods are still not available practically.

From the viewpoint of industrialization of the chemical method for the purpose of mass production, it is relatively difficult to operate the process automatically or sequentially, due to the requirement of different procedures for the separation of the D-form and L-form and also to the requirement of divergent procedures in a batch system for fractional recrystallization. In this respect, the chemical method is inferior to the above-mentioned preferential crystallization procedures using a sequential control system, and to the continuous enzymatic resolution methods described later. However, in some cases, this procedure is clearly one of the advantageous methods for production of optically active amino acids, i.e. when the objective amino acid is basic or acidic and the amino acid itself is capable of forming salts with optically active resolving agents, or when acyl derivatives are obtained as synthetic intermediates. Resolving agents that form readily separable salts of good crystallizability are available, and are inexpensive in large quantities. Thus, further investigations of the method are warranted, especially in an attempt to discover new widely applicable and efficient resolving agents which can be provided at low cost and in large quantities. Research on a generalization of operations for the resolution procedure, and for the discovery of regularities in this procedure are also called for.

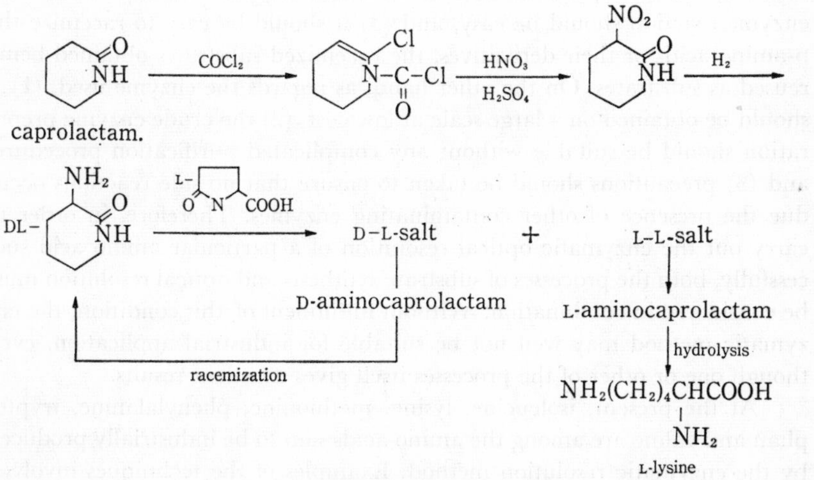

caption caprolactam $\xrightarrow{COCl_2}$ (Cl, N–C(=O)–Cl) $\xrightarrow[H_2SO_4]{HNO_3}$ (NO_2, O, NH) $\xrightarrow{H_2}$

DL-(aminocaprolactam: NH_2, O, NH) $\xrightarrow{\text{L-pyroglutamic acid (COOH)}}$ D-L-salt + L-L-salt

D-L-salt → D-aminocaprolactam → racemization (back to DL-)

L-L-salt → L-aminocaprolactam $\xrightarrow{\text{hydrolysis}}$ $NH_2(CH_2)_4CHCOOH$ (NH_2) L-lysine

Fig. 2–2. General flowsheet for L-lysine preparation from ε-caprolactam.

As an example of the efficient combination of the chemical resolution method with the chemical synthesis of DL-amino acids, Fig. 2–2 shows a general flow sheet for the preparation of L-lysine from ε-caprolactam.[89] In this procedure, ε-DL-aminocaprolactam is synthesized via ε-DL-α-nitrocaprolactam from ε-caprolactam, and resolved as the L-pyrrolidone carboxylic acid salts. The L-aminocaprolactam obtained is hydrolyzed to yield L-lysine, while the D-aminocaprolactam is racemized by heating to DL-aminocaprolactam and then reused for the resolution step.

2.3 Enzymatic methods

Enzymatic optical resolution is a procedure for the separation of DL-amino acids or their derivatives into the optical isomers by enzymatic action, such as stereospecific (asymmetric) synthesis or hydrolysis. By this method, isomers of high optical specificity can be obtained in high yield, and the method can be applied to the optical resolution of a relatively large variety of DL-amino acids. It can also handle large amounts of amino acids at one operation.

A number of different methods have been published for the enzymatic optical resolution of DL-amino acids, and these methods are available for use on a laboratory scale. However, for industrial application it is necessary that the following strict conditions be fulfilled. As regards the substrate used, (1) it should be produced on a large scale at low cost, and should be an intermediate in the amino acid synthesis or easily derived from such an intermediate, (2) the separation of both optical isomers after completion of the enzyme reaction should be easy, and (3) it should be easy to racemize the D-amino acids or their derivatives, the racemized substrates obtained being reused as substrates. On the other hand, as regards the enzyme used, (1) it should be obtained on a large scale at low cost, (2) the crude enzyme preparation should be suitable without any complicated purification procedure, and (3) precautions should be taken to ensure that no side reactions occur due the presence of other contaminating enzymes. Therefore, in order to carry out the enzymatic optical resolution of a particular amino acid successfully, both the processes of substrate synthesis and optical resolution must be considered in combination. Without fulfillment of this condition, the enzymatic method may well not be suitable for industrial application, even though one or other of the processes itself gives excellent results.

At the present, isoleucine, lysine, methionine, phenylalanine, tryptophan and valine are among the amino acids said to be industrially produced by the enzymatic resolution method. Examples of the techniques involved are described in detail in the following sections.

2.3.1 Resolution by asymmetric synthesis: the synthesis of acyl-L-amino acid anilide and phenylhydrazide

In 1937, Bergmann and Frankel-Conrat[90)] found that incubation of acyl derivatives of racemic amino acids and aniline with papain led to the formation and subsequent precipitation of the insoluble acyl-L-amino acid anilides, while the acyl-D-amino acids remained in the mother liquor. Thereafter, this method has been employed as a general optical resolution method for many kinds of amino acids. Crystallized acyl-L-amino acid anilides formed by the enzyme reaction are chemically hydrolyzed to give L-amino acids. On the other hand, the mother liquor remaining after removal of the L-anilides is acidified and extracted by organic solvent. The acyl-D-amino acids obtained can be converted to D-amino acids by chemical hydrolysis.

DL-RCH(NHCOR′)COOH + C_6H_5-NH_2

$\xrightarrow[\text{papain}]{\text{asymmetric synthesis}}$ L-RCH(NHCOR′)CONH-C_6H_5 + D-RCH(NHCOR′)COOH

L-RCH(NHCOR′)CONH-C_6H_5 $\xrightarrow{\text{chemical hydrolysis}}$ L-RCH(NH_2)COOH

D-RCH(NHCOR′)COOH $\xrightarrow{\text{chemical hydrolysis}}$ D-RCH(NH_2)COOH

In 1951, Doherty[91)] and Albertson[92)] reported fuller details of this optical resolution method. As the enzyme for the reaction, papain and similar enzymes such as bromelain and ficin may be used. The reaction rate of L-anilide formation depends on the kinds of amino acids (and of their acyl groups), and on the pH and duration of the enzyme reaction. In the case of isoleucine and lysine, the reaction must be carried out with particular care since DL-anilides are produced under certain conditions.[91)]

In place of aniline, *p*-toluidine and *p*-phenetidine have been used for the optical resolution of acetyl-DL-phenylalanine and α,ε-diacetyl-DL-lysine, respectively, and satisfactory results were recorded.[93)]

As in the case of L-anilide formation, Bergmann and Frankel-Conrat[90)] found that hydrazide derivatives of L-acylamino acids can be asymmetrically synthesized with papain by the use of phenylhydrazine instead of aniline. However, this technique has no special advantage over the anilide method.

$$\text{DL-}\underset{\text{NHCOR}'}{\text{RCHCOOH}} + \text{C}_6\text{H}_5\text{-NHNH}_2 \xrightarrow[\text{papain}]{\text{asymmetric synthesis}} \text{L-}\underset{\text{NHCOR}'}{\text{RCHCONHNH-C}_6\text{H}_5} + \text{D-}\underset{\text{NHCOR}'}{\text{RCHCOOH}}$$

$$\text{L-RCHCONHNH-C}_6\text{H}_5 \xrightarrow{\text{chemical hydrolysis}} \text{L-}\underset{\text{NH}_2}{\text{RCHCOOH}} \qquad \text{D-RCHCOOH(NHCOR}') \xrightarrow{\text{chemical hydrolysis}} \text{D-}\underset{\text{NH}_2}{\text{RCHCOOH}}$$

These methods can be applied to the optical resolution of a variety of amino acids. However, to obtain an optically active amino acid, chemical hydrolysis of the optically active amino acid derivatives is necessary subsequent to the enzyme reaction. Therefore, this method is in some respects inferior for industrial application when compared with the aminoacylase and amidase methods described later.

2.3.2 Resolution by asymmetric hydrolysis

The optical resolution method by enzymatic asymmetric hydrolysis is that procedure which comprises the asymmetric hydrolysis of ester, amide and acyl derivatives of DL-amino acids by esterase, amidase and aminoacylase, respectively. The technique has been employed for a longer period than the asymmetric synthesis method described above. At present, an optical resolution procedure using aminoacylase is carried out industrially.

A. Hydrolysis of DL-amino acid ester

In 1905, Warburg[94] found that DL-amino acid esters could be asymmetrically hydrolyzed by pancreatic enzyme, and in the next year he succeeded in obtaining L-leucine from DL-leucine propyl ester by the action of pancreatin esterase. This was the first report among the asymmetric hydrolysis methods. The L-amino acid produced by the enzyme action and the unhydrolyzed D-amino acid ester can be separated according to differences in their solubilities for solvents.

$$\text{DL-}\underset{\text{NH}_2}{\text{RCHCOOR}'} \xrightarrow[\text{esterase}]{\text{asymmetric hydrolysis}} \text{L-}\underset{\text{NH}_2}{\text{RCHCOOH}} + \text{D-}\underset{\text{NH}_2}{\text{RCHCOOR}'}$$

As the enzyme having esterase activity, crystalline chymotrypsin and crude pancreatin are employed for this method. Since the optimum pH of the enzyme reaction is generally on the alkaline side, the esters used as substrate are non-enzymatically hydrolyzed, and the optical purity of the L-amino acids obtained is lower. This suggests that the esterase method is not so advantageous in practice.

B. Hydrolysis of amino acid amides

DL-Amino acid amides are asymmetrically hydrolyzed by the action of amidase, and the L-amino acid and unhydrolyzed D-amino acid amide produced can be separated by differences in their solubilities for organic solvents.[95]

$$\text{DL-}\underset{\displaystyle NH_2}{\underset{|}{\text{RCHCONH}_2}} \xrightarrow[\text{amidase}]{\text{asymmetric hydrolysis}} \text{L-}\underset{\displaystyle NH_2}{\underset{|}{\text{RCHCOOH}}} + \text{D-}\underset{\displaystyle NH_2}{\underset{|}{\text{RCHCONH}_2}}$$

The enzyme having amidase activity is prepared from hog kidney and pancreas, mushrooms, moulds or bacteria. Since amino acid amides are more resistant to non-enzymatic hydrolysis than amino acid esters, L-amino acid in higher optical purity can be obtained by this method. In particular, the amidase method can be employed advantageously when the amino-acylase method proves difficult to apply, i.e. when chemical acylation of the amino acids is difficult, and the *N*-acylamino acids are unsusceptible to the enzyme, such as with acyl derivatives of proline,[96] *tert*-leucine,[97] dodecyline[98] and α,ε-diaminopimelic acid.[99]

C. Hydrolysis of acyl-DL-amino acids

The enzymatic method which involves hydrolysis of acylamino acids is very often used, and is one of the most advantageous of such techniques. Acyl-DL-amino acid is hydrolyzed asymmetrically by aminoacylase (*N*-acylamino acid amidohydrolase) or carboxypeptidase, and L-amino acid and unhydrolyzed acyl-D-amino acid are produced in the reaction mixture. Both compounds can easily be separated according to differences in their solubilities for solvents or by using an ion-exchange resin.

$$\text{DL-}\underset{\displaystyle NHCOR'}{\underset{|}{\text{RCHCOOH}}} \xrightarrow[\text{aminoacylase}]{\text{asymmetric hydrolysis}} \text{L-}\underset{\displaystyle NH_2}{\underset{|}{\text{RCHCOOH}}} + \text{D-}\underset{\displaystyle NHCOR'}{\underset{|}{\text{RCHCOOH}}}$$

In 1881, Schmiedeberg[100] discovered that kidney extract can hydrolyze hippuric acid to glycine and benzoic acid. Subsequently, Smorodinzev[101] found that only the L-isomer of *N*-benzoyl-α-amino-DL-butyric acid can be hydrolyzed by muscle extract. In 1924, Neuberg[102] prepared benzoyl-D-alanine from benzoyl-DL-alanine by the action of Takadiastase. For some time thereafter, the method received little attention, but in 1948 Greenstein and his co-workers began studies on hog kidney aminoacylase.[103] By 1952–53, they had been able to refine the optical resolution method by aminoacylase for common DL-amino acids. They continued detailed studies and developed the method into a widely employed and advantageous general procedure.

On the other hand, investigations have been made to isolate enzymes having aminoacylase activity from materials other than hog kidney. Neuberg and Mandl[104] carried out optical resolution of amino acids using commercial enzyme preparations such as Orthozym, Takadiastase, etc. In Japan, Michi[105] and Chibata[106] and his co-workers studied the resolution procedure with mould aminoacylase. As a result of screening possible enzyme sources, a mould displaying very high aminoacylase activity was found. At present, this mould aminoacylase is employed for industrial optical resolution. Kameda[107] and his co-workers have also studied the resolution of DL-amino acids utilizing enzyme from soil bacteria. The characteristics of the respective aminoacylases are described below.

(*i*) *Hog kidney aminoacylase*

In 1948, Greenstein and his co-workers found that rat kidney extract can specifically hydrolyze only the L-isomer of acetyl-DL-alanine. Since that time they have carried out extensive investigations on the optical resolution of DL-amino acids using hog kidney homogenate (and purified aminoacylase I and II obtained from it), and have so obtained many kinds of L-amino acids. A number of papers have been published on the purification of the enzyme,[108,109] the synthesis of the substrate and the optical resolution procedure.[108,110] Aminoacylase I has broad substrate specificity and can hydrolyze many kinds of acyl-L-amino acids except acyl-L-aspartic acid. Aminoacylase II is susceptible only to acyl-L-aspartic acid. The relative activities of hog kidney aminoacylase to different acylamino acids are shown in Table 2–1. As can be seen, hog kidney aminoacylase has relatively high activity to acylated aliphatic amino acids, but is significantly less active to acyl derivatives of aromatic amino acids such as phenylalanine, tyrosine and tryptophan. Thus, in order to resolve such aromatic amino acids, the chloroacetyl derivatives are hydrolyzed by beef pancreas carboxypeptidase instead of hog kidney aminoacylase.[113] The enzyme is also less susceptible to acyl derivatives of basic amino acids such as lysine.

From a practical industrial viewpoint, acetyl derivatives are the most suitable among acyl derivatives. However, the influence of acyl groups on the rate of the enzyme reaction has been investigated and it was found that

TABLE 2–1
Substrate specificity of aminoacylase

Amino acid	Acyl group	Relative activity†[1]	
		Kidney†[2] (hog)[111)]	Mould[112)] (*Asp. oryzae*)
Alanine	Acetyl	13	73
	Chloroacetyl	48	144
Valine	Acetyl	7	25
Leucine	Acetyl	22	31
	Chloroacetyl	69	46
Isoleucine	Acetyl	1.6	65
Alloisoleucine	Acetyl	1.1	4.7
Methionine	Acetyl	[100]	[100]
	Chloroacetyl	410	180
	Glycyl	62	26
Threonine	Chloroacetyl	3	14
Glutamic acid	Acetyl	13	37
	Chloroacetyl	53	53
Aspartic acid	Acetyl	0.1†[3]	10
	Chloroacetyl	0.5†[3]	10
Lysine	α,ε-Diacetyl	0.3	0
	α,ε-Dichloroacetyl	0.6	—
	ε-Benzoyl-α-acetyl	—	83
	ε-Benzoyl-α-chloroacetyl	0.1	158
Tryptophan	Acetyl	<0.1	74
	Chloroacetyl	<0.1	63
Phenylalanine	Acetyl	0.6	145
	Chloroacetyl	1.9	215
Tyrosine	Chloroacetyl	1.4	226

†[1] Reaction rate of acetyl methionine was taken as 100. †[2] Aminoacylase I.
†[3] Aminoacylase II.

acyl groups of higher electronegativity are more susceptible to the enzyme, and that chloroacetyl derivatives are more susceptible than acetyl ones in the case of all amino acids.[114)] Among atoms showing electronegativity, the reaction rate increases in the order I<Br<Cl<F. Although displaying higher electronegativity than monohalogeno derivatives, dihalogeno and

trihalogeno derivatives are hydrolyzed at slower rates. This may be explained by the influence of steric hindrance by these radicals. On the other hand, trifluoroacetyl derivative is hydrolyzed at a faster rate than other trihalogenoacetyl derivatives. This may be considered a result of the lesser steric hindrance of the fluorine atom due to its smaller radius. Concerning the selection of acyl groups, it is important to choose those which do not cause product inhibition after the enzyme reaction. In the enzymatic hydrolysis of acyl-DL-amino acids, most *N*-acetyl and *N*-chloroacetyl derivatives of the amino acids can be hydrolyzed at a faster rate in the presence of Co^{2+}.[115]

(*ii*) *Mould aminoacylase*

For industrial application of the aminoacylase method, the mammalian enzyme is unsatisfactory when considered from the viewpoint of supply and substrate specificity. Therefore, the possible application of microbial aminoacylase was investigated. As mentioned above, an optical resolution method using commercially available enzyme preparations such as Takadiastase has also been reported.[104]

The aminoacylase activity of various moulds was screened for the purpose of finding a more advantageous enzyme source. As a result, it was shown that various moulds can produce aminoacylase, and in particular moulds belonging to the genera *Aspergillus* and *Penicillium* displayed higher activity.[116] The susceptibilities of various acyl amino acids towards water extract from wheat-bran culture and partially purified preparation from *Asp. oryzae* are included in Table 2–1. The reaction rates are different according to the kind of amino acid, but the differences are less than in the case of hog kidney aminoacylase. The fact that mould aminoacylase has broader substrate specificity to the various acylamino acids shown indicates that the optical resolution of a variety of aliphatic and aromatic amino acids may be achieved with this single enzyme preparation.[112]

Since the reaction rate of mould aminoacylase, as well as of hog kidney enzyme, is enhanced in the presence of Co^{2+}, addition of Co^{2+} at a concentration of 5×10^{-4} –10^{-3}M is made to the enzyme reaction mixture. The effect of the acyl groups on the rate of hydrolysis of the mould enzyme is similar to that in the case of the hog kidney enzyme.

Mould aminoacylase is thus widely employed for the industrial production of optically active amino acids, due mainly to its broad substrate specificity, ease of supply, etc. Examples of such enzymatic optical resolution suitable for industrial application are described in the following sections.

(*a*) *Production of* L-*tryptophan*[112,117]

When acetyl-DL-tryptophan synthesized as shown in Fig. 2–3 is incu-

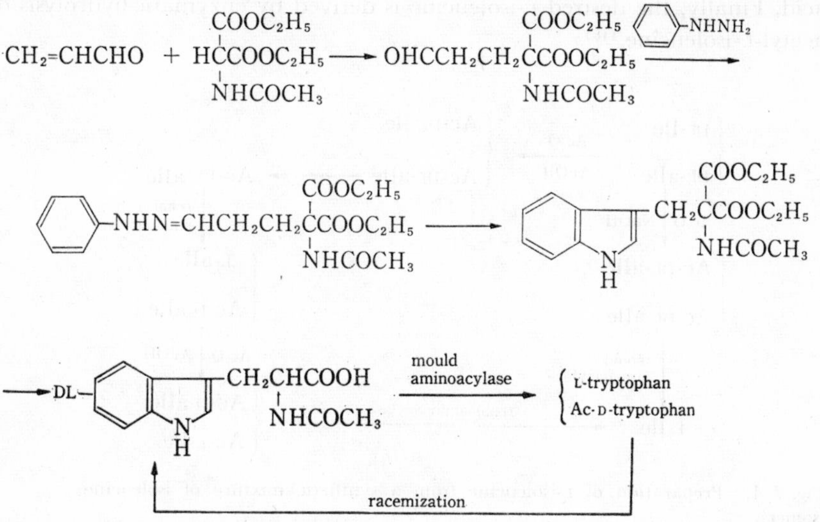

$$CH_2{=}CHCHO + HC(COOC_2H_5)(COOC_2H_5)(NHCOCH_3) \longrightarrow OHCCH_2CH_2C(COOC_2H_5)(COOC_2H_5)(NHCOCH_3) \xrightarrow{C_6H_5NHNH_2}$$

$$C_6H_5NHN{=}CHCH_2CH_2C(COOC_2H_5)(COOC_2H_5)(NHCOCH_3) \longrightarrow \text{(indol-3-yl)}CH_2C(COOC_2H_5)(COOC_2H_5)(NHCOCH_3)$$

$$\longrightarrow \text{DL-(indol-3-yl)}CH_2CH(NHCOCH_3)COOH \xrightarrow[\text{aminoacylase}]{\text{mould}} \begin{cases} \text{L-tryptophan} \\ \text{Ac-D-tryptophan} \end{cases}$$

(Ac-D-tryptophan → racemization → back to DL-acetyltryptophan)

Fig. 2–3. Preparation of L-tryptophan.

bated with mould aminoacylase, only acetyl-L-tryptophan is hydrolyzed and converted to L-tryptophan. After termination of the enzyme reaction, the reaction mixture is concentrated and the crystallized L-tryptophan easily separated. The acetyl-D-tryptophan in the mother liquor is racemized by the action of acetic anhydride, and this racemized acetyl tryptophan is then reused for the resolution process.

(*b*) *Production of* L-*isoleucine*

Isoleucine, one of the essential amino acids, is produced by the fermentative and chemical synthetic methods due to the difficulty of isolating it from protein hydrolyzates. In the case of chemical synthesis, four kinds of stereoisomers are obtained since the amino acid has two asymmetric carbon atoms. In order to isolate the nutritionally effective L-isoleucine from such mixtures, the method using hog kidney aminoacylase necessitates a very complicated procedure. That is to say, when acetyl derivatives of the mixture of isoleucine isomers are incubated with hog kidney aminoacylase, both L-isoleucine and L-alloisoleucine are formed in the reaction mixture, acetyl-L-isoleucine and acetyl-L-alloisoleucine being hydrolyzed by the enzyme with almost the same velocity (Table 2–1). Accordingly, as shown in Fig. 2–4, the acetyl-DL-isoleucine and acetyl-DL-alloisoleucine must be isolated before enzymatic resolution. After enzymatic hydrolysis of the acetyl-DL-alloisoleucine, the acetyl-D-alloisoleucine obtained is epimerized with acetic anhydride in acetic acid. Finally, the desired L-isoleucine is derived by enzymatic hydrolysis of acetyl-L-isoleucine.[118]

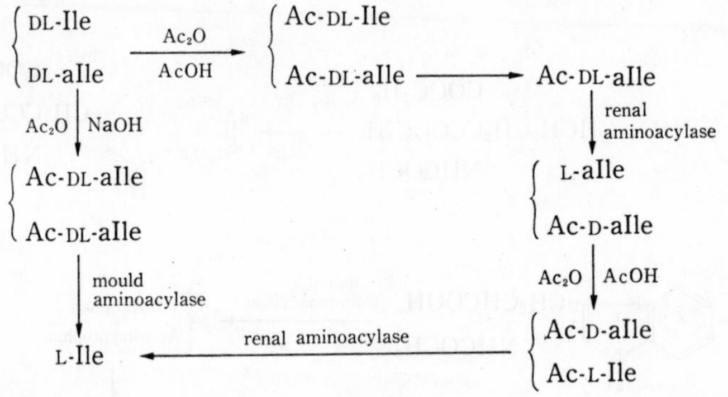

{DL-Ile, DL-aIle} —(Ac_2O, AcOH)→ {Ac-DL-Ile, Ac-DL-aIle} ⟶ Ac-DL-aIle

{DL-Ile, DL-aIle} ↓ Ac_2O | NaOH

{Ac-DL-aIle, Ac-DL-aIle}

↓ mould aminoacylase

L-Ile

Ac-DL-aIle ↓ renal aminoacylase {L-aIle, Ac-D-aIle}

↓ Ac_2O | AcOH

{Ac-D-aIle, Ac-L-Ile} —renal aminoacylase→ L-Ile

Fig. 2–4. Preparation of L-isoleucine from a synthetic mixture of isoleucine isomers.

On the other hand, mould aminoacylase is able to hydrolyze acetyl-L-isoleucine at 15 times the rate for acetyl-L-alloisoleucine (Table 2–1). Thus, by the action of mould aminoacylase, L-isoleucine can easily be obtained from the acetyl derivatives of a mixture of the four stereoisomers obtained by chemical synthesis.[119] However, in this case, reuse of the isomers other than the L-isoleucine after racemization is very difficult when compared with the case of amino acids having only one asymmetric carbon.

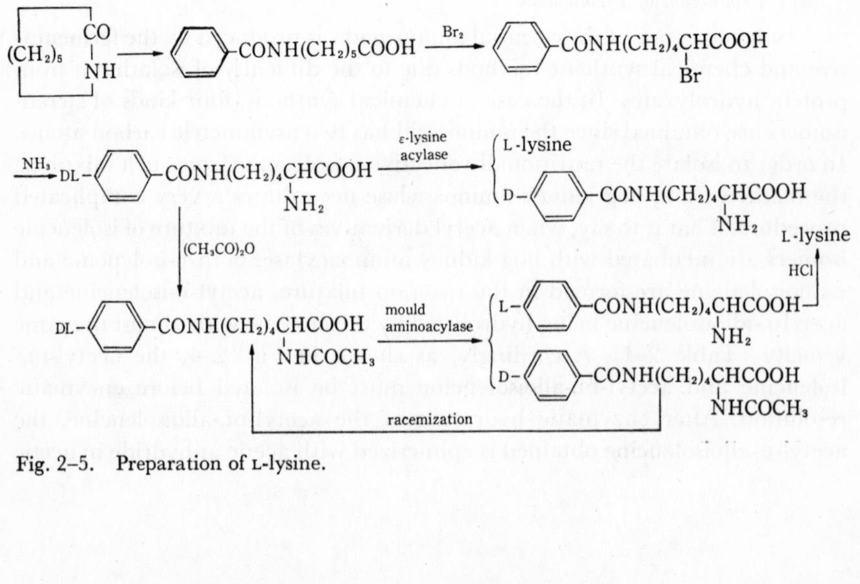

$(CH_2)_5$<CO/NH> ⟶ ⬡-$CONH(CH_2)_5COOH$ —Br_2→ ⬡-$CONH(CH_2)_4CH(Br)COOH$

—NH_3→ DL-⬡-$CONH(CH_2)_4CH(NH_2)COOH$ —ε-lysine acylase→ {L-lysine, D-⬡-$CONH(CH_2)_4CH(NH_2)COOH$}

↓ $(CH_3CO)_2O$

DL-⬡-$CONH(CH_2)_4CH(NHCOCH_3)COOH$ —mould aminoacylase→ {L-⬡-$CONH(CH_2)_4CH(NH_2)COOH$ —HCl→ L-lysine, D-⬡-$CONH(CH_2)_4CH(NHCOCH_3)COOH$}

D-⬡-$CONH(CH_2)_4CH(NHCOCH_3)COOH$ —racemization→ DL-⬡-$CONH(CH_2)_4CH(NHCOCH_3)COOH$

Fig. 2–5. Preparation of L-lysine.

(*c*) *Production of* L-*lysine*

For the preparation of L-lysine, the following method has been reported. ε-Benzoyl-DL-lysine is synthesized from ε-caprolactam, as shown in Fig. 2–5, and acetylated. The ε-benzoyl-α-acetyl-DL-lysine obtained is then asymmetrically hydrolyzed easily by mould aminoacylase. As the resulting ε-benzoyl-L-lysine is sparingly soluble, it crystallizes out during the reaction and is readily separated by filtration after completion of the reaction. Hydrolysis of the ε-benzoyl-L-lysine by hydrochloric acid gives L-lysine. After removal of the ε-benzoyl-L-lysine, the ε-benzoyl-α-acetyl-D-lysine in the mother liquor is racemized and reused for the enzymatic resolution step.[120,121)]

(*iii*) *Other aminoacylases*

In order to find other advantageous sources of microbial aminoacylase, extensive screening of yeasts,[122)] actinomycetes[123)] and bacteria[124)] has been carried out. Kameda and his co-workers studied the aminoacylase activity of soil bacteria. They found that bacteria[107)] producing L-aminoacylase acted on the L-isomer but that bacteria producing D-aminoacylase acted only on the D-isomer, and applied these enzymes for the optical resolution of DL-amino acids.[125,126)] Further, aminoacylase was found in certain glutamic acid-producing bacteria.[127)] Bacteria are thus expected to be a useful economic source of aminoacylase. However, moulds can be produced by solid wheat-bran culture as well as by liquid culture, and crude aminoacylase preparations having high enzyme activity are obtained by water extraction of the culture. This method has therefore been widely employed in industrial applications. Aminoacylase is also distributed in plants,[128)] but this may not be a suitable source for application in practice.

For the enzymatic resolution of lysine, a simpler method than that mentioned above using mould aminoacylase has been reported; namely, ε-*N*-acyl-L-lysine amidohydrolase (acyl-lysine deacylase) is employed to hydrolyze ε-acyl-L-lysine specifically. As shown above in Fig. 2–5, when ε-benzoyl-DL-lysine synthesized from ε-caprolactam is incubated with this enzyme, only the ε-benzoyl-L-lysine is hydrolyzed and converted to L-lysine. The unhydrolyzed ε-benzoyl-D-lysine is racemized, and then reused for the resolution step. Although this method has the minor disadvantage that the ε-benzoyl-DL-lysine used as substrate is less soluble in water and the rate of the enzyme reaction is relatively slow, it does have the advantage that both the processes of α-*N*-acetylation and chemical ε-*N*-deacylation can be omitted.

The occurrence of this ε-lysine acylase has been reported in rat kidney,[129)] *Pseudomonas*[130,131)] and *Achromobacter pestifer*. Among these enzymes, that of *Achromobacter pestifer* shows the highest activity.[132,133] The enzyme acts only on *N*-acyl-L-lysines and not on the other lysine derivatives, δ-*N*-acyl-ornithine and ζ-*N*-acyl-homolysine. It has been reported that the enzyme reaction rate is affected by the electronegativity and bulkiness of the acyl group.[134]

(*iv*) *Immobilized aminoacylase*

Generally, the above enzyme reactions are carried out by batch processes, incubating the substrates with soluble enzymes. Accordingly, in order to isolate L-amino acid from the reaction mixture, it is necessary as a preliminary step to remove the enzyme-protein and contaminant proteins by pH and/or heat treatment, even if the enzyme activity remains in the reaction mixture, since no efficient method to recover the enzyme has yet been developed. Taking economy of enzyme use into consideration, this procedure is therefore very disadvantageous. Furthermore, the removal process of unwanted protein and coloring materials lowers the yield of L-amino acids.

If active and stable immobilized enzyme preparations can be prepared, many of the disadvantages inherent in the conventional enzyme procedures may be overcome. Industrially advantageous application of continuous enzyme reactions also becomes possible when columns packed with immobilized enzyme are used. Based on this approach, studies on continuous enzyme reactions employing immobilized aminoacylase have been carried out, and this method is now utilized industrially for the optical resolution of certain DL-amino acids.

Techniques for the preparation of immobilized enzymes can be conveniently classified into the following three categories: (1) the binding or adsorption of enzymes to a water-insoluble carrier, (2) the cross-linking of enzymes with bifunctional reagents, (3) the entrapping of enzymes into a semi-permeable gel lattice. Of these methods, immobilization was found to be easily carried out in the case of aminoacylase by ionic binding of the native enzyme to basic ion-exchange derivatives of polysaccharides.[135] Concerning the properties of such immobilized aminoacylase, no marked differences were observed between the two enzyme preparations as regards their substrate specificity.[136] However, it was found that stability to heat treatment, proteases and protein denaturants was markedly enhanced by the conversion to the immobilized form.[136] Owing to this stability, the immobilized enzyme could be used for long period operation without appreciable loss of activity.[137,138] By employing immobilized enzyme, continuous production of L-amino acid is feasible by employing the flow system shown in Fig. 2–6.

When acetyl-DL-amino acid is passed through a column packed with

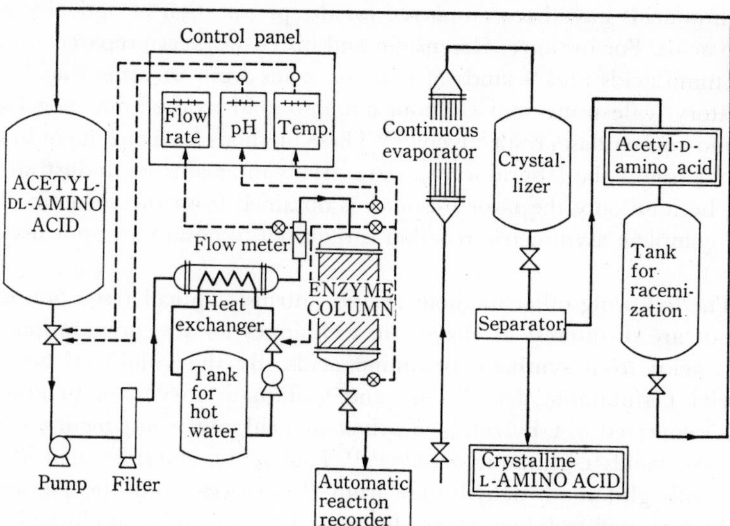

Fig. 2–6. Flow diagram for continuous production of L-amino acid by immobilized aminoacylase.

DEAE–Sephadex–aminoacylase complex at a specified temperature (usually 50°C) and flow rate (varied according to the kind of amino acid), asymmetric hydrolysis proceeds within the column. In contrast to the batch process using native enzyme, enzyme-proteins and coloring materials do not contaminate the column effluent, and crystalline L-amino acid can readily be obtained by continuous evaporation of the effluent without a complicated purification step. The acetyl-D-amino acid in the mother liquor is racemized and used for the resolution process.[137] The enzyme column is very stable, and more than 60% of the initial activity is retained after a long period of operation of 32 days. Furthermore, deteriorated enzyme columns can easily be regenerated by the addition of native aminoacylase in an amount corresponding to the deterioration in activity.[138]

Thus, by the continuous enzymatic resolution method using immobilized aminoacylase, larger amounts of L-amino acid can be produced per unit enzyme when compared with the conventional batch method using soluble enzyme. Further, this method can easily be operated automatically, and is so very advantageous for the industrial production of L-amino acids.

2.3.3 Resolution by other enzymes

Besides the methods of enzymatic asymmetric synthesis and hydrolysis mentioned above, enzymatic asymmetric oxidation and decarboxylation of

DL-amino acids have been employed for the preparation of optically active amino acids. For instance, Greenstein and his co-workers prepared 13 kinds of D-amino acids and 9 kinds of L-amino acids from DL-amino acids on a laboratory scale using snake venom L-amino acid oxidase and hog kidney D-amino acid oxidase, respectively.[139] These methods are convenient in some cases for use in the laboratory, but are disadvantageous for industrial purposes, because only the L- or D-isomer is obtained from the DL-amino acids and a complete asymmetric reaction is required to obtain an optically pure product.

The following other methods are not intrinsic optical resolution methods, but are of interest as enzymatic techniques for the preparation of L-amino acids from synthetic DL-amino acids. By the combined action of bacterial L-glutamate dehydratase and glutamate racemase, DL-glutamic acid is converted to L-pyrrolidonecarboxylic acid, and subsequent chemical hydrolysis yields L-glutamic acid.[140,141] That is, DL-glutamic acid is incubated with glutamate dehydratase from *Pseudomonas*, and the L-isomer is converted to L-pyrrolidonecarboxylic acid. The remaining D-glutamic acid is converted to DL-glutamic acid by the action of glutamate racemase from *Lactobacillus fermenti*, and this DL-glutamic acid is then converted to L-pyrrolidonecarboxylic acid by the action of L-glutamate dehydratase. By repetition of the procedure, all of the DL-glutamic acid may be converted to L-pyrrolidonecarboxylic acid, which gives L-glutamic acid on chemical hydrolysis.

The enzymatic isomerization of DL-amino acids or D-amino acids to L-amino acids is also an interesting method. D-Amino acid is converted to the α-keto acid by D-amino acid oxidase, and further enzymatic transamination or reductive amination of the α-keto acid then gives L-amino acid. For example, DL-phenylalanine[142,143] and DL-tryptophan[144,145] have been converted to the respective L-isomers by using microorganisms belonging to the genera *Pseudomonas* and *Flavobacterium*, respectively. At present, application of this isomerization technique is limited to certain specific amino acids. However, the method may well be developed as an advantageous method for the production of optically active amino acids in the future.

2.4 Biological methods

The biological method utilizes the fact that, generally, only the L-amino acids of DL-amino acids can be metabolized by living cells. Initially, Pasteur reported the resolution of DL-tartaric acid by employing living cells of various yeasts and moulds. In the case of amino acids, the following methods have been reported: (1) microorganisms are made to consume L-amino acid by cultivation in a medium containing DL-amino acid, and (2) DL-amino acid is administered to animals and excreted isomer recovered from the urine. In these biological methods, only D-isomer is obtained, so the techniques are not suitable for industrial purposes. Nevertheless, they do have some significance for physiological studies of animal metabolism.

2.5 Racemization

L-Amino acids can be derived from synthetic DL-amino acids by utilizing the various optical resolution methods described above. As a consequence, equivalent amounts of D-amino acids are produced at the same time as the L-amino acids. From an economic viewpoint, it is wasteful to discard these D-amino acids. That is to say, the synthetic DL-amino acids should be completely converted to L-amino acids by racemization of the D-amino acids to DL-amino acids and reuse of the latter for optical resolution. Accordingly, for the efficient industrial production of optically active amino acids, the three steps of DL-amino acid synthesis, optical resolution and racemization must be considered in close conjunction with one another.

It is widely recognized that free amino acids are difficult to racemize, but racemization does occur easily by heating amino acids in a concentrated alkaline solution. It is also known that racemization occurs when the amino acids are converted to their *N*-acyl derivatives, hydantoins, diketopiperazines or peptides. In this chapter, however, since the emphasis is on the synthetic production of optically active amino acids, more importance is attached to racemization procedures for converting the undesired opposite antipodes obtained by optical resolution into racemic modifications which can be reused for the optical resolution step.

2.5.1 Racemization by acids and alkalis

It has long been known that amino acids are racemized by heating in the presence of acids or alkalis. For instance, it has been reported that aspartic acid[146)] is racemized by heating at 170–80°C with acids such as hydrochloric acid, and that phenylalanine[147)] and leucine[148)] are completely racemized by heating with barium oxide at 160°C. Following these reports, many investigations were carried out on this kind of racemization, and almost all amino acids have been racemized by this method, although the yields were variable. It has also been reported that racemization occurs by heating with organic acids, such as acetic acid and formic acid, instead of mineral acids. For instance, asparagine, phenylalanine, tryptophan, methionine, etc. have all been racemized in this way.[149)]

Some degree of decomposition of the amino acids often occurs with this process. Therefore, the method is not advantageous as a general racemization procedure, although it can be applied practically to those amino acids which are not readily decomposed under the required conditions. Also, it may be attempted after other racemization methods have proved ineffective.

2.5.2 Racemization by acetic anhydride

Most amino acids which have been converted to acyl forms such as *N*-acetyl or *N*-benzoyl derivatives are completely racemized by heating at 100°C with acetic anhydride in acetic acid solution.[150] The acylamino acids are also racemized by treatment with excess acetic anhydride in aqueous solutions at room temperature, and even at 0°C in slightly acidic solutions in the presence of a base such as sodium acetate or pyridine.[151,152] The phenomenon of racemization in this case has been explained as follows. The acylamino acids are first converted to saturated azlactones and then, by a tautomeric change to the enol form, racemization occurs.

This racemization method is considered practical and convenient since it can be applied to most amino acids, and is useful for the racemization of D-acylamino acids obtained by the asymmetric hydrolysis of DL-acylamino acids with aminoacylase.

2.5.3 Racemization by catalysts

The aim of this method is to racemize amino acids under the same moderate conditions that accompany the action of amino acid racemase, and the principle of the method is based on the fact that in the reaction mechanism of amino acid racemase, pyridoxal phosphate acts as coenzyme. Taking this racemase as a model case, when amino acids were treated with pyridoxal in the presence of a metal such as copper, iron or aluminium, it was found that nonenzymatic racemization occurred.[153,154] As pyridoxal analogues, salcylaldehyde and *p*- or *o*-nitrosalcylaldehyde have also been found to accelerate racemization in the presence of a metallic ion, though the activity is low. However, on economic grounds, industrial application of these reactions is not yet feasible, since the catalysts involved are expensive and recovery of the soluble catalyst from the reaction mixture is difficult. Immobilization of of these catalysts, i.e. resin catalysts, for racemization has therefore been attempted.[155,156] For example, the CH_3 group of *o*-cresol formaldehyde resin was oxidized to –CHO group, and the resulting resin, which has a partially analogous structure to salcylaldehyde in that it possesses –OH and –CHO groups in the *ortho* position, showed racemization activity in the presence of metallic ions. Continuous racemization was accomplished by passing the amino acid solution containing metallic ions at 80°C through a column packed with this resin catalyst.

When an α-amino acid is heated with the corresponding keto acid at 50–100°C in the presence of metallic ions, a transamination reaction occurs and the α-amino acid newly produced is in the DL-form, i.e. a kind of racemization occurs.[157]

With the further development of studies on racemization catalysts, the above-mentioned racemization procedure under mild conditions is expected to constitute a practical method in the future.

2.5.4 Racemization by heating in water

This technique is more practically advantageous than any of the other related methods described above, since the procedure is very simple and requires neither conversion of the amino acids to their derivatives nor the use of special catalysts. Although the mechanism of racemization is unknown, many optically active amino acids or their salts can be racemized in aqueous solutions by heating in a sealed vessel at 150–250°C without a catalyst.[158,159] Generally speaking, the rate of reaction is faster in neutral solutions than in acidic or basic solutions, although in the case of some amino acids, the presence of an acid or base is preferred. However, the amount of base or acid required in such cases is not as great as that used in the acid/alkali racemization techniques described in section 2.5.1, so the two methods are clearly distinguishable. Racemization of lysine and glutamic acid has been carried out by this method, and it plays an important role in the industrial production of optically active amino acids by "chemical synthesis–direct resolution–racemization" processes.

2.5.5 Racemization by enzymes

Although 10 or more kinds of amino acid racemases have been found since the initial discovery of alanine racemase,[160] it still seems an impractical approach to undertake only the racemization step by an enzymatic reaction. Some promise can therefore be attached to the method of converting the whole DL-form into optically active form by coupling the racemase reaction with other enzymatic reactions. In this way, one optical isomer of the DL-amino acid would be converted to a derivative, while the other antipode would be converted simultaneously into DL-form by racemase. For instance, whole DL-glutamic acid can be completely converted to L-glutamic acid by the combined action of L-glutamic acid dehydrogenase and glutamic acid racemase.[140,141]

2.6 Conclusion

From the standpoint of producing optically active amino acids, fermentation has the merit that it yields natural L-amino acids. However, in the fermentative process, the raw materials available and kinds of amino acids produced are limited. Furthermore, the production of D-amino acids or unnatural amino acids is very difficult by fermentation. On the other hand, in chemical synthetic processes, the DL-form is generally produced and optical resolution is necessary in order to obtain optically active amino acids. This constitutes one of the characteristics but chief demerits of the synthetic method. However, this method does have the advantage that large amounts of various raw materials are readily available and it is more readily adaptable to mass production. Consequently, the synthetic method (i.e. the optical resolution of synthetic products) is expected to develop not only for the production of L-amino acids as yet unobtainable by fermentative processes, and of D-amino acids and unnatural amino acids, but also for the production of some L-amino acids where fermentative production is already established, as in the case of glutamic acid. Besides this, certain procedures for the production of various other kinds of L-amino acids, such as threonine, tryptophan, lysine, phenylalanine, methionine, valine, isoleucine and serine, have been carried out on an industrial scale by utilizing the chemical synthesis–optical resolution process.

Among the various techniques for optical resolution, automatic and continuous operation of the physicochemical method and enzymatic method is considered the most useful for industrial application. Namely, as shown in the case of the industrial production of L-glutamic acid, continuously operated optical resolution by the preferential crystallization procedure is very advantageous, especially when the resolution process can be effectively combined with the synthetic and racemization processes by the development of suitable equipment and chemical engineering techniques. The enzymatic resolution method that utilizes the high optical specificity of enzymes can be operated continuously in the case of columns packed with immobilized aminoacylase. This continuous enzymatic resolution technique has in fact already been applied to the industrial production of various L-amino acids and is considered to be a very promising method.

For the economical production of large amounts of optically active amino acids, the synthetic method can thus be regarded as an essential technique. It is therefore hoped that the techniques for optical resolution and racemization will advance together with improved methods for the chemical synthesis of DL-amino acids. On the other hand, if an excellent practical method for the asymmetric synthesis of amino acids can be developed, the process of optical resolution will in such cases become unnecessary, and the synthetic method will so expand its utility and applicability for the production of optically active amino acids.

References

1. J. P. Greenstein and M. Winitz, *Chemistry of the Amino Acids*, vol. 1, Wiley, 1961.
2. R. M. Secor, *Chem. Rev.*, **63**, 297 (1963).
3. L. Pasteur, *Ann. Chim. Phys.*, **24**, 442 (1848).
4. A. Piutti, *Compt. Rend.*, **103**, 134 (1886).
5. B. B. Dowling, *U. S. Patent* No. 2898358 (1959).
6. International Minerals and Chemical Co., *British. Patent* No. 838924 (1960).
7. I. M. Le Baron and G. L. Samsel, *U. S. Patent* No. 2992728 (1961).
8. A. Werner, *Chem. Ber.*, **47**, 2171 (1914).
9. H. W. B. Roozeboom, *Z. Physik. Chem.*, **28**, 494 (1899).
10. M. Tsuboi and T. Takenishi, *Bull. Chem. Soc. Japan*, **32**, 726 (1959).
11. D. Gernez, *Compt. Rend.*, **63**, 843 (1866).
12. D. Gernez, *Ann. Chem.*, **143**, 376 (1867).
13. R. Duschinsky, *Chem. Ind.*, **53**, 10 (1934).
14. F. Kögl, H. Erxleben and G. J. van Veersen, *Z. Physiol. Chem.*, **277**, 260 (1943).
15. T. Akashi, *Nippon Kagaku Kaishi* (Japanese), **83**, 417 (1962).
16. L. Velluz and G. Amiard, *Bull. Soc. Chim. France*, **20**, 903 (1953).
17. S. Tatsumi, I. Sasaji and K. Ohno, *Japanese Patent* No. 37–18470 (1962).
18. M. Hara, T. Akashi and K. Ohno, *Japanese Patent* No. 39–1409 (1964).
19. K. Harada, *Nature,* **206**, 1354 (1965).
20. T. Haga, M. Sato and K. Miura, *Japanese Patent* No. 42–3290 (1967).
21. I. Sasaji, K. Ohno and J. Kato, *Japanese Patent* No. 38–6183 (1963).
22. M. Shibasaki and T. Fukuro, *Japanese Patent* No. 39–24440 (1964).
23. N. Sugimoto, I. Chibata, S. Yamada and M. Yamamoto, *Japanese Patent* No. 42–19449 (1967); *U.S. Patent* No. 3440279 (1969).
24. I. Chibata, S. Yamada and M. Yamamoto, *Netherlands Patent Appl.* No. 7003706 (1970).
25. I. Chibata, S. Yamada, M. Yamamoto and M. Wada, *Experientia*, **24**, 638 (1968).
26. I. Chibata, S. Yamada and M. Yamamoto, *Japanese Patent* No. 43–24410 (1968); *U.S. Patent* No. 3523969 (1970).
27. I. Chibata, S. Yamada and M. Yamamoto, *British Patent* No. 1218158 (1969).
28. I. Chibata, S. Yamada and M. Yamamoto, *Japanese Patent* No. 45–29963 (1970); *British Patent* No. 1191100 (1969).
29. T. Uzuki and N. Sato, *Japanese Patent* No. 45–852 (1970).
30. K. Ohno and S. Tatsumi, *Japanese Patent* No. 38–7972 (1963).
31. I. Sasaji, K. Ohno and J. Kato, *Japanese Patent* No. 38–17336 (1963).
32. M. Hara, T. Akashi and K. Ohno, *Japanese Patent* No. 39–1410 (1964).
33. T. Uzuki, N. Sato and T. Akashi, *Japanese Patent* No. 43–3085 (1968).
34. N. Sugimoto, I. Chibata, S. Yamada and M. Yamamoto, *Japanese Patent* No. 44–32562 (1969).
35. T. Ogawa, T. Akashi, T. Sato and A. Yamamoto, *Japanese Patent* No. 31–423 (1956).
36. K. Ito, T. Akashi and S. Tatsumi, *Japanese Patent* No. 36–17710 (1961).

37. K. Ito, N. Mizoguchi and M. Dazai, *Japanese Patent* No. 42–9322 (1967).
38. N. Mizoguchi, M. Hara, K. Ito, T. Akashi, K. Ohno and J. Kato, *Japanese Patent* No. 37–9971 (1961).
39. T. Watanabe, H. Kurokawa, T. Koga, Y. Kawauchi and G. Noyori, *Kogyo Kagaku Zasshi* (Japanese), **70**, 330 (1967).
40. H. Wakamatsu, *Food Eng.*, **40**, (11), 92 (1968).
41. J. L. Purvis, *Japanese Patent* No. 33–6760 (1958); *U.S. Patent* No. 2790001 (1957).
42. H. L. Fike, *U.S. Patent* No. 2937200 (1960).
43. S. Suzuki, H. Tatemichi, H. Watanabe and M. Ogasawara, *Japanese Patent* No. 41–5447 (1966).
44. B. D. Sarma and J. C. Bailar Jr., *J. Am. Chem. Soc.*, **78**, 895 (1956).
45. M. Shibata, Y. Fujita, M. Naito and K. Hori, *Bull. Chem. Soc. Japan*, **36**, 485 (1963).
46. K. Harada, *Nature*, **205**, 590 (1965).
47. H. Krebs and J. A. Wagner, *German Patent* No. 1013657 (1957).
48. H. Krebs and J. Diewald, *German Patent* No. 1013655 (1957).
49. D. L. Gramaise and J. Colucci, *U.S. Patent* No. 2957886 (1960).
50. H. Suda, *Kobunshi* (Japanese), **17**, 139 (1968); *Japanese Patent* No. 45–1829 (1970).
51. A. G. Inventa, *British Patent* No. 1007665 (1961).
52. C. W. Roberts and H. Haigh, *J. Org. Chem.*, **27**, 3375 (1962).
53. L. Pasteur, *Compt. Rend.*, **37**, 162 (1853).
54. E. Fischer, *Chem. Ber.*, **32**, 2451 (1899).
55. F. H. Radke, R. B. Fearing and S. W. Fox, *J. Am. Chem. Soc.*, **76**, 2801 (1954).
56. K. Kanao, S. Hori and J. Kato, *Japanese Patent* No. 32–5419 (1957).
57. J. Cuvillier, *German Patent* No. 1118794 (1961).
58. F. L. Pyman, *J. Chem. Soc.*, **99**, 1386 (1911).
59. R. D. Emmick, *U.S. Patent* No. 2556907 (1951).
60. A. O. Rogers, *U.S. Patent* No. 2657230 (1953).
61. M. C. Backer, *U.S. Patent* No. 2942024 (1960).
62. K. Togo, M. Hara and K. Ohno, *Japanese Patent* No. 40–11606 (1965).
63. M. Brenner and H. R. Rickenbacher, *Helv. Chim. Acta*, **41**, 181 (1958).
64. International Minerals and Chemical Co., *British Patent* No. 773653 (1957).
65. M. Fields and M. A. Stevens, *U.S. Patent* No. 2865928 (1958).
66. T. N. Coker, W. L. Kohlhase, M. Field, A. O. Rogers and M. A. Stevens, *J. Org. Chem.*, **27**, 850 (1962).
67. J. Kato and Y. Tsuchiya, *Agr. Biol. Chem.* (*Tokyo*), **26**, 467, 473 (1962).
68. H. D. Dewitt and A. W. Ingersoll, *J. Am. Chem. Soc.*, **73**, 5782 (1951).
69. I. Sollin, *U.S. Patent* No. 2945879 (1960).
70. K. Volger and P. Lanz, *Helv. Chim. Acta*, **49**, 1348 (1966).
71. K. Okawa, K. Hori, K. Hirose and Y. Nakagawa, *Nippon Kagaku Kaishi* (Japanese), **89**, 998 (1968).
72. L. Velluz, G. Amiard and R. Heymès, *Bull. Soc. Chim. France*, **20**, 904 (1953); **21**, 1015 (1954); **22**, 201 (1955).
73. T. Perlotto and M. Vignolo, *Farmaco* (*Pavia*) *Ed. Sci.*, **21**, (1), 30 (1966).
74. L. R. Overby and A. W. Ingersoll, *J. Am. Chem. Soc.*, **73**, 3363 (1951).
75. W. A. H. Huffman and A. W. Ingersoll, *ibid.*, **73**, 3366 (1951).
76. G. P. Wheeler and A. W. Ingersoll, *ibid.*, **73**, 4604 (1951).
77. L. R. Overby, *J. Org. Chem.*, **23**, 1393 (1958).

78. J. E. Nelemans, A. H. Peeasse, W. Pesch and U. Verstrigden, *U.S. Patent* No. 3105067 (1963).
79. Stamicarbon N. V., *Netherlands Patent Appl.* No. 300552 (1965); *Japanese Patent* No. 41–21222 (1966).
80. Y. Suzuki and N. Nagashima, *Bull. Chem. Soc. Japan*, **39**, 1701 (1966).
81. Y. Suzuki, T. Akashi and M. Nakamura, *Japanese Patent* No. 44–20983 (1969).
82. Australian National University and Monsanto Chemicals, *French Patent* No. 1360884 (1964).
83. D. F. Holmes and R. Adams, *J. Am. Chem. Soc.*, **56**, 2093 (1934).
84. S. Berlingozzi, G. Adembri and G. Bucci, *Gazz. Chim. Ital.*, **84**, 383 (1954).
85. S. Terashima, K. Achiwa and S. Yamada, *Chem. Pharm. Bull.* (*Tokyo*), **13**, 1399 (1965).
86. K. Okuda, I. Kuniyoshi and K. Saito, *Japanese Patent* No. 43–5727 (1968).
87. E. Shapiro and R. F. Newton, *J. Am. Chem. Soc.*, **65**, 777 (1943).
88. E. Gil-Av, R. Charles-Sigler, G. Fischer and D. Nurok, *J. Gas Chromatog.*, **4**, (2), 51 (1966).
89. H. H. Ottenheim and P. J. Jenneskens, *J. Agr. Food. Chem.*, **18**, 1010 (1970).
90. M. Bergmann and H. Frankel-Conrat, *J. Biol. Chem.*, **119**, 707 (1937).
91. D. G. Doherty and E. A. Popenoe Jr., *ibid.*, **189**, 447 (1951).
92. N. F. Albertson, *J. Am. Chem. Soc.*, **73**, 452 (1951).
93. S. Kuno, T. Murachi and S. Uchino, *Symp. Enzyme Chem. Japan*, **12**, 121 (1957).
94. O. Warburg, *Chem. Ber.*, **38**, 187 (1905).
95. M. Bergmann, J. S. Fruton and H. Pollok, *Science*, **85**, 410 (1937).
96. D. Hamer and J. P. Greenstein, *J. Biol. Chem.*, **193**, 81 (1951).
97. N. Izumiya, S-C. J. Fu, S. M. Birnbaum and J. P. Greenstein, *ibid.*, **205**, 221 (1953).
98. C. G. Baker and A. Meister, *J. Am. Chem. Soc.*, **73**, 1336 (1951).
99. E. Work, S. M. Birnbaum, M. Winitz and J. P. Greenstein, *ibid.*, **77**, 1916 (1955).
100. O. Schmiedeberg, *Arch. Exptl. Pathol. Pharmakol.*, **14**, 379 (1881).
101. I. A. Smorodinzev, *Z. Physiol. Chem.*, **124**, 123 (1922).
102. C. Neuberg and K. Linhardt, *Biochem. Z.*, **147**, 372 (1924).
103. V. E. Price and J. P. Greenstein, *J. Biol. Chem.*, **175**, 969 (1948).
104. C. Neuberg and I. Mandl, *Enzymologia*, **14**, 128 (1950).
105. K. Michi and H. Nonaka, *J. Agr. Chem. Soc. Japan*, **28**, 343 (1954).
106. S. Yamada, I. Chibata and S. Yamada, *J. Pharm. Soc. Japan*, **75**, 113 (1955).
107. Y. Kameda, E. Toyoura, H. Yamazoe, Y. Kimura and Y. Yasuda, *Nature*, **170**, 888 (1952).
108. J. P. Greenstein, L. Levintow, C. G. Baker and J. White, *J. Biol. Chem.*, **188**, 647 (1951).
109. S. M. Birnbaum, *Methods in Enzymology* (ed. S. P. Colowick, N. O. Kaplan), II, p. 115, Academic Press, 1955.
110. J. P. Greenstein, *ibid.*, III, p. 554, Academic Press, 1957.
111. W. S. Fones and M. Lee, *J. Biol. Chem.*, **210**, 227 (1954).
112. I. Chibata, T. Ishikawa and S. Yamada, *Bull. Agr. Chem. Soc. Japan*, **21**, 304 (1957).
113. S-C. J. Fu, S. M. Birnbaum and J. P. Greenstein, *J. Am. Chem. Soc.*, **76**, 6054 (1954).
114. W. S. Fones and M. Lee, *J. Biol. Chem.*, **201**, 847 (1953).

115. K. R. Rao, S. M. Birnbaum, R. B. Kingsley and J. P. Greenstein, *ibid.*, **198**, 507 (1952).
116. I. Chibata, T. Ishikawa and S. Yamada, *Bull. Agr. Chem. Soc. Japan*, **21**, 300 (1957).
117. I. Chibata and S. Yamada, *ibid.*, **21**, 58 (1957).
118. J. P. Greenstein, L. Levintow, C. G. Baker and J. White, *J. Biol. Chem.*, **188**, 647 (1951).
119. I. Chibata, S. Yamada and H. Itoh, *Yakugaku Kenkyu* (Japanese), **32**, 247 (1960); *Japanese Patent* No. 37–12261 (1962).
120. I. Chibata, S. Yamada and S. Yamada, *Bull. Agr. Chem. Soc. Japan*, **20**, 174 (1956).
121. I. Chibata, A. Watanabe and S. Yamada, *ibid.*, **21**, 296 (1957).
122. I. Chibata and T. Ishikawa, *ibid.*, **22**, 218 (1958).
123. S. Kudo, K. Kushiro, Y. Noguchi and Y. Yokote, *Japanese Patent* No. 43–24456 (1968).
124. I. Chibata, M. Kisumi and S. Yamada, *Bull. Agr. Chem. Soc. Japan*, **22**, 24 (1957).
125. Y. Kameda, E. Toyoura and K. Matsui, *J. Pharm. Soc. Japan*, **78**, 202 (1958).
126. Y. Kameda, E. Toyoura and Y. Kimura, *Nature*, **181**, 1225 (1958).
127. K. Mineura and M. Tanaka, *J. Agr. Chem. Soc. Japan*, **28**, 343 (1954).
128. I. Chibata and T. Tosa, *Bull. Agr. Chem. Soc. Japan*, **23**, 370 (1959).
129. W. K. Paik, L. Bloch-Frankenthal, S. M. Birnbaum, M. Winitz and J. P. Greenstein, *Arch. Biochem. Biophys.*, **69**, 56 (1957).
130. Y. Kameda, E. Toyoura, Y. Kimura and K. Matsui, *Chem. Pharm. Bull.* (*Tokyo*), **6**, 394 (1958).
131. S. Wada, *J. Biochem.*, **46**, 445 (1959).
132. I. Chibata, T. Ishikawa and T. Tosa, *Nature*, **195**, 80 (1962).
133. I. Chibata, T. Ishikawa and T. Tosa, *Methods in Enzymology* (ed. G. E. Perlmann, L. Lorand), XIX, p. 756, Academic Press, 1970.
134. I. Chibata, T. Tosa and T. Ishikawa, *Arch. Biochem. Biophys.*, **104**, 231 (1964).
135. T. Tosa, T. Mori, N. Fuse and I. Chibata, *Enzymologia*, **31**, 214 (1966).
136. T. Tosa, T. Mori and I. Chibata, *Agr. Biol. Chem.*, **33**, 1053 (1969).
137. T. Tosa, T. Mori, N. Fuse and I. Chibata, *Biotech. Bioeng.*, **9**, 603 (1967).
138. T. Tosa, T. Mori, N. Fuse and I. Chibata, *Agr. Biol. Chem.*, **33**, 1047 (1969).
139. J. R. Parikh, J. P. Greenstein, M. Winitz and S. M. Birnbaum, *J. Am. Chem. Soc.*, **80**, 953 (1958).
140. M. Tanaka, Y. Nagano and S. Kinoshita, *J. Agr. Chem. Soc. Japan*, **34**, 737 (1960).
141. M. Tanaka, Y. Kato and S. Kinoshita, *ibid.*, **36**, 237 (1962).
142. I. Chibata, T. Tosa and R. Sano, *Appl. Microbiol.*, **13**, 618 (1965).
143. I. Chibata, T. Tosa and R. Sano, *Amino Acid and Nucleic Acid* (Japanese), **11**, 163 (1965).
144. J. R. Martin and N. N. Durham, *Biochem. Biophys. Res. Comm.*, **14**, 388 (1964).
145. J. R. Martin and N. N. Durham, *Can. J. Microbiol.*, **12**, 1269 (1966).
146. A. Michael and J. F. Wing, *Chem. Ber.*, **17**, 2984 (1884); *Am. Chem. J.*, **7**, 278 (1885).
147. E. Schulze, *Z. Physiol. Chem.*, **9**, 63 (1885).
148. E. Schulze and E. Bosshard, *ibid.*, **10**, 134 (1886).
149. I. Sakieki and M. Mitsuno, *J. Chem. Soc. Japan*, **80**, 1035 (1959).

150. M. Bergmann and L. Zervas, *Biochem. Z.*, **203**, 280 (1929).
151. V. du Vigneaud and R. R. Sealock, *J. Biol. Chem.*, **96**, 511 (1932).
152. V. du Vigneaud and C. E. Meyer, *ibid.*, **98**, 295 (1932).
153. J. Olivard, D. E. Metzler and E. E. Snell, *ibid.*, **199**, 669 (1952).
154. D. E. Metzler, M. Ikawa and E. E. Snell, *J. Am. Chem. Soc.*, **76**, 648 (1954).
155. K. Toi, T. Izumi and S. Akabori, *Bull. Chem. Soc. Japan*, **35**, 1422 (1962).
156. K. Toi, *ibid.*, **36**, 739 (1963).
157. H. Ogasawara, H. Tatemichi, H. Ito and S. Suzuki, *Japanese Patent* No. 42–13446 (1967).
158. I. Sasaji, M. Hara, S. Tatsumi, K. Seki, T. Akashi and K. Ohno, *U.S. Patent* No. 3213106 (1965).
159. K. Kageyama, H. Okamoto, M. Iida and K. Ohno, *U.S. Patent* No. 3185703 (1965).
160. W. A. Wood and I. C. Gunsalus, *J. Biol. Chem.*, **190**, 403 (1951).

CHAPTER 3

Asymmetric Synthesis

Y. Izumi*

In general, synthetic methods for the production of amino acids are technically well-developed, and further expansion of the industry is more dependent on the opening-up of new end-uses for amino acids than on technical process development. However, methods for asymmetric synthesis based on the chemistry of natural products still remain to be explored. Although many reports have been published on asymmetric synthesis, less than 100 are of industrial significance.

The nature and scope of asymmetric synthesis will be discussed in the following sections in relation to industrial requirements.[1)]

3.1 Classification and Definition of Asymmetric Synthesis

Asymmetric synthetic reactions have been newly classified by us[1)] on the basis of the reaction process by which the new chiral center is introduced. Thus, two classifications are introduced at the initial stage of the reaction, namely diastereoselective and enantioselective syntheses.

* Yoshiharu Izumi, Institute for Protein Research, Osaka University, Suita-shi, Osaka 565, Japan

3.1.1 Diastereoselective synthesis

A molecule which consists of more than two carbon atoms and which contains one asymmetric center may have a geometric or configurational plane such that the two sides of the plane are sterically unsymmetrical. The two sides are then said to be in a diastereotopic relationship, and if the plane contains a prochiral center, a pair of diastereomers is produced by the attack of a reagent on the prochiral center. Since the diastereotopic sides are not symmetrical, the reagent will attack the two sides unequally, producing the diastereomers in different proportions irrespective of the structure of the reagent. Reactions proceeding in this way are called diastereoselective.

Examples of configurational and conformational diastereoselective syntheses are shown in Fig. 3–1 and 3–2, respectively. Fig. 3–1 shows a diketopiperazine derivative (I) with the ethyl and methyl groups below and above the plane (O) respectively. Since catalytic hydrogenation proceeds preferentially from the less hindered methyl side, the reaction shown will predominate.

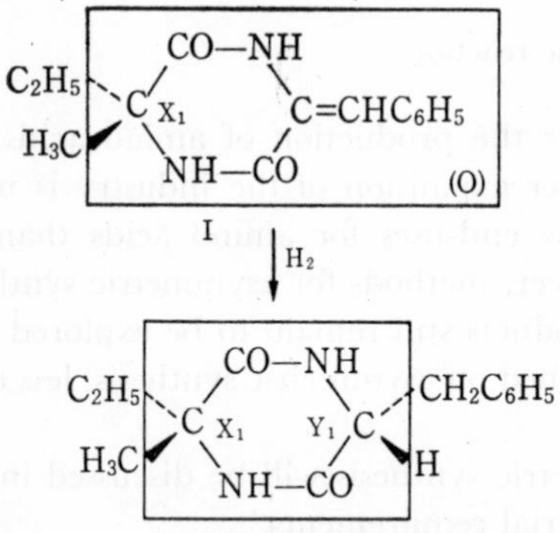

Fig. 3–1. Configurational diastereoselective synthesis.

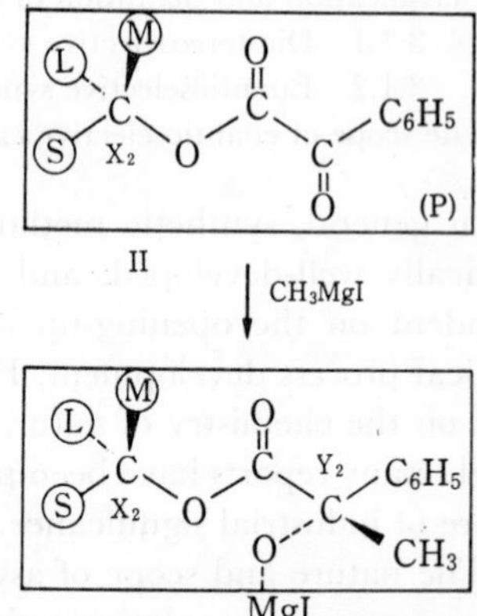

Fig. 3–2. Conformational diastereoselective synthesis.

Fig. 3–2 shows α-ketoglutarate (II). It is assumed that the conformation in which the two anticoplanar carbonyl groups, the central carbon atom, the smallest substituent Ⓢ and the oxygen atom of the chiral group lie in one plane (P) is preferred, since it is the least sterically hindered. Thus, in reaction with Grignard reagents, attack occurs preferentially from the side with the medium-size substituent Ⓜ, so that the product with the methyl group on the opposite side to the largest substituent Ⓛ will be produced more readily.

This type of synthesis produces diastereoisomers in proportions that depend on the diastereoselectivity of the reagents or catalyst. Almost all reagents and catalysts possess such selectivity to some extent. Enantiomers are not produced.

Inductive asymmetric syntheses are special cases, most of which are designed finally to obtain enantiomers through diastereoselective syntheses. If optically active substrates are used, the final reaction products will contain one enantiomer preferentially. As regards the preferential formation of one enantiomer, enantioselective and diastereoselective syntheses seem to be similar types of reactions, and confusion has often occurred in the literature. However, diastereoselective syntheses are substantially different from enantioselective syntheses.

In diastereoselective synthesis, at least one chiral center is required for the formation of a new chiral center, and several moles of optically active compound may be necessary for the formation of one mole of the new chiral center. This is therefore not an economically useful method.

3.1.2 Enantioselective synthesis

Compounds which are two-dimensionally non-disymmetric, as shown in Fig. 3–3, yield *R* or *S* enantiomers, depending on which side of the molecular plane (Q) the reagent attacks. The two sides of this plane are said to be in an enantiotopic relationship.

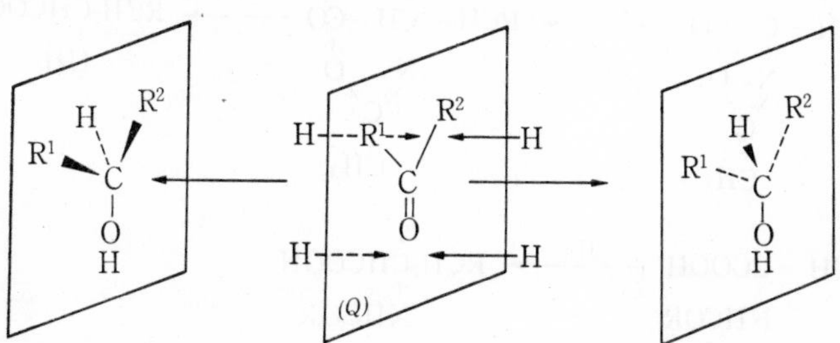

Fig. 3–3. Enantioselective synthesis.

Normal reagents attack on both sides of the plane in equal proportions, but enantioselective reagents or catalysts, which can differentiate between the two sides, produces a preponderance of either the *R* or *S* enantiomer. This type of reaction is referred to as enantioselective synthesis.

This category includes most enzymatic reactions, hydrogenation with asymmetrically modified Raney nickel catalysts[1)] and the use of chiral organometallic catalysts. In this type of synthesis, one chiral center in the catalyst can lead to the formation of thousands of molecules of optically active product, so catalytic reactions of this type are economically and industrially attractive.

3.2 THE SCOPE OF ENANTIOSELECTIVE CATALYTIC REACTIONS

Enantioselective catalytic reactions appear to be industrially desirable for the production of amino acids, but some restrictions on the structure of substrates and some special reaction conditions are necessary. For instance,

$$\text{① RCHO} \xrightarrow{\text{HCN}} \text{R}\overset{*}{\text{C}}\text{H(OH)CN} \xrightarrow{\text{NH}_3} \text{RCH(NH}_2\text{)CN} \longrightarrow \text{RCH(NH}_2\text{)COOH}$$

$$\text{② RCHO} + \text{R}'\text{CONH}_2 + \text{CO} \longrightarrow \text{R}\overset{*}{\text{C}}\text{H(NHCOR}'\text{)COOH}$$

$$\text{③ RC(=NR}'\text{)COOH} \xrightarrow{\text{H}_2} \text{R}\overset{*}{\text{C}}\text{H(NH}_2\text{)COOH}$$

$$\text{④ RCH=}\underset{\text{N=C(CH}_3\text{)–O}}{\text{C–CO}} \xrightarrow{\text{H}_2} \text{RCH}_2\text{–}\underset{\text{N=C(CH}_3\text{)–O}}{\overset{*}{\text{C}}\text{H–CO}} \longrightarrow \text{RCH}_2\text{CH(NH}_2\text{)COOH}$$

$$\text{⑤ RCH=C(NHCOR}'\text{)COOH} \xrightarrow{\text{H}_2} \text{RCH}_2\overset{*}{\text{C}}\text{H(NHCOR}'\text{)COOH}$$

$$\text{⑥ RC(N(CO)}_2\text{R}'\text{)=C=O} \xrightarrow{\text{R}''\text{OH}} \text{R}\overset{*}{\text{C}}\text{H(N(CO)}_2\text{R}'\text{)COOR}'' \longrightarrow \text{RCH(NH}_2\text{)COOH}$$

$$\text{⑦ HOOCCH=CHCOOH} \xrightarrow{\text{NH}_3} \text{HOOCCH}_2\overset{*}{\text{C}}\text{H(NH}_2\text{)COOH}$$

⑧ RCHO $\xrightarrow{HCN}$ $R\overset{*}{C}HCN$ (OH) $\xrightarrow{(NH_4)_2CO_3}$ R—CH—CO (N, NH, C, H ring) ⟶ RCHCOOH (NH_2)

⑨ RCH = C—CO (N, NH, C, H ring) $\xrightarrow{H_2}$ $RCH_2\overset{*}{C}H$—CO (N, NH, C, H ring) ⟶ $RCH_2CHCOOH$ (NH_2)

the carbon atom which will become the α-carbon of the amino acid should be a constituent of a double bond in the starting material. The following synthetic methods for amino acids, ① through ⑨, satisfy the necessary conditions for asymmetric catalytic synthesis.

The first enantioselective catalytic synthesis (method ①) was performed by Bredig,[2] yielding cyanhydrin from benzaldehyde and hydrogen cyanide in the presence of alkaloids. Albers[3,4] and Prelog[5] proposed that the hydrogen cyanide–quinine salt (B) enantioselectively bound to the aldehyde and that another molecule of hydrogen cyanide–quinine reacts with the resulting complex.

$$B\langle^{H^+}_{CN^-} + \overset{O}{H}\!\!>\!CC_6H_5 \longrightarrow B\langle^{H—O}_{CN—CHC_6H_5}$$

Tsuboyama[6-8] obtained benzcyanhydrin in an optical yield of 19.6% by a catalytic reaction employing polymers containing chiral centers, and studied the relationship between the structure of the polymer and the enantioselectivity. However, enantioselective synthesis of amino acids by the Strecker reaction has not yet been attempted.

The research group of the present author has enantioselectively synthesized glutamic acid and phenylalanine by the hydrogenation of ethyl oximinoglutarate (method ③), benzylidenazlactone (method ④) and α-acetamidocinnamic acid (method ⑤) using a silk-palladium catalyst, obtaining optical yields of 6.1%, 35.6% and 26.3%, respectively.[9] Alanine was also obtained in 0.5% optical yield by the hydrogenation of ethyl acetoximinopyruvate with a histidine-modified Raney nickel catalyst (method ③).[10]

Knowles and co-workers[11] at Monsanto Co. have succeeded in the enantioselective synthesis of phenylalanine derivatives from α-acylaminocinnamic acids using rhodium chiral phosphine complexes known as Wil-

$$RhCl_3 \cdot (CH_3O\text{-}C_6H_4\text{-}\overset{\displaystyle C_6H_{11}}{\overset{|}{P}}\text{-}Me)_3$$

III

TABLE 3–1
Asymmetric synthesis of Dopa with Wilkinson's complexes (RhL_3CL)

Chiral ligand, L $R^1\text{-}P(R^2)\text{-}R^3$				Substrate $R^4CH{=}C(NHCOR^5)\text{-}COOH$		Product $R^4CH_2CH(NHCOR^5)COOH$
R^1	R^2	R^3	Optical purity (%)	R^4	R^5	Optical purity (%)
o-Anisyl	Me	Ph	95	3-MeO-4-(OH)C_6H_3	Ph	58
Me	Ph	Pr^n	90	3-MeO-4-(OH)C_6H_3	Ph	28
Me	Ph	Pr^i	90	3-MeO-4-(OH)C_6H_3	Ph	28
m-Anisyl	Me	Ph	80	3-MeO-4-(OH)C_6H_3	Ph	1
o-Anisyl	Cyclohexyl	Me	95	3-MeO-4-(OH)C_6H_3	Ph	87
o-Anisyl	Cyclohexyl	Me	95	3-MeO-4-(OH)C_6H_3	Ph	90
Cyclohexyl	Me	Ph	75	3-MeO-4-(OH)C_6H_3	Ph	32
o-Anisyl	Ph	Pr^i	80	3-MeO-4-(OH)C_6H_3	Ph	1
o-Anisyl	Me	Ph	95	3-MeO-4-(AcO)C_6H_3	Me	55
o-Anisyl	Me	Pr^n	95	3-MeO-4-(AcO)C_6H_3	Me	20
o-Anisyl	Cyclohexyl	Me	90	3-MeO-4-(AcO)C_6H_3	Me	77
o-Anisyl	Cyclohexyl	Me	95	3-MeO-4-(AcO)C_6H_3	Me	85
o-Anisyl	Cyclohexyl	Me	95	3-MeO-4-(AcO)C_6H_3	Me	88
o-Anisyl	Cyclohexyl	Me	95	Ph	Me	85
o-Anisyl	Cyclohexyl	Me	95	Ph	Ph	58
o-Anisyl	Cyclohexyl	Me	95	H	Me	60

Anisyl = $CH_3OC_6H_4CH_2$–
Me = CH_3–
MeO = CH_3O–
Ph = C_6H_5–
Cyclohexyl = C_6H_{11}–
AcO = CH_3COO–
Pr^n = $CH_3CH_2CH_2$–
Pr^i = $(CH_3)_2CH$–

kinson's complexes (III) (method ⑤). The results are shown in Table 3–1. These catalysts are capable of the economical industrial synthesis of amino acids, but this method is limited to the synthesis of aromatic amino acids at present.

The synthesis of amino acids via ketenes by enantioselective catalysis using optically active amines such as brucine (method ⑥) was investigated by Pracejus.[12)] Though it was not necessary to isolate the ketenes, the optical yield was not sufficient for practical use. Isoda[13)] carried out enantioselective reductive amination by using Raney nickel treated with an optically active amino acid and obtained glutamic acid from α-ketoglutarate in 2.5–7% optical yield (method ③). Reductive amination could become of major importance if a more effective enantioselective catalyst and synthetic process for α-keto acids could be developed.

Recently, Wakamatsu[14] developed a catalyst for the synthesis of acylamino acids from aldehydes, acylamides and carbon monoxide (method ②). Although this has not yet been applied practically, it appears to be a potentially valuable process, particularly since acylamino acids may be nutritionally superior to amino acids.

The development of enantioselective synthesis of aspartic acid by the addition of ammonia to maleic or fumaric acid (method ⑦) is not industrially useful at present, since L-aspartic acid is already produced industrially from fumaric acid and ammonia by enzymatic methods.

Methods ⑧ and ⑨ above are not of interest, because racemization of the amino acid will occur during the alkaline hydrolysis of the hydantoin derivatives.

Thus, although enantioselective synthesis is scientifically and technically interesting, it is economically unsatisfactory since the number of intermediates suitable for enantioselective reactions is rather limited. Also, even with very efficient enantioselective syntheses, optical purification will be required, together with racemization and optical resolution of the by-produced antipode, on economic grounds to maximize the yield of the desired enantiomer, unless the optical yield of synthesis is nearly 100%. Unless this resolution step can be avoided, utilization of enantioselective synthesis in the amino acid industry will provide no economic advantages over other synthetic methods.

Many amino acids can easily be optically resolved by preferential crystallization or by enzymatic action, and are easily racemized by heating an aqueous solution of the amino acid or a fatty acid solution of an acylamino acid. Thus, by a combination of optical resolution and racemization of the

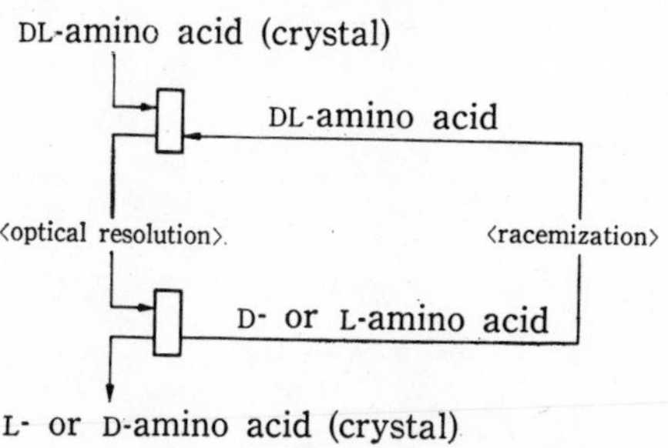

Fig. 3–4. Indirect asymmetric synthesis of amino acids.

unwanted enantiomer, as shown in Fig. 3–4, optically active amino acid can be obtained continuously from the racemic compound in high yield. This may be referred to as indirect asymmetric synthesis, and may be used in conjunction with any economically acceptable method of synthesis of amino acids.

Although enantioselective synthesis may be of great utility in the synthesis of diasymmetric amino acids with a β-asymmetric center such as isoleucine or threonine, which are hard to racemize, other methods such as preferential crystallization or the development of racemization catalysts and particle-bound "captive" enzymes for optical resolution, have much greater economic and industrial value for the synthesis of optically active amino acids.

References

1. Y. Izumi, *Angew. Chem. Intern. Ed. Engl.*, **10**, 871 (1971).
2. G. Bredig, *Zurich Chem. Ztg.*, **35**, 324 (1912).
3. H. Albers and E. Albers, *Z. Naturforsch.*, **9b**, 122 (1954).
4. H. Albers and E. Albers, *ibid.*, **9b**, 133 (1954).
5. V. Prelog and M. Wilhelm, *Helv. Chim. Acta*, **37**, 1634 (1954).
6. S. Tsuboyama, *Bull. Chem. Soc. Japan*, **35**, 1004 (1962).
7. S. Tsuboyama, *ibid.*, **38**, 354 (1965).
8. S. Tsuboyama, *ibid.*, **39**, 698 (1966).
9. S. Akabori, Y. Izumi, Y. Fujii and S. Sakurai, *Nature*, **178**, 323 (1956).
10. Y. Izumi, H. Takizawa, K. Nakagawa, R. Imamura, M. Imaida, T. Ninomiya and S. Yajima, *Bull. Chem. Soc. Japan*, **43**, 1792 (1970).
11. W. S. Knowles and M. J. Sabacky, *Chem. Comm.*, **1968**, 1445.
12. G. Pracejus, *Ann. Chem.*, **622**, 10 (1959).
13. T. Isoda, A. Ichikawa and T. Shimamoto, *Riken Hokoku* (Japanese), **34**, 134 (1958).
14. H. Wakamatsu, J. Uda and N. Yamakami, *Chem. Comm.*, **1971**, 1540.

CHAPTER **4**

Synthetic Methods for Individual Amino Acids

T. Shirai*[1] et al.

*1 Tadashi SHIRAI, Pharmaceutical Division, Research Laboratories, Nippon Kayaku Co., Ltd., Kita-ku, Tokyo 115, Japan

*2 Koji TOI, Central Research Laboratories, Ajinomoto Co., Inc., Kawasaki-shi, Kanagawa-ken 210, Japan

*3 Ichiro CHIBATA, Research Laboratory of Applied Biochemistry, Tanabe Seiyaku Co., Ltd., Higashi-Yodogawa-ku, Osaka 532, Japan

*4 Ryonosuke YOSHIDA, Product Development Laboratories, Ajinomoto Co., Inc., Kawasaki-shi, Kanagawa-ken 210, Japan

*5 Takeo KANEKO, Shiseido Laboratories, Shiseido Co., Ltd., Kohoku-ku, Yokohama-shi, Kanagawa-ken 222, Japan

*6 Katsuyuki TOKI, Sumitomo Chemical Co., Ltd., Takarazuka-shi, Hyogo-ken 665, Japan

4.1 Alanine

Alanine is an unusual amino acid in that its synthesis preceded its discovery in nature. In 1850, Strecker synthesized alanine, demonstrating its presence after treatment of acetaldehyde with ammonia and hydrogen cyanide in an attempt to prepare lactic acid. This amino acid was first found in hydrolyzates of silk fibroin by Weyl in 1888, about 40 yr after the above synthesis.

Since L-alanine could be prepared only with difficulty by direct fermentation until recently, while it is easily synthesized because of its simple structure, synthetic DL-alanine has been predominantly utilized. However, it is clear that further improvement in the quality of food additives will be required in the future, and demand for L-alanine preparations containing no D-isomer, which cannot be utilized directly by the human body, is expected to increase.

L-Alanine is prepared by enzymatic β-decarboxylation of L-aspartic acid at present. However, other methods such as asymmetric synthesis and optical resolution, as well as direct fermentation yielding L-alanine, should be reconsidered to cut the cost of production.

4.1.1 Chemical synthesis of alanine

The manufacture of DL-alanine on an industrial scale is based on the methods of Strecker and of Bucherer at present. Since post-treatment is common to both methods, and the separation technique is more important than the synthetic process itself because of the high solubility of this amino acid in water, both methods will be described together.

Alanine (III) is synthesized via aminopropionitrile (II) derived from acetaldehyde (I), ammonia and hydrogen cyanide in the Strecker method.

$$\underset{\text{I}}{CH_3CHO} + NH_3 + HCN \longrightarrow \underset{\text{II}}{CH_3\underset{|}{C}HCN \atop \quad NH_2} \xrightarrow[H^+]{} \underset{\text{III}}{CH_3\underset{|}{C}HCOOH \atop \quad NH_2}$$

This method was further developed through several modifications[1-4] into a new synthetic process, the Bucherer-Berg method, in 1934 by Bucherer *et al.*[5,6] who synthesized alanine via 5-methylhydantoin (V) prepared from lactonitrile (IV) or aminopropionitrile (II).

Later, Gaudry[7] obtained alanine (III) in 80% yield by hydrolyzing the hydantoin (V) with barium hydroxide. The yields of the hydantoin and

$$\underset{\underset{\text{X}}{|}}{CH_3CHCN} + (NH_4)_2CO_3 \longrightarrow \begin{array}{c} CH_3CH{-}CO \\ |\quad\ \ | \\ HN\quad NH \\ \diagdown\ \diagup \\ C \\ \| \\ O \\ \text{V} \end{array} \xrightarrow[OH^-]{} \begin{array}{c} CH_3CHCOOH \\ | \\ NH_2 \\ \text{III} \end{array}$$

$$\begin{array}{ll} X = NH_2 & \text{II} \\ X = OH & \text{IV} \end{array}$$

of alanine from the hydantoin were 90% and 89%, respectively. On the other hand, White[8,9] synthesized alanine in 70% yield from acetaldehyde. In this case, the hydantoin was prepared from 1 mole of acetaldehyde, 2 moles of ammonium bicarbonate and 1 mole of sodium cyanide, and was directly hydrolyzed with sodium hydroxide. The process of separating the hydantoin from the reaction mixture was omitted in this method.

For a considerable period after the development of the Strecker method, the use of acid, especially hydrochloric acid, had been recommended for the hydrolysis of aminopropionitrile, an intermediate in alanine synthesis.[4] However, in 1933, Farbenindustrie A.-G.[10] showed that the hydrolysis can be carried out using a metal hydroxide. Since then, alkaline hydrolysis has been widely applied, and the use of a nonvolatile, non-corrosive alkaline agent is recommended and practiced at present.

In preparing alanine, the technique for separating alanine from other inorganic compounds in the alkaline hydrolyzate of aminopropionitrile or 5-methylhydantoin is more important than the synthetic process itself because of the high solubility of alanine in water. However, since ion-exchange resins have been used in the manufacture of this amino acid, the yield has been improved appreciably and the process simplified. For example, Tokime and Mishima[11] hydrolyzed aminopropionitrile in a saturated ammonia solution containing an equivalent number of moles of sodium hydroxide, removed the excess ammonia *in vacuo* or by boiling, passed the solution through a column of weakly acidic ion-exchange resin to remove sodium ions, concentrated the eluate, and finally obtained alanine in 89.2% yield from aminopropionitrile.

White[12] hydrolyzed 5-methylhydantoin with sodium hydroxide, mixed the solution with a suspension of strongly acidic ion-exchange resin, adjusted the pH to 5, removed the resin by filtration, concentrated the filtrate under reduced pressure, and obtained alanine in 85.5% yield.

If a hydroxide of an alkaline earth metal such as calcium or barium is used in the hydrolyzing process, metal ions can be mostly removed as the insoluble carbonate salt after the hydrolysis, and this process reduces the amount of ion-exchange resin required.

A flowsheet for alanine synthesis is illustrated in Fig. 4–1. Although this scheme is essentially based on alkaline hydrolysis of an intermediate, aminopropionitrile or 5-methylhydantoin, and desalting with an ion-exchange resin, the detailed features of each process are not clear because commercial manufacturers have not made them public.

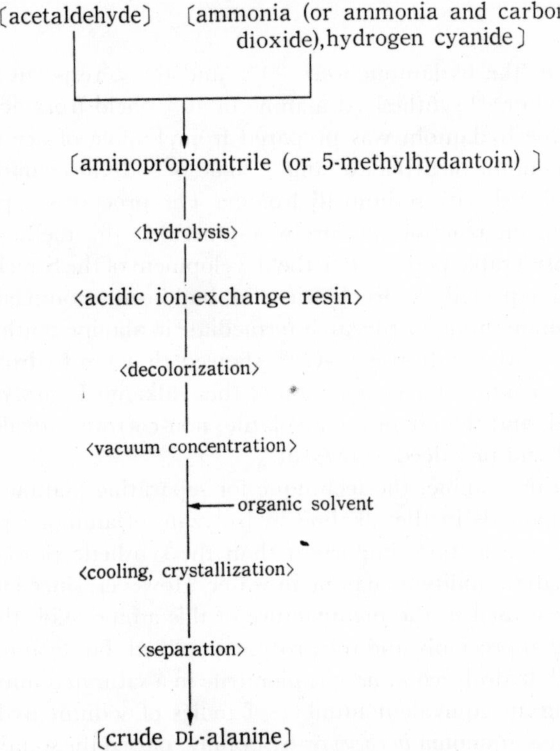

Fig. 4–1. General flowsheet for DL-alanine synthesis.

Several interesting methods of intermediate preparation have been reported recently. One example is the continuous process described in 1962 by Kato *et al.*,[13] who synthesized alanine in 85.1% yield. Acetaldehyde was passed through a reaction tower filled with McMahon packing, and a mixture of hydrogen cyanide and ammonia (prepared over a platinum-rhodium catalyst net from a mixture of methane:ammonia:air, 12.5:11.3:76.2) was blown in. The effluent was further treated with ammonia to carry out the Strecker reaction and the resulting intermediate was hydrolyzed with sodium hydroxide. This work is a valuable step towards the development of the reaction or simplification of the process, in view of the expected increase in demand for alanine.

4.1.2 Enzymatic preparation of L-alanine from aspartic acid

The preparation of L-alanine by direct fermentation is not easy compared to that of other amino acids. The difficulty is due to the coexistence of alanine racemase in the fermentation system, and the product is usually DL-alanine, with few exceptions.

Watanabe *et al.*[14] have however reported on the L-aspartate β-decarboxylase produced by *Xanthomonas oryzae* No. 531 in 1965. L-Aspartic acid could be converted to L-alanine in cell suspensions of this strain at pH 4–6 and 40°C. The reaction proceeded at high substrate concentration, and was accelerated by surface-active agents. The optimum pH and temperature for activity of the enzyme in the cell extract were 5.3 and 40°C. Later, Osawa *et al.*[15] using the same strain, prepared L-alanine in 95% yield from 15% L-aspartic acid solution. Chibata *et al.*[16] using *Pseudomonas dacunhae* or *Achromobacter pestifer*, and Yoshinaga *et al.*[17] using *Pseudomonas* No. 618, obtained L-alanine in yields of 88–92% and 98.2% (in solution, and 84% for the crude crystal), respectively.

4.1.3 Resolution of DL-alanine

DL-Alanine can be resolved by enzymatic, chemical, and physicochemical techniques. For example, asymmetric hydrolysis of acyl-DL-alanine by aminoacylase from hog kidney[18] or fungi,[19,20] chemical resolution of benzoyl-DL-alanine with brucine or strychnine,[21] and resolution by fractional crystallization of alanine benzene sulfonate[22] have been reported. Except for cases where D-alanine is desired, resolution is not practiced commercially because L-alanine can be prepared by an enzymatic method from L-aspartic acid.

References

1. N. Zelinsky and G. Stadnikov, *Chem. Ber.*, **41**, 2061 (1908).
2. E. C. Kendall and B. F. Mckenzie, *Org. Syn. Coll.*, vol. 1, p. 21, 1941.
3. A. L. Barker and G. S. Skinner, *J. Am. Chem. Soc.*, **46**, 403 (1924).
4. W. Cocker and A. Lapworth, *J. Chem. Soc.*, **1931**, 1391.
5. H. T. Bucherer and W. Steiner, *J. Prakt. Chem.*, **140**, 291 (1934).
6. H. T. Bucherer and A. Lieb, *ibid.*, **141**, 5 (1934).
7. R. Gaudry, *Can. J. Res.*, **26B**, 773 (1948).
8. H. C. White, *U.S. Patent* No. 2557920 (1951).

9. H. C. White, *U.S. Patent* No. 2642459 (1953).
10. I. G. Farbenindustrie A.-G., *French Patent* No. 746641 (1933).
11. M. Tokime and Y. Mishima, *Japanese Patent* No. 29–8677 (1954).
12. H. C. White, *U.S. Patent* No. 2700054 (1955).
13. J. Kato, Y. Iwanaga, I. Noda and K. Tawara, *Japanese Patent* No. 37–4312 (1962); *British Patent* No. 908735 (1962).
14. S. Watanabe, S. Isshiki, T. Osawa and S. Yamamoto, *Hakko Kogaku Zasshi* (Japanese), **43**, 697 (1965).
15. T. Osawa, S. Watanabe and S. Isshiki, *Japanese Patent* No. 42–11995 (1967).
16. I. Chibata, T. Kamimoto and J. Kato, *Appl. Microbiol.*, **13**, 638 (1965); *British Patent* No. 1059668 (1967).
17. F. Yoshinaga, Y. Yoshihara and S. Okumura, *Japanese Patent* No. 44–1192 (1969).
18. S. M. Birnbaum, L. Levintow, R. B. Kingsley and J. P. Greenstein, *J. Biol. Chem.*, **194**, 455 (1952).
19. K. Michi and H. Tsuda, *Bull. Agr. Chem. Soc. Japan*, **21**, 235 (1957).
20. I. Chibata, M. Kisumi and S. Yamada, *ibid.*, **22**, 24 (1958).
21. E. Pacsu and J. W. Hullen, *J. Biol. Chem.*, **136**, 335 (1940).
22. I. Chibata and T. Ishikawa, *Bull. Agr. Chem. Soc. Japan*, **22**, 218 (1958).

4.2 β-ALANINE

β-Alanine was first synthesized in 1870 by Heintz from β-iodopropionic acid and ammonia. This amino acid, unlike other common amino acids, is not detected in hydrolyzates of protein, but is biologically important since it occurs in carnosine (I) in animal muscle and in pantothenic acid (II), a component of vitamin B complex.

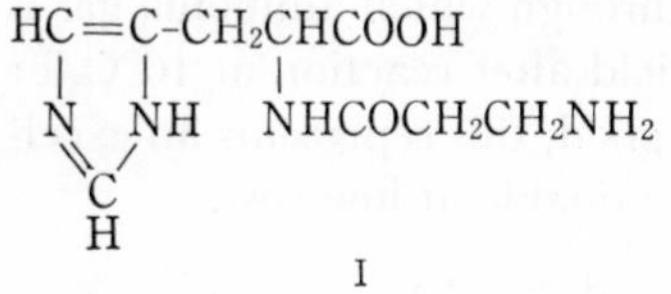

$$\mathrm{HC{=}C\text{-}CH_2CH(NHCOCH_2CH_2NH_2)COOH}\ \text{(imidazole ring: N, NH, CH)}\quad \text{I}$$

$$\mathrm{HOCH_2C(CH_3)_2CH(OH)CONHCH_2CH_2COOH}\quad \text{II}$$

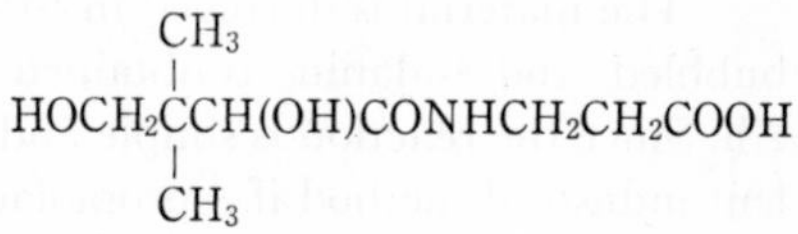

As β-alanine does not possess an asymmetric carbon atom, it is produced only by synthetic methods. Following the development of the petrochemical industry, new methods of synthesis from cheap and easily obtainable materials have been devised, and mass production of β-alanine has been undertaken.

Although many methods for β-alanine synthesis have been reported, the most important raw materials for synthesis are: (1) ethylenecyanohydrin, (2) β-propiolactone, (3) derivatives of acrylic acid, and (4) cyanoacetic acid. The commercial advantages of each of these materials depend on the specificity of the particular plant for each raw material for β-alanine synthesis. Therefore, methods for synthesis starting from several interesting raw materials will be described below.

4.2.1 Synthesis from ethylenecyanohydrin

Paden *et al.*[1] accomplished both amination and hydrolysis in a single process, treating β-alkoxypropionitrile with ammonia water at high temperature (above 150°C).

$$\mathrm{ROCH_2CH_2CN} \xrightarrow{\mathrm{NH_3}} \underset{\text{III}}{\mathrm{H_2NCH_2CH_2COOH}}$$

According to a patent by Röhm and Hass,[2] β-alanine (III) can be prepared in 71% yield, by hydrolysis (with 50% sulfuric acid) of a polymer formed on autoclaving ethylenecyanohydrin at 175°C in the presence of a metal oxide catalyst. This method is industrially advantageous, but the yield should be further improved.

4.2.2 Synthesis from β-propiolactone

Gresham *et al.*[3] reported an interesting method using β-propiolactone as a starting material.

$$\underbrace{CH_2CH_2CO}_{\lfloor\_\_\_O\rfloor} \xrightarrow{NH_3} \underset{\text{III}}{H_2NCH_2CH_2COOH}$$

The material is dissolved in *tert*-butanol through which ammonia gas is bubbled, and β-alanine is obtained in 79% yield after reaction at 16°C for 2 h. Since the reaction is simple and the yield good, this represents an excellent industrial method if β-propiolactone is obtainable at low cost.

4.2.3 Synthesis from derivatives of acrylic acid

Various methods of synthesis starting from acrylic acid, acrylamide, and acrylonitrile have been reported. Since acrylonitrile has been obtainable at very low cost recently, following the development of the Sohio method, synthesis of β-alanine from acrylonitrile has become the principal commercial process.

A. Synthesis from acrylic acid

This method was first examined by Wender *et al.* Later, Babcock *et al.*[4] treated methyl acrylate with ammonia water at 140–145°C for 5 h in an autoclave. After completion of the reaction, β-alanine was obtained in 68% yield by concentration followed by the addition of an organic solvent.

$$CH_2{=}CHCOOX \xrightarrow{NH_3} \underset{\text{III}}{H_2NCH_2CH_2COOH}$$

X = alkyl or metal

Provided that better methods for separation and purification of the product are developed, this process may be more advantageous. It is an interesting industrial method, but depends on the cost of acrylic acid.

B. Synthesis from acrylamide

Matlack *et al.*[5] reported a method using acrylamide as the raw material for synthesis of β-alanine.

$$\underset{\text{IV}}{CH_2{=}CHCONH_2} \longrightarrow {+}NHCH_2CH_2CO{+}_n \xrightarrow[H^+]{} \underset{\text{III}}{H_2NCH_2CH_2COOH}$$

Acrylamide (IV) is converted to a polymer by strongly basic catalysts in the presence of quinones, which are polymerization inhibitors. The resulting polymer can be hydrolyzed with 50% sulfuric acid to give β-alanine in 90% yield. Since the amide (IV) can also be prepared from acrylonitrile by treatment with sulfuric acid, this method may be applicable for the commercial production of β-alanine.

C. Synthesis from acrylonitrile

This method is based on amination of acrylonitrile (V) followed by hydrolysis, and proceeds as illustrated below.

$$\underset{\text{V}}{CH_2{=}CHCN} \xrightarrow{NH_3} \underset{\text{VI}}{H_2NCH_2CH_2CN} \xrightarrow[H^+ \text{ or } OH^-]{H_2O} \underset{\text{III}}{H_2NCH_2CH_2COOH}$$

$$\underset{\text{VI}}{H_2NCH_2CH_2CN} \xrightarrow{\text{V}} \underset{\text{VII}}{HN(CH_2CH_2CN)_2} \xrightarrow{\text{V}} \underset{\text{VIII}}{N(CH_2CH_2CN)_3}$$

Use of excess ammonia is advantageous since secondary (VII) and tertiary (VIII) amino compounds may be formed as by-products in the

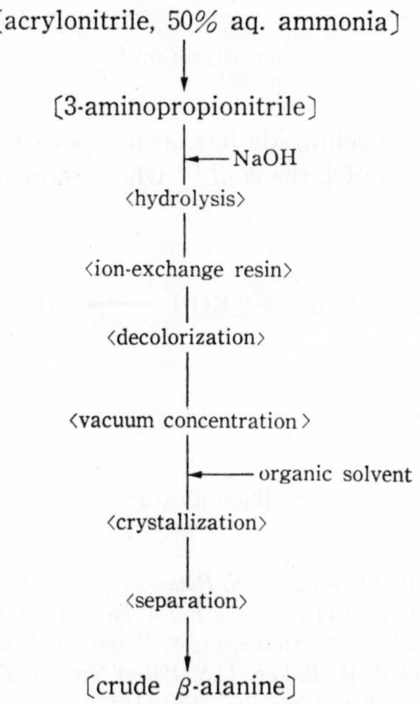

Fig. 4–2. General flowsheet for β-alanine synthesis from acrylonitrile.

amination of acrylonitrile (V). Although several methods have been reported by Hoffmann *et al.*,[6] Buc *et al.*,[7] Whitmore,[8] Ford *et al.*[9] and Carlson *et al.*,[10] the reported yields are below 60%.

Recently, Szlompeck-Nesteruk[11] prepared aminonitrile (VI) in 81% yield, by adding diphenylamine to the reaction system and substituting methanol for water. A continuous reaction method using the tubular reactor proposed by Smolin *et al.*[12] may be interesting for industrial use.

There have been many investigations on acid or alkaline hydrolysis of aminonitrile, and the yields have been reported to be as good as 80–95%. Griffith *et al.*,[13] utilizing cationic and anionic ion-exchange resins in post-treatment, obtained β-alanine in 93.4% yield.

A flowsheet for β-alanine synthesis from acrylonitrile is illustrated in Fig. 4–2.

4.2.4 Synthesis from cyanoacetic acid and other materials

These methods are at a more experimental stage when compared with the methods already described. Syntheses from cyanoacetic acid have been reported by Merck[14] and Products Roche Soc.,[15] and the yield in the latter case was 74%.

$$\mathrm{NCCH_2COOK} \xrightarrow[\substack{\text{Raney nickel,}\\ \text{saturated } \mathrm{NH_3/CH_3OH}\\ \text{100 atm, 80°C}}]{\mathrm{H_2}} \underset{\mathrm{III}}{\mathrm{H_2NCH_2CH_2COOH}}$$

A synthesis from succinimide has been reported by Hoogewerff,[16] and was reexamined later by Clarke *et al.*[17] who obtained β-alanine in 41–45% yield.

$$\begin{matrix}\mathrm{CH_2CO}\diagdown \\ | \qquad\quad \mathrm{NH} \\ \mathrm{CH_2CO}\diagup\end{matrix} + \mathrm{KOBr} + 2\,\mathrm{KOH} \longrightarrow \underset{\mathrm{III}}{\mathrm{H_2NCH_2CH_2COOH}}$$

References

1. J. H. Paden and P. M. Kirk, *U.S. Patent* No. 2335605 (1943).
2. Röhm and Haas G.m.b.H., *German Patent* No. 1105426 (1961).
3. T. L. Gresham and F. W. Shaver, *U.S. Patent* No. 2525794 (1950).
4. S. H. Babcock and B. R. Baker, *U.S. Patent* No. 2376334 (1945).
5. A. S. Matlack, *U.S. Patent* No. 2672480 (1954).
6. U. Hoffmann and B. Jacobi, *U.S. Patent* No. 1992615 (1935).

7. S. R. Buc, J. H. Ford and E. C. Wise, *J. Am. Chem. Soc.*, **67**, 92 (1945).
8. J. Whitmore, *U.S. Patent* No. 2742491 (1945).
9. J. H. Ford, S. R. Buc and J. W. Greiner, *J. Am. Chem. Soc.*, **69**, 844 (1947).
10. G. H. Carlson and C. N. Hotchkiss, *U.S. Patent* No. 2377401 (1945).
11. Danuta Szlompek-Nesteruk, *Przemysl Chem.*, **44**, 85 (1965).
12. E. M. Smolin and L. C. Beegle, *Ind. Eng. Chem.*, **50**, 1115 (1958).
13. R. Griffith, W. A. Disalvo, R. Kapp and L. T. Rosenberg, *U.S. Patent* No. 2819303 (1958).
14. F. E. Merck, *German Patent* No. 597305 (1934).
15. Products Roche Soc. Avon, *Belgian Patent* No. 446511 (1942).
16. Hoogewerff and Van Dorp, *Rec. Trav. Chim.*, **10**, 5 (1891).
17. H. T. Clarke and L. D. Behr, *Org. Syn. Coll.*, vol. 2, p. 540, 1943.

4.3 Arginine

In 1886, Schulze and Steiger isolated a crystalline nitrate salt of a basic compound with the formula, $C_6H_{14}N_2O_2 \cdot HNO_3 \cdot \frac{1}{2}H_2O$, and named it "arginine." Schulze and Winterstein found that arginine yields ammonia, carbon dioxide, and a new basic compound when hydrolyzed. They assumed the presence of guanidino group in the arginine molecule, since its dibenzoyl derivative was identified as dibenzoyl ornithine, and proved their assumption by synthesizing an identical compound from ornithine and cyanamide. In 1910, Sörensen proved that the guanidino group is at the δ-position of ornithine. Isolation of arginine from protein, on the other hand, was first achieved in 1895 by Hedin who isolated it from acid hydrolyzates of horn in the form of the nitrate salt, $C_6H_{14}N_2O_2 \cdot HNO_3 \cdot \frac{1}{2}H_2O$, reported by Schulze and Steiger.

L-Arginine is prepared commercially by extraction and isolation from acid hydrolyzates of gelatin, defatted soybeans, etc. The synthetic method using L-ornithine which is produced by fermentation, is also advantageous. A method for direct fermentation of L-arginine has been established.

Among the methods of synthesis of L-arginine, an industrially practical process is that guanidizing the δ-amino group of L-ornithine which is obtainable in large quantities by fermentation. Since the material is the L-isomer, and racemization does not occur in the synthetic process, the resulting arginine is also the L-isomer. There are several guanidizing agents such as cyanamide, *S*-methylisothiourea, and *O*-methylisourea.

4.3.1 Synthesis from L-ornithine and cyanamide

Schulze and Winterstein[1,2] synthesized arginine (III) from free ornithine (I) and cyanamide (II) by thermal reaction in an aqueous solution. In this process, α,δ-diguanidinovaleric acid (IV) is also formed subsidiarily.

$$\underset{\text{I}}{H_2NCH_2CH_2CH_2\underset{\displaystyle NH_2}{\underset{|}{C}}HCOOH} + \underset{\text{II}}{HN{=}C{=}NH}$$

$$\longrightarrow \underset{\text{III}}{H_2N\underset{\displaystyle NH}{\underset{\|}{C}}NHCH_2CH_2CH_2\underset{\displaystyle NH_2}{\underset{|}{C}}HCOOH} + \underset{\text{IV}}{H_2N\underset{\displaystyle NH}{\underset{\|}{C}}NHCH_2CH_2CH_2\underset{\displaystyle NH\underset{\displaystyle NH}{\underset{\|}{C}}NH_2}{\underset{|}{C}}HCOOH}$$

Odo,[3] examining the conditions of the reaction in detail, improved the yield to 42%. The subsidiary formation of α,δ-diguanidinovaleric acid was inhibited by heat treatment of the aqueous solution of ornithine hydrochloride with sodium hydroxide and cyanamide. Since cyanamide is inexpensive when compared with other guanidizing agents, its use may be advantageous in industry if the subsidiary production of diguanidinovaleric acid can be effectively repressed and the yield further improved.

Sörensen *et al.*[4] obtained arginine hydrochloride by hydrolyzing α-benzoylarginine with hydrochloric acid. The α-benzoylarginine was prepared by treating cyanamide with α-benzoylornithine derived from ornithine in which the α-amino group was masked.

Fujii *et al.*[5] greatly improved the yield of L-arginine by treating L-ornithine with cyanamide in the presence of a catalytic amount of methylmercaptan. L-Ornithine solution containing 0.1–0.5 mole of methylmercaptan was treated with a solution containing 1.0–1.3 moles of cyanamide at pH 9.5, and L-arginine was obtained in 90% yield. In this process, it is thought that the cyanamide first reacts with methylmercaptan to form *S*-methylisothiourea, which guanidizes the L-ornithine, and that methylmercaptan is then regenerated and utilized again in the reaction.

4.3.2 Synthesis from L-ornithine and *S*-methylisothiourea

Odo[3] synthesized L-arginine (III) in 68% yield by treating L-ornithine (I) with *S*-methylisothiourea (V) under weakly alkaline conditions.

$$\underset{\text{I}}{H_2NCH_2CH_2CH_2\underset{\displaystyle NH_2}{\underset{|}{C}}HCOOH} + \underset{\text{V}}{HN{=}\overset{\displaystyle SCH_3}{\overset{|}{C}}NH_2} \longrightarrow$$

$$\underset{\text{III}}{H_2N\underset{\displaystyle NH}{\underset{\|}{C}}NHCH_2CH_2CH_2\underset{\displaystyle NH_2}{\underset{|}{C}}HCOOH} + CH_3SH$$

Samejima *et al.*[6] examined the reaction, and proved that L-ornithine could be converted to L-arginine without any subsidiary reactions at pH 10–12. Also, when L-ornithine is treated with *S*-methylisothiourea after its α-amino group has been masked as a copper salt, only the δ-amino group is guanidized. L-Arginine is then obtained by removing copper from the salt by treatment with hydrogen sulfide.[7–9]

4.3.3 Synthesis from L-ornithine and *O*-methylisourea

Kurtz[8] and Turba and Schuster,[10] treating the copper salt of L-ornithine with *O*-methylisourea instead of *S*-methylthiourea, obtained L-arginine in 94% yield after removal of the copper from the salt with hydrogen sulfide. Although it is possible to treat L-ornithine directly with *O*-methylisourea, the yield is then poor.

4.3.4 Resolution of DL-arginine

There are several methods for resolving DL-arginine. These include asymmetric hydrolysis of the acetylated derivative by aminoacylase,[11] and recovery of D-arginine by treating DL-arginine with L-amino acid oxidase.[12] However, resolution is not yet practiced commercially since the profitable methods for L-arginine production, such as extraction from protein hydrolyzates and synthesis from the L-ornithine obtained by fermentation, do not require a resolution step.

References

1. E. Schulze and E. Winterstein, *Chem. Ber.*, **32**, 3191 (1899).
2. E. Schulze and E. Winterstein, *Z. Physiol. Chem.*, **34**, 128 (1901).
3. K. Odo, *Nippon Kagaku Kaishi* (Japanese), **74**, 1, 744 (1953).
4. S. P. L. Sörensen, M. Hoyrup and A. C. Anderson, *Z. Physiol. Chem.*, **76**, 44 (1911).
5. K. Fujii, S. Fujisawa and K. Nakano, *Japanese Patent* No. 42–8012 (1967).
6. H. Samejima, Y. Nagano, H. Teranishi and T. Naruse, *Japanese Patent* No. 39–21232 (1964).
7. F. Turba and K. H. Schustev, *Naturwissenschaften*, **33**, 370 (1946).
8. A. C. Kurtz, *J. Biol. Chem.*, **180**, 1253 (1949).
9. C. Simons and H. T. Young, *British Patent* No. 672684 (1952).
10. F. Turba and K. H. Schuster, *Z. Physiol. Chem.*, **283**, 27 (1948).
11. S. M. Birnbaum, M. Winitz and J. P. Greenstein, *Arch. Biochem. Biophys.*, **60**, 496 (1956).
12. J. R. Parilch, J. P. Greenstein, M. Winitz and S. M. Birnbaum, *J. Am. Chem Soc.*, **80**, 953 (1958).

4.4 Asparagine

Asparagine was first found in 1806 by Vauquelin and Robiquet, who isolated it from asparagus juice. The name asparagine was proposed by Duloug in 1826. Later, Liebig presented its exact structural formula. In 1932, Damodaran, isolating asparagine from edestin hydrolyzate prepared by the action of pepsin, trypsin and yeast peptidase, demonstrated that asparagine is a constituent of protein.

4.4.1 Synthesis from aspartic acid

There are few reported examples of asparagine synthesis. Bergmann *et al.*[1] obtained L-asparagine (V) after treating *N*-benzyloxycarbonyl-L-aspartic acid anhydride (I) with benzyl alcohol to prepare the α-benzyl ester of *N*-benzyloxycarbonyl-L-aspartic acid (II). This product (II) was converted with ammonia to the amide (IV), which was then reduced with palladium to give L-asparagine (V).

$$\underset{\text{I}}{\begin{matrix} C_6H_5CH_2OCONHCHCO \\ | \\ CH_2CO \end{matrix}\!\!>\!O} \xrightarrow{C_6H_5CH_2OH} \underset{\text{II}}{\begin{matrix} C_6H_5CH_2OCONHCHCOOCH_2C_6H_5 \\ | \\ CH_2COOH \end{matrix}}$$

$$\longrightarrow \underset{\text{III}}{\begin{matrix} C_6H_5CH_2OCONHCHCOOCH_2C_6H_5 \\ | \\ CH_2COCl \end{matrix}} \xrightarrow{NH_3} \underset{\text{IV}}{\begin{matrix} C_6H_5CH_2OCONHCHCOOCH_2C_6H_5 \\ | \\ CH_2CONH_2 \end{matrix}}$$

$$\xrightarrow[Pd]{H_2} \underset{\text{V}}{\begin{matrix} H_2NCHCOOH \\ | \\ CH_2CONH_2 \end{matrix}}$$

Bovarnik[2] treated *N*-tosyl-DL-aspartic acid anhydride (VI) with benzyl alcohol to prepare α- and β-benzyl esters. The β-benzyl ester of *N*-tosyl-DL-aspartic acid (VII) was separated from the α-isomer and aminated with ammonia water. DL-Asparagine (V) was then obtained on detosylation of the resulting *N*-tosylaspartic acid (VIII) with liquid ammonia and sodium.

$$\underset{\text{VI}}{\begin{matrix} CH_3C_6H_4SO_2NHCHCO \\ | \\ CH_2CO \end{matrix}\!\!>\!O} \xrightarrow{C_6H_5CH_2OH} \underset{\text{VII}}{\begin{matrix} CH_3C_6H_4SO_2NHCHCOOH \\ | \\ CH_2COOCH_2C_6H_5 \end{matrix}}$$

$$\xrightarrow{NH_3} \underset{\text{VIII}}{\underset{CH_2CONH_2}{CH_3C_6H_4SO_2NHCHCOOH}} \xrightarrow[\text{liq. ammonia}]{Na} \underset{\text{V}}{\underset{CH_2CONH_2}{H_2NCHCOOH}}$$

Weygand, Klinke and Eigen,[3] treating *N*-trifluoroacetyl-L-aspartic acid anhydride (IX) with ethanol, prepared the ethyl ester of *N*-trifluoroacetyl-L-aspartic acid (X). Its β-carboxylic group was converted to the acid chloride giving (XI), and the ethyl ester of *N*-trifluoroacetyl-L-asparagine (XII) was then prepared by treatment of (XI) with ammonia. L-Asparagine (V) was obtained from the ester (XII) by deacylation with barium hydroxide followed by saponification.

$$\underset{\text{IX}}{\underset{CH_2CO}{CF_3CONHCHCO}\!\!>\!O} \xrightarrow{C_2H_5OH} \underset{\text{X}}{\underset{CH_2COOH}{CF_3CONHCHCOOC_2H_5}} \longrightarrow$$

$$\underset{\text{XI}}{\underset{CH_2COCl}{CF_3CONHCHCOOC_2H_5}} \xrightarrow{NH_3} \underset{\text{XII}}{\underset{CH_2CONH_2}{CF_3CONHCHCOOC_2H_5}} \xrightarrow[Ba(OH)_2]{}$$

$$\underset{\text{V}}{\underset{CH_2CONH_2}{H_2NCHCOOH}}$$

The syntheses described above were carried out in the course of studies on peptide synthesis and the reactions involved are complicated. For commercial purposes, the method described next is considered more advantageous.

Beecham[4] synthesized L-asparagine (V) by amidizing the β-methyl ester of L-aspartic acid (XIII) with ammonia. A solution of the β-methyl ester (XIII) in ammonia-saturated ethanol was allowed to stand for 42 h, and after the solvent had been distilled off, the residue was dissolved in warm water. On addition of methanol to this solution, L-asparagine hydrate was crystallized out giving a 72% yield.

Since L-aspartic acid is produced at low cost by fermentation and is easily convertible to the β-methyl ester in good yield by normal esterification procedures, the above method for synthesizing L-asparagine by amidizing

$$\underset{\text{XIII}}{\underset{\displaystyle |\atop NH_2}{CH_3OOCCH_2CHCOOH}} \xrightarrow{NH_3} \underset{\text{V}}{\underset{\displaystyle |\atop NH_2}{H_2NCOCH_2CHCOOH}}$$

the ester is both simple and advantageous for commercial application, and gives a good yield.

4.4.2 Synthesis from maleic acid

Frankel *et al.*[5] treated maleic acid anhydride (XIV) with ammonia to prepare maleinamic acid (XV), which was then treated with benzylamine. DL-Asparagine was obtained when the resulting *N*-benzyl-DL-asparagine (XVI) was reduced with palladium catalyst.

$$\underset{\text{XIV}}{\begin{matrix} CHCO \\ \| \\ CHCO \end{matrix}\!\!>O} \xrightarrow{NH_3} \underset{\text{XV}}{\begin{matrix} CHCOOH \\ \| \\ CHCONH_2 \end{matrix}} \xrightarrow{C_6H_5CH_2NH_2} \underset{\text{XVI}}{\underset{\displaystyle |\atop CH_2CONH_2}{C_6H_5CH_2NHCHCOOH}}$$

$$\xrightarrow[Pd]{H_2} \underset{\text{V}}{\underset{\displaystyle |\atop CH_2CONH_2}{H_2NCHCOOH}}$$

4.4.3 Resolution of DL-asparagine

Since L-asparagine is easily synthesized from L-aspartic acid, which is obtainable at low cost by fermentation, resolution of DL-asparagine is not practiced commercially. However, fractional crystallization may be a possible way of resolving DL-asparagine, when this can be produced at low cost.[6]

References

1. M. Bergmann, L. Zervas and L. Salzmann, *Chem. Ber.*, **66**, 1288 (1933).
2. M. R. Bovarnick, *J. Biol. Chem.*, **148**, 151 (1943).
3. F. Weygand, P. Klinke and I. Eigen, *Chem. Ber.*, **90**, 1896 (1957).
4. A. F. Beecham, *J. Am. Chem. Soc.*, **76**, 4615 (1953).
5. M. Frankel, Y. Liwschitz and Y. Amiel, *J. Am. Chem. Soc.*, **75**, 330 (1953).
6. A. Piutti, *Compt. Rend.*, **103**, 134 (1886).

4.5 Aspartic acid

In 1806, Vauquelin and Robiquet found asparagine (the β-amide of aspartic acid) in the juice of asparagus, as mentioned in section 4.4. In 1827, Plisson and Henry obtained the free acid by hydrolysis of asparagine, and named it "aspartic acid". In 1868, Ritthausen isolated aspartic acid from protein hydrolyzates, suggesting that this amino acid is in fact a constituent of proteins. Piutti confirmed the chemical structure of aspartic acid by total synthesis in 1887.

L-Aspartic acid can be produced in good yield by the hydrolysis of asparagine isolated from plant material such as ethiolated seedlings of lupines. It is also obtainable by isolation from protein hydrolyzates or by optical resolution of the synthesized DL-amino acid. Nowadays, however, L-aspartic acid is produced mainly by the enzymatic action of aspartase on fumaric acid, which can be supplied economically in large quantities. The optical resolution method cannot compete effectively with the enzymatic method, except in the case of producing D-aspartic acid. Therefore, it seems probable that a synthetic method for obtaining aspartic acid will not be carried out industrially unless an excellent technique for the asymmetric synthesis of L-aspartic acid can be developed. In this section, the typical methods for obtaining aspartic acid are described.

4.5.1 Synthesis from fumaric acid or maleic acid

DL-Aspartic acid can be synthesized by amination of fumaric acid, maleic acid or derivatives of these acids.

$$\underset{\text{I}}{\begin{matrix}\text{CHCOOH}\\ \|\\ \text{HOOCCH}\end{matrix}} \xrightarrow[\text{ii) HCl}]{\text{i) } NH_4OH,\ NH_4Cl} \underset{\text{II}}{\text{HOOCCH}_2\underset{\displaystyle NH_2}{\underset{|}{\text{CH}}}\text{COOH}}$$

Autoclaving of fumaric acid with ammonia and ammonium hydrochloride at 180°C under pressure (10 atm), acidification of the reaction mixture with HCl and subsequent addition of ethanol, gave DL-aspartic acid in a yield of 68%.[1)]

Fusion of diammonium fumarate and ammonium acetate without pressure led to the formation of aspartic acid in a yield of 70%.[2)]

Aspartic acid can also be synthesized from diethylfumarate (III).[3)] The reaction of the ester (IV) with ammonia in ethanol resulted in the formation of diketopiperazine diacetamide (IV). Hydrolysis of (IV) with sodium hydroxide, and subsequent neutralization with hydrochloric acid, led to the formation of aspartic acid (II). The amino acid was isolated as the insoluble copper salt, and removal of the copper with hydrogen sulfide gave the crude amino acid in a yield of 96%. Pure aspartic acid was then obtained by recrystallization (yield, 59%).

$$\underset{\text{III}}{\begin{array}{l} \quad CHCOOC_2H_5 \\ \quad \| \\ H_5C_2OOCCH \end{array}} \xrightarrow[C_2H_5OH]{NH_3} \underset{\text{IV}}{\begin{array}{l} H_2NOCCH{-}CO{-}NH \\ \qquad | \qquad\quad | \\ \quad NH{-}CO{-}CHCH_2CONH_2 \end{array}}$$

$$\xrightarrow[NaOH]{} \underset{\text{II}}{\begin{array}{l} HOOCCH_2CHCOOH \\ \qquad\quad | \\ \qquad\; NH_2 \end{array}}$$

On the other hand, reaction of one mole of maleic acid (V) with 2 moles of benzylamine (VI) gave the benzylamine salt of *N*-benzylaspartic acid (VII) quantitatively. Removal of benzylamine from the salt (VII) by the addition of sodium hydroxide led to *N*-benzylaspartic acid (VIII). Reduction with palladium catalyst then gave aspartic acid (II) in a high yield.[4)]

$$\underset{\text{V}}{\begin{array}{l} CHCOOH \\ \| \\ CHCOOH \end{array}} + \underset{\text{VI}}{C_6H_5{-}CH_2NH_2} \xrightarrow[\text{quantitative}]{} \underset{\text{VII}}{\begin{array}{l} C_6H_5{-}CH_2NH_2\cdot HOOCCH_2CHCOOH \\ \qquad\qquad\qquad\qquad\qquad\quad | \\ \qquad\qquad\qquad\qquad\qquad NHCH_2{-}C_6H_5 \end{array}}$$

$$\xrightarrow[NaOH]{} \underset{\text{VIII}}{\begin{array}{l} HOOCCH_2CHCOOH \\ \qquad\quad | \\ \qquad\; NHCH_2{-}C_6H_5 \end{array}} \xrightarrow[Pd\text{-}C]{H_2} \underset{\text{II}}{\begin{array}{l} HOOCCH_2CHCOOH \\ \qquad\quad | \\ \qquad\; NH_2 \end{array}}$$

(NaOH step: 95%; H_2, Pd-C step: quantitative)

Besides the above synthesis of racemic aspartic acid, asymmetric synthesis of this amino acid using optically active α-methylbenzylamine has been investigated.[5)] For example, L- and D-isomers of aspartic acid, at an optical purity of 13.7% and 15.4% respectively, were obtained in a yield of 86% from diethylmaleate.

4.5.2 Synthesis from aminomalonate derivatives

Aspartic acid can be synthesized from derivatives of aminomalonates, as is the case with other amino acids. Condensation of sodium diethylphthalimidomalonate (IX) with ethylchloroacetate (X), followed by hydrolysis of the condensation product (XI) with a mixture of hydrochloric acid and acetic acid, gave aspartic acid (II) in an overall yield of 46%.[6)]

$$\underset{\text{IX}}{\text{HC(COOC}_2\text{H}_5)_2\text{-N(CO)}_2\text{C}_6\text{H}_4} + \underset{\text{X}}{\text{ClCH}_2\text{COOC}_2\text{H}_5} \xrightarrow[95\sim99\%]{\text{Na}} \underset{\text{XI}}{\text{H}_5\text{C}_2\text{OOCCH}_2\text{C(COOC}_2\text{H}_5)_2\text{-N(CO)}_2\text{C}_6\text{H}_4}$$

$$\xrightarrow[42\sim43\%]{\text{HCl, CH}_3\text{COOH}} \underset{\text{II}}{\text{HOOCCH}_2\text{CH(NH}_2\text{)COOH}}$$

Ethyl formylaminomalonate[7,8] and acetamide derivatives of malonic acid[9] were also employed in place of phthalimidomalonate, and DL-aspartic acid was obtained in a yield of some 50%.

4.5.3 Resolution of DL-aspartic acid

In order to obtain optically active aspartic acid, resolution of the synthetic DL-amino acid is required. Enzymatic resolution of racemic aspartic acid has been performed by asymmetric hydrolysis of *N*-chloroacetyl-DL-aspartic acid with aminoacylase II from hog kidney.[10] DL-Aspartic acid can also be resolved by a preferential crystallization procedure. Seeding of optical isomer into an aqueous solution of ammonium formate[11] or ammonium chloride,[12] supersaturated with DL-aspartic acid (or its ammonium or hydrochloric acid salt), resulted in precipitation of one or other of the two isomers.

The L-aspartic acid presently used for drugs or foodstuffs is not manufactured by optical resolution of the synthetic product, but by the enzymatic action of microbial aspartase on fumaric acid. This acid is produced industrially on a large scale and can be supplied at low cost. The microbial aspartase catalyzes the following reaction:

$$\text{HOOCCH}{=}\text{CHCOOH} + \text{NH}_3 \underset{}{\overset{\text{aspartase}}{\rightleftharpoons}} \text{L-HOOCCH}_2\text{CH(NH}_2\text{)COOH}$$

Preparation of L-aspartic acid using resting yeast cells was previously reported in 1928,[13] and research on industrial enzymatic procedures has become very active since around 1958. For example, a one-step fermentation procedure culturing *Escherichia coli* in a medium containing fumarate as the main carbon source was developed.[14] There have also been many reports on the production of L-aspartic acid by feeding ammonium fumarate at high concentrations to a fermented broth or cell suspension of microorganisms having high aspartase activity.[15,16] Most of these microorganisms belonged to the genera *Pseudomonas*, *Bacillus* or *Escherichia*. Maleic acid has also been used in place of fumaric acid.[17]

References

1. Y. Tutiya, *J. Agr. Chem. Soc. Japan*, **17**, 706 (1941).
2. T. Maki and T. Mori, *J. Pharm. Soc. Japan*, **75**, 604 (1955).
3. M. S. Dunn and S. W. Fox, *J. Biol. Chem.*, **101**, 493 (1933).
4. M. Frankel, Y. Liwschitz and Y. Amiel, *J. Am. Chem. Soc.*, **75**, 330 (1953).
5. K. Harada and K. Matsumoto, *J. Org. Chem.*, **31**, 2985 (1966).
6. M. S. Dunn and B. W. Smart, *Org. Syn. Coll.*, vol. 4, p. 55, 1963.
7. A. Galat, *J. Am. Chem. Soc.*, **69**, 965 (1947).
8. A. Cohen, E. G. Hughes and J. A. Silk, *British Patent* No. 621706 (1948).
9. F. H. McMillan and N. F. Albertson, *J. Am. Chem. Soc.*, **70**, 3778 (1948).
10. S. M. Birnbaum, L. Levintow, R. B. Kingsley and J. P. Greenstein, *J. Biol. Chem.*, **194**, 455 (1952).
11. K. Harada, *Bull. Chem. Soc. Japan*, **38**, 1552 (1965).
12. T. Haga, M. Sato and K. Miura, *Japanese Patent* No. 42–3290 (1967).
13. Y. Sumiki, *Bull. Jap. Soc. Ferment.*, **23**, 33 (1928).
14. M. Kisumi, Y. Ashikaga and I. Chibata, *Bull. Agr. Chem. Soc. Japan*, **24**, 296 (1960).
15. S. Kinoshita, K. Nakayama and S. Kitada, *J. Ferment. Assoc. Japan*, **16**, 517 (1958).
16. K. Kitahara, S. Fukui and M. Misawa, *J. Agr. Chem. Soc. Japan*, **34**, 44 (1960).
17. Y. Takamura, I. Kitamura, M. Ikura, K. Kono and A. Ozaki, *Agr. Biol. Chem.*, **30**, 338 (1966).

4.6 Citrulline

Citrulline (δ-carbamoyl ornithine) is not a constituent of proteins but occurs widely in the free state in nature. It is contained in large amounts in melons and onions, and in animal liver. In 1914, Koga and Otake isolated a compound with the chemical formula $C_6H_{13}N_3O_3$ from water-melon juice. Wada again isolated the same compound in 1930. He found that its benzoyl derivative was identical to α-benzoyl-δ-carbamoyl ornithine, a derivative of ornithine, and finally identified the compound occurring in water-melons as δ-carbamoyl ornithine.

4.6.1 Synthesis from ornithine

Synthesis of citrulline from L-ornithine, which is now produced in large quantities by fermentation, is commercially profitable. In 1949, Kultz[1)] obtained L-citrulline (IV) from L-ornithine (I) in 60–70% yield. He converted the L-ornithine to the copper complex (II) in order to mask its α-amino group. On heating an alkaline solution of the copper complex (II) with an excess of urea, the water-insoluble copper complex of L-citrulline (III) was formed. The copper was then removed from the complex (III) by treatment with hydrogen sulfide. Crystalline L-citrulline was obtained on addition of alcohol to the concentrated filtrate.

$$\underset{\text{I}}{H_2NCH_2CH_2CH_2\underset{\displaystyle NH_2}{\underset{|}{C}}HCOOH} \xrightarrow[OH^-]{Cu^{2+}} \underset{\text{II}}{\left[H_2NCH_2CH_2CH_2CH\begin{matrix} COO \searrow \\ NH_2 \nearrow \end{matrix} \right]_2 Cu}$$

$$\xrightarrow{NH_2CONH_2} \underset{\text{III}}{\left[H_2NCONHCH_2CH_2CH_2CH\begin{matrix} COO \searrow \\ NH_2 \nearrow \end{matrix} \right]_2 Cu}$$

$$\xrightarrow[H^+]{H_2S} \underset{\text{IV}}{H_2NCONHCH_2CH_2CH_2\underset{\displaystyle NH_2}{\underset{|}{C}}HCOOH}$$

Since the method of Kultz is simple and does not involve any racemization process, it is particularly advantageous, especially in the respect that the

L-citrulline is synthesized from L-ornithine while maintaining the steric configuration in an unchanged state. As carbamoylating reagents other than urea, urethane,[2] nitrourea[1] and potassium cyanate[3] have been used. They are however inferior for practical purposes.

4.6.2 Synthesis from arginine

Fox[4] synthesized DL-citrulline from L-arginine. He isolated a copper complex of DL-citrulline in 40% yield from a solution of L-arginine hydrochloride refluxed with sodium hydroxide. This method appears rather simple, but is insufficiently reliable. Because the heating process is carried out in strongly alkaline conditions, DL-citrulline is obtained as a result of complete racemization.

4.6.3 Resolution of DL-citrulline

Due to the fact that L-citrulline is produced by synthesis from L-ornithine, which is produced by fermentation or by on enzymatic process from L-arginine, a commercial method for the resolution of DL-citrulline has not yet been reported.

References

1. A. C. Kultz, *J. Biol. Chem.*, **180**, 1253 (1949).
2. M. Wada, *Biochem. Z.*, **224**, 420 (1930).
3. L. H. Smith, *J. Am. Chem. Soc.*, **77**, 6691 (1955).
4. S. W. Fox, *J. Biol. Chem.*, **123**, 687 (1938).

4.7 Cystine and Cysteine

Cystine was the first amino acid to be isolated. Wallaston found it in a urinary calculus in 1810 and Berzelius named it in 1932. Morner isolated it from hydrolyzates of protein (horn) in 1899. The relationship between cystine and cysteine was recognized in 1883, and their structures were elucidated by synthesis from benzoylserine and phosphorus sulfide.

Cystine is converted to cysteine on reduction with tin or zinc in acidic solution, while cysteine is converted to cystine by atmospheric oxidation in weakly alkaline solution in the presence of a catalytic amount of ferric ions. Although it has been ascertained that cysteine occurs in proteins, acid hydrolysis of proteins gives only cystine. Cystine is found in many proteins but its content is usually low. Cystine-rich proteins include the keratins of hair and horn.

Cystine is obtained from hydrolyzate of human hair, based on its solubility, which is the lowest of all the amino acids in the hydrolyzate. Chemical synthesis has not been undertaken industrially, and no fermentation technique has been reported for cystine production.

Methods for the chemical synthesis of cystine include: (1) condensation of acetamidomalonic ester with $RSCH_2Cl$ (R is a masking group which is easily removed); (2) the Strecker method, or Bucherer hydantoin synthesis; (3) routes via thiazoline; (4) use of serine as a starting material; and (5) introduction of a thiol group on the β-carbon atom of acetamidoacrylate or α-chloroacrylate. Although these methods involve many problems which have still to be resolved, methods which may be useful in practice on an industrial scale from the viewpoint of starting material and ease of operation will be described.

4.7.1 The Bucherer method[1)]

The aldehyde (III) is prepared by acid treatment of the acetal (II) derived from the bromoacetal (I) by condensation with benzylmercaptan in the presence of sodium alcoholate. *S*-Benzyl-DL-cysteine (V) is synthesized via the hydantoin intermediate (IV) derived from the aldehyde (III) by Bucherer's hydantoination process. Cysteine (VI) is obtained when *S*-benzyl-DL-cysteine (V) is reduced with metallic sodium in liquid ammonia to remove the benzyl group. Cysteine is converted to cystine (VII) by atmospheric oxidation. The yield at each step is comparatively high. L-Cystine can also be prepared by resolution of *S*-benzyl-DL-cysteine followed by debenzylation. However, the processes of bromoacetal (I) synthesis and *S*-benzylcysteine reduction should be further improved for mass production.

$$\underset{\text{I}}{BrCH_2CH(OC_2H_5)_2} \xrightarrow[C_2H_5ONa]{C_6H_5CH_2SH} \underset{\text{II}}{C_6H_5CH_2SCH_2CH(OC_2H_5)_2} \xrightarrow[HCl]{}$$

$$\underset{\text{III}}{C_6H_5CH_2SCH_2CHO} \xrightarrow{NH_4CN,\ CO_2} \underset{\text{IV}}{C_6H_5CH_2SCH_2\text{—}CH\text{—}CO \ (\text{NH—CO—NH ring})} \longrightarrow$$

$$\underset{\text{V}}{C_6H_5CH_2SCH_2CH(NH_2)COOH} \xrightarrow[\text{liq. ammonia, Na}]{} \underset{\text{VI}}{\left(HSCH_2CH(NH_2)COOH\right)}$$

$$\xrightarrow[Fe^{3+}]{O_2} \underset{\text{VII}}{HOOCCH(NH_2)CH_2SSCH_2CH(NH_2)COOH}$$

4.7.2 Synthesis from serine

In addition to the historic method[2] of cystine synthesis from serine, in an attempt to identify cystine and cysteine, there is another method of converting the hydroxyl group of serine to a mercapto group by using thioacetic acid.[3]

$$\underset{\text{VIII}}{HOCH_2CH(NH_2)COOH} \longrightarrow HOCH_2CH(NHCOC_6H_5)COOH \xrightarrow{P_2S_5} HSCH_2CH(NHCOC_6H_5)COOH$$

$$\longrightarrow \underset{\text{VI}}{HSCH_2CH(NH_2)COOH} \xrightarrow[FeCl_3,\ NH_4OH]{O_2} \underset{\text{VII}}{\left(HOOCCH(NH_2)CH_2S\text{—}\right)_2}$$

$$\underset{\text{VIII}}{HOCH_2CH(NH_2)COOH} \xrightarrow[\text{pyridine}]{CH_3COSH,\ (CH_3CO)_2O} \underset{\text{IX}}{CH_3COSH_2\text{—}CH\text{—}CO \ (\text{N=C(CH}_3\text{)—O ring})}$$

$$\xrightarrow{HCl} \underset{\text{VI}}{HSCH_2CH(NH_2)COOH} \xrightarrow[78\%]{} \underset{\text{VII}}{\left(HOOCCH(NH_2)CH_2S\text{—}\right)_2}$$

When serine (VIII) and thioacetic acid in acetic anhydride are heated in the presence of a small amount of pyridine, *N,S*-diacetylcysteineazlactone

(IX) is formed. Its hydrolysis yields cysteine, which can be converted to cystine by oxidation. As the yield is comparatively good and the process is short, it may represent an industrially useful method provided that serine, the starting material, is obtainable at low cost.

4.7.3 Synthesis from α-chloroacrylic ester[4]

Hydrogen chloride is bubbled into an alcoholic solution of thiourea, and methyl α-chloroacrylate (X) is added. Hydrolysis of the addition product gives *S*-(β-chloro-β-carboxyethyl) isothiourea (XI) which is aminated and cyclized simultaneously by treatment with ammonia to yield 2-aminothiazoline-4-carboxylic acid (XII). Cysteine is obtained when the 2-aminothiazoline-4-carboxylic acid (XII) is reductively decyclized with sodium amalgam or by electrolysis. Cystine is prepared by oxidizing the cysteine.

$$\underset{\text{X}}{CH_2{=}\underset{\substack{|\\Cl}}{C}{-}COOCH_3} \xrightarrow[\text{ii) HCl}]{\text{i) }NH_2CSNH_2,\ C_2H_5OH\text{-}HCl} \underset{\text{XI}}{\begin{matrix}H_2N^+\\H_2N\end{matrix}\!\!\gg\! C{-}S{-}CH_2\underset{\substack{|\\Cl}}{C}HCOO^-}$$

$$\xrightarrow{NH_3} \underset{\text{XII}}{\begin{matrix}CH_2{-}CH{-}COOH\\ |\qquad\ \ |\\ S\qquad N\\ \diagdown\ \diagup\!\!\!\!\!/\\ C\\ |\\ NH_2\end{matrix}} \xrightarrow{H_2} \underset{\text{VI}}{HSCH_2\underset{\substack{|\\NH_2}}{C}HCOOH}$$

The yield at each step of the process is about 70%, so that further development may be required for a commercial method, both from the viewpoint of the raw material and the number of steps involved.

4.7.4 Synthesis of cysteine from cystine

L-Cystine, obtained from natural materials or by synthesis, is converted to L-cysteine by reduction. Methods of reduction include treatment with an excess of thiol reagents such as thioglycolic acid in neutral or alkaline solution,[5] reduction with metallic sodium in liquid ammonia,[6] treatment with tin in hydrochloric acid[7] and electrolytic reduction.[8,9] Of these methods, electrolytic reduction is advantageous for practical use on an industrial scale.

There are two methods for the electrolytic reduction of cystine. In one method, a solution of cystine in ammonia water is the catholyte,[9] and in the other, a solution of cystine in hydrochloric acid is the catholyte. The latter method is preferred.

The catholyte is a solution of L-cystine hydrochloride in dilute hydrochloric acid and the anolyte is 10% hydrochloric acid. A lead plate is used as the cathode and a graphite rod as the anode. Electrolysis is carried out at 5–6 V and 1.7–7 A/dm². After electrolysis is complete, L-cysteine hydrochloride is obtained almost quantitatively from the concentrated catholyte.

$$\begin{array}{l} SCH_2CHCOOH \\ \quad\quad | \\ \quad\quad NH_2\cdot HCl \\ SCH_2CHCOOH \\ \quad\quad | \\ \quad\quad NH_2\cdot HCl \end{array} + 2H_2O + 2e^- \longrightarrow \begin{array}{l} HSCH_2CHCOOH \\ \quad\quad\quad | \\ \quad\quad\quad NH_2\cdot HCl \end{array} + 2OH^-$$

Not only is the current efficiency high, but also the purity of the product is good with this reduction method. It is used to produce cysteine from the cystine extracted from hydrolyzates of human hair.

4.7.5 Resolution of cystine and cysteine

Cystine prepared by usual synthetic methods contains DL- and *meso*-isomers, and is optically inactive. By utilizing the action of aminoacylase, L-, D- and *meso*-isomers are separable from the mixture of acetylated compounds.[10] Resolution of cystine is generally accmplished at the level of cysteine derivatives, and optically active cysteine is easily oxidized to cystine. For example, *N*-formyl-*S*-benzylcysteine is resolved by brucine,[11] and *S*-benzyl-L-cysteine can be separated from *S*-benzyl-*N*-acetyl-DL-cysteine hydrolyzed asymmetrically by aminoacylase.[12] *S*-Benzyl-L-cysteine is converted to L-cysteine, releasing the benzyl group on treatment with metallic sodium in liquid ammonia. L-Cystine is prepared from L-cysteine by oxidation with air in the presence of a catalytic amount of ferric chloride.

References

1. G. Nadeau and R. Gaudry, *Can. J. Res.*, **27**, 421 (1949).
2. E. Erlenmeyer Jr., *Chem. Ber.*, **36**, 2720 (1903); E. Erlenmeyer Jr. and F. Stoop, *Ann. Chem.*, **337**, 236 (1904).
3. P. Rambacher, *Chem. Ber.*, **101**, 2595 (1968).
4. H. Behringer and P. Zillikens, *Ann. Chem.*, **574**, 140 (1951).
5. D. Goddard and L. Michaelis, *J. Biol. Chem.*, **112**, 361 (1935).
6. V. du Vigneaud, L. F. Audrieth and H. S. Loring, *J. Am. Chem. Soc.*, **52**, 4500 (1930).
7. *Ed.* S. Akabori and S. Mizushima, *Tampakushitsu Kagaku* (Japanese), vol. I, p. 389, Kyoritsu Shuppan, 1954.

8. Y. Hosaka, *Japanese Patent* No. 37–2022 (1962).
9. Y. Hosaka, *Japanese Patent* No. 37–3712 (1962).
10. R. Marshall, M. Winitz, S. M. Birnbaum and J. P. Greenstein, *J. Am. Chem Soc.*, **79**, 4538 (1957).
11. J. L. Wood and V. du Vigneaud, *J. Biol. Chem.*, **130**, 109 (1939).
12. S. M. Birnbaum and J. P. Greenstein, *Arch. Biochem. Biophys.*, **39**, 108 (1952).

4.8 3,4-Dihydroxyphenylalanine (Dopa)

3,4-Dihydroxyphenylalanine (Dopa) was first isolated from the broad bean, *Vicia faba*, in 1913 by Torquati. Guggenheim isolated it at almost the same time and elucidated the structure of its *N,O,O*-tribenzoyl derivative. While enzymatic methods have been examined for preparing Dopa from catechol and amino acids such as tyrosine, cysteine and serine, various methods of chemical synthesis have also been known from early times. The principal method of chemical synthesis involves protocatechualdehyde (3,4-dihydroxybenzaldehyde) or its derivatives with masked hydroxyl groups such as piperonal, vanillin, or veratraldehyde as starting materials.

4.8.1 Synthesis from vanillin

Harington and McCartney[1)] synthesized ethyl α-benzamido-3-methoxy-4-hydroxycinnamate (IV) by alkaline treatment of the azlactone (III)

$$\underset{\text{I}}{\text{HO-}C_6H_3(OCH_3)\text{-CHO}} + \underset{\text{II}}{CH_2(NHCOC_6H_5)COOH} \xrightarrow[75\%]{(CH_3CO)_2O,\ CH_3COONa} \underset{\text{III}}{CH_3COO\text{-}C_6H_3(OCH_3)\text{-}CH{=}C\text{—}CO\ (N{=}C(C_6H_5)\text{—}O)}$$

$$\xrightarrow[OH^-,\ 65\%]{C_2H_5OH} \underset{\text{IV}}{\text{HO-}C_6H_3(OCH_3)\text{-}CH{=}C(NHCOC_6H_5)COOC_2H_5} \xrightarrow[\text{red P},\ CH_3COOH]{HI} \underset{\text{V}}{(HO)_2C_6H_3\text{-}CH_2CH(NH_2)COOH}\ (50\%)$$

derived from vanillin (I) and hippuric acid (II) by the Erlenmeyer reaction. Refluxing the ester (IV) with concentrated hydrogen iodide and acetic acid in the presence of red phosphorus to reduce the double bond and liberate the masking group, they obtained Dopa (V).

Prior to this work, Fromherz and Hermanns[2)] synthesized Dopa (V) from α-benzamido-3-methoxy-4-hydroxycinnamic acid (VI) which was reduced with sodium amalgam to α-benzamido-β-(3-methoxy-4-hydroxy-

$$\underset{\text{VI}}{\text{HO-}C_6H_3(OCH_3)\text{-}CH{=}C(NHCOC_6H_5)COOH} \xrightarrow[50\%]{Na/Hg} \underset{\text{VII}}{\text{HO-}C_6H_3(OCH_3)\text{-}CH_2CH(NHCOC_6H_5)COOH}$$

$$\xrightarrow{HI} \text{HO-}\underset{\text{HO}}{\bigcirc}\text{-}CH_2\underset{NH_2}{CH}COOH \quad \text{(V)}$$

phenyl) propionic acid (VII) and then heated in hydrogen iodide.

Shaw *et al.*[3] treated vanillin with acetylglycine (VIII), and then treated the condensation product (the azlactone (IX)) with alkali and reduced the product with sodium amalgam to obtain 3-methoxy-4-hydroxyphenylalanine (X).

$$\underset{\text{I}}{\text{HO-}\underset{CH_3O}{\bigcirc}\text{-CHO}} + \underset{\text{VIII}}{CH_2(NHCOCH_3)COOH} \xrightarrow[69\%]{(CH_3CO)_2O,\ CH_3COONa} \underset{\text{IX}}{CH_3COO\text{-}\underset{CH_3O}{\bigcirc}\text{-}CH{=}C\text{—}CO \;(N{=}C(CH_3)\text{—}O)}$$

$$\xrightarrow{\text{i) }OH^-\ \text{ii) }Na/Hg} \underset{\text{X}}{\text{HO-}\underset{CH_3O}{\bigcirc}\text{-}CH_2\underset{NH_2}{CH}COOH} \quad 76\%$$

Britton and White[4] treated vanillin (I) with hydantoin (XI) and reduced the condensation product (5-vanillidene hydantoin (XII)) to 5-

$$\underset{\text{I}}{\text{HO-}\underset{CH_3O}{\bigcirc}\text{-CHO}} + \underset{\text{XI}}{CH_2\text{—}CO \;(NH\ NH)\; CO} \xrightarrow[CH_3OH]{NH(CH_2CH_2OH)_2} \underset{\text{XII}}{\text{HO-}\underset{CH_3O}{\bigcirc}\text{-}CH{=}C\text{—}CO \;(HN\ NH)\; CO}$$

$$\xrightarrow{H_2} \underset{\text{XIII}}{\text{HO-}\underset{CH_3O}{\bigcirc}\text{-}CH_2CH\text{—}CO \;(NH\ NH)\; CO} \xrightarrow{H^+} \underset{\text{V}}{\text{HO-}\underset{HO}{\bigcirc}\text{-}CH_2\underset{NH_2}{CH}COOH}$$

vanillyl hydantoin (XIII) which was then hydrolyzed to Dopa (V) with hydrogen bromide.

Prior to the above scheme, Deulofeu and Mendivelzua[5] treated vanillin with hydantoin in the presence of acetic anhydride and sodium acetate, and reduced the condensation product (XII) with sodium amalgam to obtain vanillyl hydantoin (XIII).

Hirai[6] treated vanillin (I) with glycine anhydride (XIV) in the presence of acetic anhydride to prepare di-(3-methoxy-4-acetoxybenzylideneglycine) anhydride (XV). He then obtained Dopa (V) by heating the conden-

CH_3O, HO–⟨ring⟩–CHO (I) + CH_2(CONH)(NHCO)CH_2 (XIV) $\xrightarrow{(CH_3CO)_2O}$

CH_3O, CH_3COO–⟨ring⟩–CH=C(CONH)(NHCO)C=CH–⟨ring⟩–$OCOCH_3$, OCH_3 (XV)

$\xrightarrow[\text{red P}]{\text{HI}}$ HO, HO–⟨ring⟩–$CH_2CHCOOH$ with NH_2 (V)

sation product (XV) with hydrogen iodide containing red phosphorus to remove the masking groups on the hydroxyl groups of the aromatic rings, to reduce the double bonds and to hydrolyze the amide bonds simultaneously. This method has been reexamined by Losse *et al.*[7]

4.8.2 Synthesis from piperonal

Mori[8] reduced piperonylidene acetone (XVII), which was formed by condensation of piperonal (XVI) with acetone, to prepare piperonyl acetone (XVIII), which was then hydroxyaminated by treatment with butyl nitrite. The resulting α-hydroxyiminopiperonyl acetone (XIX) was oxidized with sodium hypobromite to α-hydroxyimino-β-(3,4-methylenedioxyphenyl)propionic acid (XX), which was then catalytically reduced to 3,4-methylenedioxyphenylalanine (XXI). Dopa was obtained on hydrolysis of the product (XXI) with hydrogen iodide.

$$\text{XVI} \xrightarrow{CH_3COCH_3} \text{XVII (3,4-methylenedioxyphenyl)}-CH{=}CHCCH_3(=O) \xrightarrow[\text{Raney nickel}]{H_2}$$

$$\text{XVIII (3,4-methylenedioxyphenyl)}-CH_2CH_2CCH_3(=O) \text{ (73\% based on XVI)} \xrightarrow{C_4H_9ONO} \text{XIX (3,4-methylenedioxyphenyl)}-CH_2CCCH_3(=NOH)(=O) \text{ (84\%)} \xrightarrow{NaOBr}$$

$$\text{XX (3,4-methylenedioxyphenyl)}-CH_2CCOOH(=NOH) \text{ (80.5\%)} \xrightarrow[\text{Raney nickel}]{H_2} \text{XXI (3,4-methylenedioxyphenyl)}-CH_2CH(NH_2)COOH \text{ (82\%)} \xrightarrow{HI}$$

$$\text{V (3,4-dihydroxyphenyl)}-CH_2CH(NH_2)COOH \text{ (81\%)}$$

Barry *et al.*[9] prepared α-hydroxyimino-β-(3,4-methylenedioxyphenyl)-propionic acid (XX), by hydroxyiminating 3,4-methylenedioxybenzylmalonic acid (XXIII) and 3,4-methylenedioxybenzyl acetoacetic ester (XXII), derived from piperonal (XVI), with butyl nitrite. The hydroxyimino compound (XX) was catalytically reduced to 3,4-methylenedioxy-

$$\text{XVI (piperonal, -CHO)} \rightarrow \text{XXII (3,4-methylenedioxyphenyl)}-CH_2CH(COOC_2H_5)COCH_3 \xrightarrow[62\%]{C_4H_9ONO} \text{XX (3,4-methylenedioxyphenyl)}-CH_2CCOOH(=NOH)$$

$$\text{XVI} \rightarrow \text{XXIII (3,4-methylenedioxyphenyl)}-CH_2CH(COOH)COOH \xrightarrow[89\%]{C_4H_9ONO} \text{XX}$$

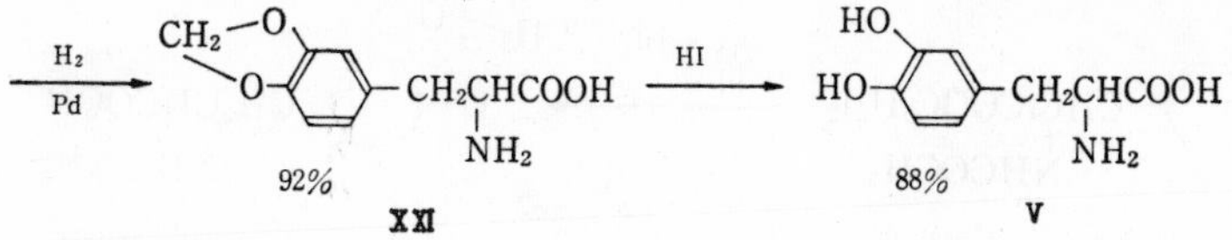

H₂/Pd → XXI (92%) → HI → V (88%)

phenylalanine (XXI), which was then hydrolyzed with hydrogen iodide to obtain Dopa (V).

Yamada *et al.*[10] treated the azlactone (XXIV), a condensation product of piperonal with acetylglycine,[11] with sodium-*l*-menthol. Partially asymmetric reduction of the resulting unsaturated ester (XXV) in methanol or benzene in the presence of palladium-carbon gave *l*-menthyl α-acetamido-β-(3,4-methylenedioxyphenyl)propionate (XXVI). Optically active 3,4-methylenedioxyphenylalanine (XXI) was obtained by hydrolyzing the ester (XXVI) with hydrogen iodide-acetic anhydride in the presence of red phosphorus.

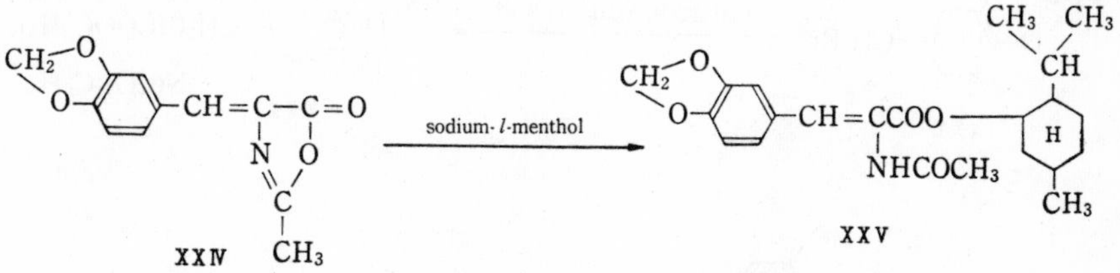

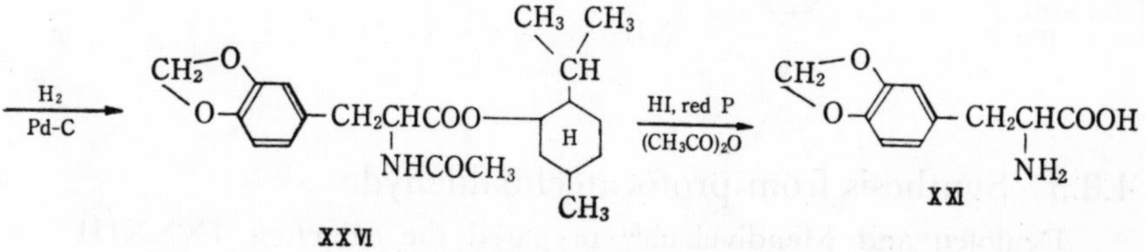

XXIV → sodium-*l*-menthol → XXV

H₂/Pd-C → XXVI → HI, red P, $(CH_3CO)_2O$ → XXI

Yamada *et al.*[12] prepared 3,4-methylenedioxybenzyl chloride (XXVIII) from 3,4-methylenedioxybenzyl alcohol (XXVII), derived from piperonal. They obtained 3,4-methylenedioxyphenylalanine (XXI) by saponifying and decarboxylating acetamido-3,4-methylenedioxybenzylmalonic acid ethyl ester (XXIX), the condensation product of the 3,4-methylenedioxybenzyl chloride (XXVIII) with ethyl acetamidomalonate.

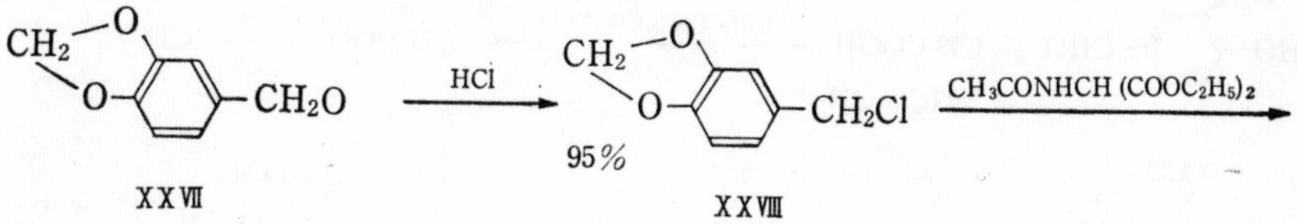

XXVII → HCl → XXVIII (95%) → $CH_3CONHCH(COOC_2H_5)_2$ →

$$\text{CH}_2\text{O}_2\text{C}_6\text{H}_3\text{-CHC(COOC}_2\text{H}_5)_2(\text{NHCOCH}_3) \xrightarrow[\text{ii) HCl}]{\text{i) NaOH}} \text{CH}_2\text{O}_2\text{C}_6\text{H}_3\text{-CH}_2\text{CH(NH}_2)\text{COOH}$$

64% XXIX 87% XXI

Matsuda and Matsumoto[13] have performed a similar synthesis using ethyl acetaminocyanoacetate instead of ethyl acetaminomalonate.

Stephens and Weizmann,[14] treating piperonyl bromide (XXX) with sodium salt of ethyl phtalimidomalonate, obtained Dopa (V) from the condensation product (XXXI) by hydrolysis with barium hydroxide followed by heating in a mixture of acetic acid and hydrochloric acid.

$$\text{CH}_2\text{O}_2\text{C}_6\text{H}_3\text{-CH}_2\text{Br} \xrightarrow{\text{C}_6\text{H}_4\text{(CO)}_2\text{NCH(COOC}_2\text{H}_5)_2} \text{CH}_2\text{O}_2\text{C}_6\text{H}_3\text{-CH}_2\text{C(COOC}_2\text{H}_5)_2\text{N(CO)}_2\text{C}_6\text{H}_4$$

XXX XXXI

$$\xrightarrow[\text{ii) H}^+]{\text{i) OH}^-} (\text{HO})_2\text{C}_6\text{H}_3\text{-CH}_2\text{CH(NH}_2)\text{COOH}$$

V

4.8.3 Synthesis from protocatechualdehyde

Deulofeu and Mendivelzua[5] prepared the azlactone (XXXIII) by condensing protocatechualdehyde (XXXII) with hippuric acid (II), and then converted it to the acrylate derivative (XXXIV) by acid treatment in alcohol. Dopa was obtained when the derivative (XXXIV) was heated in hydrogen iodide in the presence of red phosphorus to reduce the double bond and release the substituting group.

$$(\text{HO})_2\text{C}_6\text{H}_3\text{-CHO} + \text{CH}_2\text{(NHCOC}_6\text{H}_5)\text{COOH} \xrightarrow{\text{(CH}_3\text{CO)}_2\text{O, CH}_3\text{COONa}} (\text{CH}_3\text{COO})_2\text{C}_6\text{H}_3\text{-CH=C-CO(N=C(C}_6\text{H}_5)\text{-O)}$$

XXXII II 65% XXXIII

$\xrightarrow[C_2H_5OH]{H^+}$ HO, HO-C₆H₃-CH=C(NHCOC₆H₅)COOC₂H₅ (XXXIV, 50%) $\xrightarrow[\text{Raney nickel}]{H_2}$ HO, HO-C₆H₃-CH₂CH(NHCOC₆H₅)COOC₂H₅ (XXXV, 90%) $\xrightarrow{78\%}$ HO, HO-C₆H₃-CH₂CH(NH₂)COOH (V)

XXXIV $\xrightarrow{\text{HI, red P, } CH_3COOH}$ V

Matsuda and Matsumoto,[15] on the other hand, synthesized Dopa (V) by hydrolyzing α-benzoylamino-β-(3,4-dihydroxyphenyl)propionic acid ester (XXXV), which was derived from the acrylic acid ester derivative (XXXIV) by catalytic reduction.

Deulofeu and Mendivelzua[5] condensed protocatechualdehyde (XXXII) with glycine anhydride (XIV) by means of acetic anhydride and sodium acetate, obtaining bis-(3,4-diacetoxybenzylidene)diketopiperazine (XXXVI). Dopa (V) was obtained when the condensation product (XXXVI) was heated in hydrogen iodide in the presence of red phosphorus.

HO, HO-C₆H₃-CHO (XXXII) + CH_2(CONH)(NHCO)CH_2 (XIV) $\xrightarrow{(CH_3CO)_2O,\ CH_3COONa}$

CH_3COO, CH_3COO-C₆H₃-CH=C(CONH)(NHCO)C=CH-C₆H₃-$OCOCH_3$, $OCOCH_3$ (XXXVI, 71%)

$\xrightarrow[\text{red P}]{HI}$ HO, HO-C₆H₃-CH₂CH(NH₂)COOH (V)

Harington and Randall[16] synthesized Dopa (V) from protocatechualdehyde (XXXII) and acetylglycine (VIII). Treatment of (XXXII) and (VIII) in the presence of acetic anhydride and sodium acetate gave an azlactone derivative, which was then treated with water without any purification process. The resulting α-acetamino-β-(3,4-diactoxyphenyl)acrylic acid (XXXVII) was catalytically reduced to α-acetamino-β-(3,4-diacetoxy) propionic acid (XXXVIII) and the product treated with brucine for resolution of the optical isomers. Optically active Dopa (V) was obtained on hydrolysis of the resulting product with hydrochloric acid.

HO, HO–C₆H₃–CHO (XXXII) + CH_2COOH / $NHCOCH_3$ (VIII) $\xrightarrow{\text{i) } (CH_3CO)_2O,\ CH_3COONa;\ \text{ii) } H_2O}$

CH_3COO, CH_3COO–C₆H₃–$CH{=}CCOOH$ / $NHCOCH_3$ (XXXVII) $\xrightarrow{H_2}$ CH_3COO, CH_3COO–C₆H₃–$CH_2CHCOOH$ / $NHCOCH_3$ (XXXVIII)

$\xrightarrow{H^+}$ HO, HO–C₆H₃–$CH_2CHCOOH$ / NH_2 (V)

4.8.4 Synthesis from tyrosine

Waser and Lewandowski[17] and Sealock[18] synthesized L-Dopa from L-tyrosine. L-Tyrosine (XXXIX) was nitrated to form 3-nitro-L-tyrosine (XL) which was then reduced with hydrochloric acid and tin. The resulting 3-amino-L-tyrosine (XLI) was diazotized with barium nitrite in dilute sulfuric acid, and the solution boiled with copper sulfate. Dopa (V) was obtained in 30–40% yield.

HO–C₆H₄–$CH_2CHCOOH$ / NH_2 (XXXIX) $\xrightarrow{HNO_3}$ NO_2, HO–C₆H₃–$CH_2CHCOOH$ / NH_2 (XL)

$\xrightarrow{H_2}$ NH_2, HO–C₆H₃–$CH_2CHCOOH$ / NH_2 (XLI) $\xrightarrow{\text{i) } HNO_2;\ \text{ii) } CuSO_4}$ HO, HO–C₆H₃–$CH_2CHCOOH$ / NH_2 (V)

Since $[\alpha]_D$ for the synthesized product was $-12.7°$ (4% HCl), i.e identical with that of Dopa isolated from naturally occurring materials, the optical stereostructure of natural Dopa was elucidated.

4.8.5 Miscellaneous methods of synthesis

Funk,[19] treating 3,4-carbonyldioxybenzaldehyde (XLII) with hippuric acid (II) in the presence of acetic anhydride and sodium acetate,

XLII $-CHO + CH_2COOH(NHCOC_6H_5)$ (II) $\xrightarrow{(CH_3CO)_2O,\ CH_3COONa}$ XLIII ($-CH=C-CO$, N, O, C, C_6H_5)

$\xrightarrow{OH^-}$ HO, HO$-$ $-CH=CCOOH$ ($NHCOC_6H_5$) XLIV $\xrightarrow{Na/Hg}$ HO, HO$-$ $-CH_2CHCOOH$ ($NHCOC_6H_5$) XLV

$\xrightarrow{H^+}$ HO, HO$-$ $-CH_2CHCOOH$ (NH_2) V

prepared a condensation product, the azlactone (XLIII), which was converted to α-benzoylamino-3,4-dihydroxycinnamic acid (XLIV) by treatment with alkali. The resulting product (XLIV) was reduced with sodium amalgam to *N*-benzoyl Dopa (XLV) from which Dopa (V) was obtained by hydrolysis with hydrochloric acid.

Deulofeu and Mendivelzua[5] treated veratraldehyde (XLVI) with hippuric acid (II), and converted the resulting azlactone (XLVII) by alkali treatment to the acrylate derivative (XLVIII), which was reduced to form the intermediate (XLIX) and then hydrolyzed with alkali to yield 3,4-dimethoxyphenylalanine (L).

CH_3O, CH_3O- $-CHO + CH_2COOH$ ($NHCOC_6H_5$) XLVI, II $\xrightarrow{(CH_3CO)_2O,\ CH_3COONa}$ CH_3O, CH_3O- $-CH=C-CO$ (N, O, C, C_6H_5) 71% XLVII

$$\xrightarrow{OH^-} \text{3,4-}(CH_3O)_2C_6H_3-CH{=}C(NHCOC_6H_5)COOH \ (82\%,\ \text{XLVIII}) \xrightarrow{Na/Hg} \text{3,4-}(CH_3O)_2C_6H_3-CH_2CH(NHCOC_6H_5)COOH \ (60\%,\ \text{XLIX})$$

$$\xrightarrow{OH^-} \text{3,4-}(CH_3O)_2C_6H_3-CH_2CH(NH_2)COOH \ (75\%,\ \text{L})$$

Substituting hydantoin (XI) for hippuric acid, they synthesized 3,4-dimethoxyphenylalanine (L) through reduction and alkaline hydrolysis of the condensation product, 3,4-dimethoxybenzylidenehydantoin (LI).

$$\text{3,4-}(CH_3O)_2C_6H_3-CHO \ (\text{XLVI}) + \text{hydantoin} \ (CH_2-CO-NH-CO-NH,\ \text{XI}) \xrightarrow{(CH_3CO)_2O,\ CH_3COONa} \text{3,4-}(CH_3O)_2C_6H_3-CH{=}C\text{(hydantoin)} \ (71\%,\ \text{LI})$$

$$\xrightarrow{Na/Hg} \text{3,4-}(CH_3O)_2C_6H_3-CH_2\text{-}CH\text{(hydantoin)} \ (90\%,\ \text{LII}) \xrightarrow{OH^-} \text{3,4-}(CH_3O)_2C_6H_3-CH_2CH(NH_2)COOH \ (72\%,\ \text{L})$$

4.8.6 Resolution of DL-Dopa

An enzymatic method for L-Dopa production using *Erwinia herbicola*, which accumulates L-Dopa from pyrocatechol and serine, has been reported.[20]

There are two principal methods of obtaining optically active Dopa by resolution. One is to resolve either DL-Dopa or a simple derivative of Dopa, and the other is to synthesize Dopa after a chemically stable synthetic intermediate has been resolved. An enzymatic method of resolution by asymmetrically hydrolyzing acyl derivatives of intermediates in Dopa synthesis has been reported. Specifically, the *N*-acetyl derivative of 3,4-methylenedioxyphenylalanine was resolved using Takadiastase.[12] Several methods for chemical resolution have been reported, e.g. resolution of the ethyl ester of Dopa by using dibenzoyl-D-tartarate as a resolving agent,[21] and resolution of *N*-acyl derivatives of intermediates in Dopa synthesis, such as 3,4-methylenedihydroxyphenylalanine,[12] 3,4-dimethoxyphenylalanine,[22] 3,4-diacetoxyphenylalanine[16] and 3-methoxy-4-hydroxyphenylalanine,[23] by using optically active bases such as cinchonine, ephedrine or brucine as resolving agent.

One of the resolution methods advantageous for large-scale production is fractional crystallization. It is directly applicable to free DL-Dopa. Krieger *et al.*[24] achieved resolution using dilute hydrochloric acid as a solvent. This is advantageous, since the solubility and stability of DL-Dopa are greater in dilute hydrochloric acid, while the solubility in water is low. Use of the ammonium salt or *n*-dibutylamine salt of *N*-acetyl-3,4-methylenedihydroxyphenylalanine, an intermediate of Dopa synthesis, for resolution by fractional crystallization[25,26] is considered to be advantageous, permitting high recovery in fractional crystallization and in racemization of the undesired antipode.

References

1. C. R. Harington and W. McCartney, *Biochem. J.*, **21**, 852 (1927).
2. K. Fromherz and L. Hermanns, *Z. Physiol. Chem.*, **91**, 221 (1914).
3. K. N. F. Shaw, A. McMillan and M. D. Armstrong, *J. Org. Chem.*, **23**, 27 (1958).
4. E. C. Britton and H. C. White, *U.S. Patent* No. 2605282 (1952).
5. V. Deulofeu and G. Mendivelzua, *Z. Physiol. Chem.*, **219**, 233 (1933).
6. K. Hirai, *Biochem. Z.*, **114**, 67 (1921).
7. G. Losse, A. Barth and K. Jasche, *J. Prakt. Chem.*, **21**, 32 (1963).
8. K. Mori, *Nippon Kagaku Kaishi* (Japanese), **81**, 464 (1960).
9. R. H. Barry, A. M. Mattocks and W. H. Hartung, *J. Am. Chem. Soc.*, **70**, 693 (1948).
10. S. Yamada, T. Shioiri and T. Fujii, *Chem. Pharm. Bull.*, **10**, 688, 693 (1962).
11. T. Okuda and Y. Fujii, *Bull. Chem. Soc. Japan*, **30**, 698 (1957).
12. S. Yamada, T. Fujii and T. Shioiri, *Chem. Pharm. Bull.*, **10**, 680 (1962).
13. Y. Matsuda and I. Matsumoto, *Yakugaku Kenkyu* (Japanese), **29**, 508 (1957).
14. H. Stephens and C. Weizmann, *J. Chem. Soc.*, **105**, 1152 (1914).
15. Y. Matsuda and I. Matsumoto, *Japanese Patent* No. 35–916 (1960).
16. C. R. Harington and S. S. Randall, *Biochem. J.*, **25**, 1029 (1931).
17. E. Waser and M. Lewandowski, *Helv. Chim. Acta*, **4**, 657 (1921).
18. R. R. Sealock, *J. Biol. Chem.*, **166**, 1 (1946).
19. C. Funk, *J. Chem. Soc.*, **99**, 554 (1911).
20. H. Enei, H. Matsui, S. Okumura and H. Yamada, *Biochem. Biophys. Res. Comm.*, **43**, 1345 (1971).
21. G. Losse, A. Barth and W. Langenbeck, *Chem. Ber.*, **94**, 2271 (1961).
22. H. Nakamoto, M. Aburatani and M. Inagaki, *J. Med. Chem.*, **14**, 1021 (1971).
23. K. Volger and H. Baumgartner, *Helv. Chim. Acta*, **35**, 1776 (1952).

24. K. H. Krieger, J. Lago, M. Hill and J. A. Wantuck, *U.S. Patent* No. 3405159 (1968).
25. S. Yamada, M. Yamamoto and I. Chibata, *Nihon Hakkokogakkai Taikai Koen-yoshishu* (Japanese), p. 67, 1971.
26. K. Toi, T. Uzuki, M. Yuda, N. Nakayama and N. Sato, *Swiss Patent* No. 511774 (1971).

4.9 Glutamic acid

In 1866, Ritthausen isolated glutamic acid from a hydrolyzate of gliadin, an alcohol-soluble component of wheat gluten, suggesting the name "glutamic acid" on the basis of gluten being the source. The structure of glutamic acid was confirmed by the following experiments: (1) α-hydroxyglutaric acid was obtained on treatment of glutamic acid with nitrous acid, and this hyroxy acid was reduced by hydroiodic acid to glutaric acid; (2) natural and synthetic glutamic acid from levulinic acid were treated separately with nitrous acid, and the α-hydroxyglutaric acids obtained were found to be identical. In addition, it is now well known that glutamic acid is contained in numerous kinds of proteins, although it is especially abundant in wheat gliadin (43.7%) and soy-bean protein (19.5%).

In 1908, Ikeda discovered that L-glutamic acid is the essential taste substance of "*konbu*" (kelp, *Laminaria japonica*). Since then, monosodium L-glutamate (MSG) has been widely used as a seasoning to enhance the natural flavor of foodstuffs. MSG is now the amino acid with the largest individual industrial production.

In early times, glutamic acid was manufactured by extraction from natural sources, i.e. from proteins which contain glutamic acid as a major peptide (such as wheat gluten and soy-bean protein), or from Steffen's waste. In the latter process, pyrrolidonecarboxylic acid was separated from Steffen's waste, a transformation product of beet glutamine formed in the processing stage of sugar refining.

In 1957, Kinoshita and Asai independently reported new methods for the production of glutamic acid by direct fermentation. Glutamic acid was obtained in cultures of *Micrococcus glutamicus* or *M. varians* using a medium containing glucose and ammonium salts. The initial yield of glutamic acid was very high (20–30% from glucose). This fermentation method thus caused a revolution in the glutamic acid-producing industry. Subsequently, the yield of glutamic acid by this method has been greatly enhanced. The fermentation process requires starch or glucose as raw material, and the cost of producing L-glutamic acid is thus influenced by fluctuations in the price of these substances. Recently, this weak point has been overcome by shifting to the use of hydrocarbons or acetic acid, which can be stably supplied at low cost, as raw materials. On the other hand, in 1962, a new synthetic process for glutamic acid production was started commercially, as described below.

The old processes of extraction from natural sources have thus gradually been replaced by these new methods, and the fermentation process is fast becoming the main one of the MSG industry.

Many synthetic methods for glutamic acid production have been published.[1] They may be classified as follows: (1) applications of Strecker's reaction or of Bucherer's reaction via hydantoin, (2) condensation of a compound containing a carbon atom with attached amino and carboxylic groups (i.e. $H_2N{-}\overset{|}{\underset{|}{C}}{-}COOH$) with a 3-carbon compound (e.g. condensation of ethyl acetamidomalonate with acrylonitrile), (3) reduction of α-ketoglutaric acid in the presence of ammonia, and (4) other miscellaneous methods. However, these synthetic techniques were suitable only on a laboratory scale since the starting materials and reagents used are unsatisfactory from an industrial standpoint, and the processes are too complicated and the yields too low. In the 1950's, methods for synthesis of glutamic acid on an industrial scale were thoroughly studied. Ajinomoto Co., Inc. in Japan chose an oxo reaction of acrylonitrile as the best process for glutamic acid synthesis. Production was started at the plant in Yokkaichi City, Mie Pref., at an initial rate of 300 tons per month as monosodium L-glutamate monohydrate. The plant capacity was later raised to 1000 tons per month.

Recently, a new synthetic method from acrolein has been proposed, whereby acrolein is reacted with hydrogen cyanide to give α-hydroxyglutarodinitrile, which is then converted to glutamic acid by amination and hydrolysis. This method may be useful for industrial application.

All the above synthetic processes produce racemic glutamic acid. However, since the main use of glutamic acid is as a seasoning agent, for which only L-glutamic acid is suitable, it is necessary to resolve the racemic glutamic acid into its L- and D-isomers. L-Glutamic acid is isolated and converted to MSG, whereas the unwanted D-isomer is racemized and subjected again to optical resolution. Thus, further research is required to develop a practical method for obtaining L-glutamic acid by asymmetric synthesis, whereby the optical resolution step and racemization of unwanted D-isomer can be omitted.

4.9.1 Synthesis of glutamic acid from acrylonitrile

Acrylonitrile was selected as starting material because of its ready availability. (It is produced on a large scale for the manufacture of synthetic fibres.) The other raw materials of the process are carbon monoxide, hydrogen, methane and ammonia, and glutamic acid is obtained at a high rate of utilization of the carbon, nitrogen and hydrogen contained in these substances. The whole process has been adapted to automatically controlled continuous flow, so reducing the labor requirement.[2,3]

β-Cyanopropionaldehyde (II) is synthesized from acrylonitrile (I) by an oxo reaction, in which the reaction of acrylonitrile with carbon monoxide and hydrogen is carried out at high pressure in the presence of a soluble cobalt catalyst. Normally, oxo reactions involving simple olefins give a mixture of isomeric aldehydes; however, in this case, β-cyanopropionaldehyde is predominantly formed and the laborious process of isomer separation can be omitted. Moreover, the solvent and catalyst are both recovered and reused. The β-cyanopropionaldehyde (II) obtained is subjected to Strecker's reaction, giving the intermediate, α-aminoglutarodinitrile (III), which

$$\underset{\text{I}}{CH_2{=}CHCN} \xrightarrow{\text{oxo reaction}} \underset{\text{II}}{OHCCH_2CH_2CN} \xrightarrow{\text{Strecker's reaction}} \underset{\text{III}}{NCCH_2CH_2\underset{\displaystyle NH_2}{\underset{|}{C}}HCN}$$

$$\xrightarrow{\text{hydrolysis}} \underset{\text{IV}}{HOOCCH_2CH_2\underset{\displaystyle NH_2}{\underset{|}{C}}HCOOH}$$

contains two cyano groups and one amino group. The two nitrogen atoms of the cyano groups are recovered as ammonia in the following hydrolysis step, which yields glutamic acid (IV). This ammonia is recycled for use in the production of hydrogen cyanide. However, since the conversion rate to hydrogen cyanide is not quantitative, additional ammonia must also be supplied.

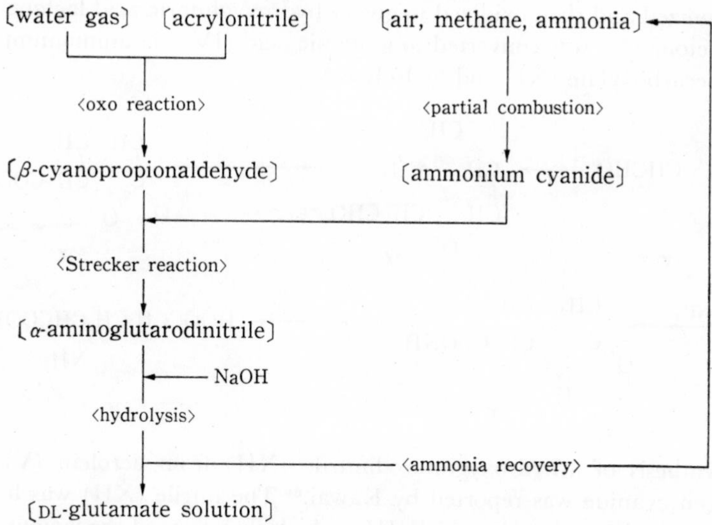

Fig. 4–3. General flowsheet for DL-glutamate synthesis from acrylonitrile.

DL-glutamic acid is crystallized out from the alkaline hydrolyzate, after neutralization with recycled glutamic acid solution containing an excess of sulfuric acid. A simple scheme for the synthetic process is given in Fig. 4–3.

4.9.2 Synthesis of glutamic acid from acrolein

The synthesis of glutamic acid from acrolein has been investigated using several different routes. One of the earliest methods was that using Strecker's reaction, reported in 1925 by Keimatsu.[4)] Synthesis of the crucial aldehyde intermediate, succinic semialdehyde (VIII), was effected from acrolein (V) via the intermediates β-chloropropionaldehyde acetal (VI) and β-cyanopropionaldehyde acetal (VII). The aldehyde (VIII) was then converted to the pertinent aminonitrile by Strecker's reaction, and this hydrolyzed to give glutamic acid (IV).

$$\underset{\text{V}}{CH_2{=}CHCHO} \xrightarrow{HCl,\ C_2H_5OH} \underset{\text{VI}}{ClCH_2CH_2CH(OC_2H_5)_2} \xrightarrow{KCN} \underset{\text{VII}}{NCCH_2CH_2CH(OC_2H_5)_2}$$

$$\xrightarrow[H^+]{} \underset{\text{VIII}}{HOOCCH_2CH_2CHO} \xrightarrow{\text{Strecker's reaction}} \underset{\text{IV}}{HOOCCH_2CH_2CH(NH_2)COOH}$$

Acrolein dimer (IX), which is easily prepared from acrolein, also proved suitable as a starting material for glutamic acid synthesis. The dimer (IX) was ozonized and then oxidized to give α-hydroxyglutaric acid lactone (X). This lactone (X) was converted to glutamic acid (IV) via ammonium pyrrolidonecarboxylate (XI) and hydrolysis.[5)]

$$\underset{\text{V}}{CH_2{=}CHCHO} \longrightarrow \underset{\text{IX}}{\text{(dihydropyran: }CH{=}CH{-}CH_2{-}CH_2{-}CH(CHO){-}O\text{, ring)}} \xrightarrow{O_3} \underset{\text{X}}{\text{(lactone: }CH_2{-}CH_2{-}C(=O){-}O{-}CH(COOH)\text{, ring)}}$$

$$\xrightarrow{NH_3} \underset{\text{XI}}{\text{(}CH_2{-}CH_2{-}C(=O){-}NH{-}CH(COONH_4)\text{, ring)}} \longrightarrow \underset{\text{IV}}{HOOCCH_2CH_2CH(NH_2)COOH}$$

Synthesis of α-hydroxyglutarodinitrile (XII) from acrolein (V) and hydrogen cyanide was reported by Kawai.[6)] The nitrile (XII) was hydrolyzed to α-hydroxyglutaric acid (XIII), which then formed the lactone (X). This was converted to glutamic acid (IV) by heating with ammonia.

$$\underset{\text{V}}{CH_2{=}CHCHO} \xrightarrow[50°C]{2HCN} \underset{\text{XII}}{NCCH_2CH_2\underset{\underset{OH}{|}}{C}HCN} \xrightarrow[H^+]{H_2O} \underset{\text{XIII}}{HOOCCH_2CH_2\underset{\underset{OH}{|}}{C}HCOOH}$$

$$\longrightarrow \underset{\text{X}}{\text{(CH}_2\text{-CH}_2\text{, C(=O)-O, CH-COOH lactone)}} \xrightarrow{NH_3} \underset{\text{XI}}{\text{(CH}_2\text{-CH}_2\text{, C(=O)-NH, CH-COONH}_4\text{ lactam)}} \longrightarrow \underset{\text{IV}}{HOOCCH_2CH_2\underset{\underset{NH_2}{|}}{C}HCOOH}$$

It was also reported that reaction of hydrogen cyanide with the double bond of acrolein in the presence of alkaline catalyst gave β-cyanopropionaldehyde (II).[7]

$$\underset{\text{V}}{CH_2{=}CHCHO} \xrightarrow[alkali]{HCN} \underset{\text{II}}{NCCH_2CH_2CHO}$$

The synthetic methods described above were not adopted industrially because of their low yield. However, in 1964, Shell Oil Co. patented a synthetic procedure for obtaining α-hydroxyglutarodinitrile from acrolein.[8] The technique was simple and the yield excellent. The reaction of acrolein with hydrogen cyanide was carried out in acetonitrile or aqueous or alcoholic media, in the presence of a catalytic amount of an alkali such as sodium cyanide or sodium hydroxide. The reaction temperature and time were generally controlled within the limits 50–60°C and 15–60 min.

$$\underset{\text{V}}{CH_2{=}CHCHO} \xrightarrow[OH^-]{2HCN} \underset{\text{XII}}{NCCH_2CH_2\underset{\underset{OH}{|}}{C}HCN} \xrightarrow{\text{i) amination, ii) hydrolysis}} \underset{\text{IV}}{HOOCCH_2CH_2\underset{\underset{NH_2}{|}}{C}HCOOH}$$

For example, 72 ml (1 mole) of acrolein and 103 ml (2.5 moles) of hydrogen cyanide were mixed into 500 ml of acetonitrile. Two grams of potassium *tert*-butoxide were cautiously added to the stirred mixture. An exothermic reaction occurred, and the temperature was kept below 36°C using a cooling bath. The temperature of the mixture was then raised to 71°C in 15 min, resulting in a very pale yellow coloration. After refluxing for 60 min, the acetonitrile was evaporated off. The yield of α-hydroxyglutarodinitrile was 72–75% as determined by gas chromatography after acetylation with acetic anhydride. The α-hydroxyglutarodinitrile thus obtained was aminated with ammonium hydroxide, and then hydrolyzed to glutamic acid.

When water or an alcoholic solution was used as a solvent for the reaction of acrolein and hydrogen cyanide, the subsequent amination and hydrolysis could be carried out without separation of the dinitrile (XII). Typical reaction conditions and yields of dinitrile are listed in Table 4–1.

This process is considered advantageous because of the low cost of the raw materials, of which acrolein is prepared from propylene and hydrogen cyanide obtained as a by-product of Sohio's acrylonitrile process. Moreover, since the reaction can be operated under relatively mild conditions, without high temperature or pressure, the cost of construction need not be great.

TABLE 4–1
Reaction conditions of acrolein and hydrogen cyanide, and yield of α-hydroxyglutarodinitrile

Cyanide	Catalyst	Solvent	Yield (%)
NaCN	—	water	37–38
HCN	NaCN	—	—
HCN	$KOC(CH_3)_3$	acetonitrile	72–75
HCN	$KOC(CH_3)_3$	*tert*-amylalcohol	62
HCN	$NaOC_2H_5$	ethanol	53
HCN	$KOC(CH_3)_3$	*tert*-butanol–water	68

4.9.3 Other miscellaneous methods of synthesis

Among the many known methods for glutamic acid synthesis, several others have been considered for possible industrial application. The details are as follows.

(1) Furfural (XIV) was converted to α-ketoglutaric acid (XV), and glutamic acid (IV) was then obtained by reductive amination of the acid (XV) via catalytic hydrogenation in the presence of aqueous ammonia.[9)]

$$\underset{\text{XIV}}{\text{(furan)–CHO}} \longrightarrow \underset{\text{XV}}{HOOCCH_2CH_2COCOOH} \xrightarrow[\text{amination}]{\text{reductive}} \underset{\text{IV}}{HOOCCH_2CH_2\underset{\displaystyle NH_2}{\underset{|}{C}}HCOOH}$$

(2) 3-Chlorocyclopentene, formed by the hydrochlorination of cyclopentadiene (XVI), was treated with ozone in a suitable solvent to give an ozonolysis intermediate, which was oxidized to α-chloroglutaric acid (XVII). The chloro acid (XVII) was then converted to glutamic acid (IV) by reaction with aqueous ammonia.[10)]

$$\text{cyclopentadiene (XVI)} \xrightarrow{HCl} \text{3-chlorocyclopentene} \xrightarrow{O_3} \underset{\text{XVII}}{HOOCCH_2CH_2CH(Cl)COOH}$$

$$\xrightarrow{NH_4OH} \underset{\text{IV}}{HOOCCH_2CH_2CH(NH_2)COOH}$$

(3) Reaction of diethyl oxalate (XVIII) with diethyl succinate (XIX) yielded diethyl α-ketoglutarate. This was then converted to glutamic acid (IV) by reduction in the presence of ammonia.[11]

$$\underset{\text{XVIII}}{C_2H_5OOCCOOC_2H_5} + \underset{\text{XIX}}{C_2H_5OOCCH_2CH_2COOC_2H_5} \xrightarrow[Na]{} C_2H_5OOCCH_2CH(COOC_2H_5)COCOOC_2H_5$$

$$\longrightarrow C_2H_5OOCCH_2CH_2COCOOC_2H_5 \longrightarrow \underset{\text{IV}}{HOOCCH_2CH_2CH(NH_2)COOH}$$

(4) Methyl β-formylpropionate (XXI) was obtained from methyl acrylate (XX) by a hydroformylation reaction. It was then subjected to Strecker's reaction and hydrolysis to give glutamic acid (IV).[12]

$$\underset{\text{XX}}{CH_2{=}CHCOOCH_3} \xrightarrow[\text{cobalt carbonyl}]{CO,\ H_2} \underset{\text{XXI}}{OHCCH_2CH_2COOCH_3}$$

$$\xrightarrow{\text{i) Strecker's reection; ii) hydrolysis}} \underset{\text{IV}}{HOOCCH_2CH_2CH(NH_2)COOH}$$

4.9.4 Resolution of DL-glutamic acid

Many processes are available for the optical resolution of racemic glutamic acid. They include the following:

(1) *N*-Acyl-DL-glutamic acid or DL-pyrrolidonecarboxylic acid is brought in contact with an optically active base, and the diastereoisomeric salts formed are separated from each other on the basis of their difference in solubility.

(2) An asymmetric reaction of *N*-acyl-DL-glutamic acid is carried out through the action of enzymes such as acylase or papain.

(3) Asymmetric oxidation of DL-glutamic acid is carried out through the action of amino acid oxidase.

However, the above techniques are not suitable for application on an industrial scale.

The most suitable method for the industrial resolution of DL-glutamic acid is preferential crystallization. Since DL-glutamic acid forms a racemic mixture, D- and L-glutamic acid can be obtained by a selective crystallization procedure. For this purpose, DL-glutamic acid solution of a specific concentration is seeded alternately with L- and D-glutamic acid.[13)] This technique does not require the use of expensive optically active resolving agents or enzymes, or the protection of the amino group and carboxyl groups. The D-glutamic acid obtained in the resolution process is racemized to DL-glutamic acid and this then re-used. Racemization occurs when D-glutamic acid is heated at 170–190°C in 30–50% sulfuric acid, and DL-glutamic acid is recovered at 97% yield.[14)] Also, when D-glutamic acid is heated at 190–200°C for 3 h in water, DL-pyrrolidonecarboxylic acid is formed. This can then be hydrolyzed quantitatively to DL-glutamic acid.[15)] A typical flowsheet for the resolution and racemization processes is given in Chapter 2 (sect. 2.1.1. B).

References

1. C. W. Huffman and W. G. Skelly, *Chem. Rev.*, **63**, 625 (1963).
2. J. Kato, H. Wakamatsu, T. Komatsu, R. Iwanaga and T. Yoshida, *Kogyo Kagaku Zasshi* (Japanese), **64**, 2142 (1961); J. Kato, I. Hayashi, I. Noda, R. Iwanaga and T. Yoshida, *ibid.*, **65**, 552 (1962).
3. R. Iwanaga, I. Iwamoto and T. Yoshida, *Japanese Patent* No. 33–267 (1958); J. Kato, H. Wakamatsu and H. Ishihara, *Japanese Patent* No. 36–2574 (1961); J. Kato, R. Iwanaga, I. Noda and K. Tahara, *Japanese Patent* No. 37–4312 (1962); J. Kato, R. Iwanaga, Y. Akiyama, H. Wakamatsu, T. Fujii, H. Ishihara and O. Hiwatashi, *Japanese Patent* No. 37–17209 (1962).
4. S. Keimatsu and S. Sugasawa, *J. Pharm. Soc. Japan*, **531**, 369 (1925).
5. R. Kodras, *U.S. Patent* No. 2904555 (1959).
6. Y. Kawai, *Nippon Kagaku Kaishi* (Japanese), **80**, 1321 (1959); **81**, 1606 (1960).
7. D. T. Warner and O. A. Moe, *U.S. Patent* No. 2565537 (1951).
8. G. A. Kurhajee and D. S. La France, *U.S. Patent* No. 3131211 (1964).
9. J. Kato, O. Hiwatashi and S. Kanao, *Japanese Patent* No. 31–5713 (1956); 31–10218 (1956).
10. J. Purvis, *U.S. Patent* No. 2833786 (1958).
11. L. Friedman and E. Kosower, *Org. Syn.*, **26**, 42 (1946).
12. R. Iwanaga, I. Iwamoto and T. Yoshida, *Japanese Patent* No. 32–7768 (1957).
13. T. Akashi, *Nippon Kagaku Kaishi* (Japanese), **81**, 421 (1960); **83**, 417 (1962); T. Ogawa and T. Akashi, *Japanese Patent* No. 31–2972 (1956).
14. T. Akashi, *British Patent* No. 893272 (1962).
15. L. E. Arnow and J. C. Opsahl, *J. Biol. Chem.*, **134**, 649 (1940).

4.10 Glutamine

Glutamine was first found in 1868 by Scheibler, who suggested the presence in beet molasses of a compound whose carbon number was one greater than asparagine. In 1877, Schulze isolated a compound which was considered to be an amide of glutamic acid from beet-root extract, and named it "glutamine". Schulze and Basshard then prepared pure glutamine from beet-root extract in 1883. In 1932, Damodavin *et al.* isolated glutamine from hydrolyzates of edestin and gliadin treated with pepsin, trypsin and yeast peptidase, and emphasized that glutamine is a constituent of proteins.

Glutamine is obtained by extraction from natural materials, fermentation, or synthesis. For the extraction process, beet-root, which contains large quantities of glutamine, is the main starting material.[1)] For large-scale production, either fermentation or synthesis is suitable.

4.10.1 Synthesis from glutamic acid

Since the major part of the structure of the glutamine molecule is identical with that of glutamic acid, which is produced on a large scale at low cost, the latter can be used as raw material for glutamine synthesis. Glutamic acid is esterified and the resulting γ-methyl ester converted to γ-hydrazine or γ-hydroxamic acid, which is then reduced to give glutamine. In another method, the γ-methyl ester is amidated with ammonia to prepare glutamine. In this case, the α-amino group of the γ-methyl ester must be masked with a suitable group before the amidation step is carried out. If the α-amino group is not masked, amidation readily leads to intramolecular cyclization, forming pyrrolidonecarboxylic acid.

$$\underset{\text{I}}{H_3COOCCH_2CH_2\underset{\displaystyle NH_2}{\underset{|}{C}}HCOOH} \xrightarrow{NH_2NH_2} \underset{\text{II}}{H_2NHNOCCH_2CH_2\underset{\displaystyle NH_2}{\underset{|}{C}}HCOOH}$$

$$\xrightarrow[\text{Raney nickel}]{H_2} \underset{\text{III}}{\left[\left(H_2NOCCH_2CH_2\underset{\displaystyle NH_2}{\underset{|}{C}}HCOO\right)_2 Ni\right]} \xrightarrow{H_2S \text{ or dimethylglyoxime}}$$

$$\underset{\text{IV}}{H_2NOCCH_2CH_2\underset{\displaystyle NH_2}{\underset{|}{C}}HCOOH}$$

Akabori and Narita[2] and Chang,[3] after synthesizing the γ-hydrazide (II) by treatment of the γ-methyl ester of L-glutamic acid (I) with hydrazine, prepared L-glutamine by treating the hydrazide (II) with Raney nickel. In practice, however, since L-glutamine forms a nickel complex (III), it is necessary to cleave the complex or to remove nickel with dimethylglyoxime prior to isolating the L-glutamine (IV). This method is not economic because it requires a large quantity of Raney nickel, which must then be removed with dimethylglyoxime.

Roper *et al.*[4] and Gipson *et al.*[5] synthesized L-glutamine by reacting the γ-methyl ester of L-glutamic acid (I) with hydroxylamine, followed by reduction of the resulting γ-hydroxamic acid (V) with Raney nickel. The

$$\underset{\text{I}}{H_3COOCCH_2CH_2\underset{\displaystyle NH_2}{\underset{|}{C}}HCOOH} \xrightarrow{NH_2OH} \underset{\text{V}}{HOHNOCCH_2CH_2\underset{\displaystyle NH_2}{\underset{|}{C}}HCOOH}$$

$$\xrightarrow[\text{Raney nickel}]{H_2} \underset{\text{III}}{\left[\left(H_2NOCCH_2CH_2\underset{\displaystyle NH_2}{\underset{|}{C}}HCOO\right)_2 Ni\right]} \xrightarrow{H_2S} \underset{\text{IV}}{H_2NOCCH_2CH_2\underset{\displaystyle NH_2}{\underset{|}{C}}HCOOH}$$

nickel complex of L-glutamine (III) is again formed by this method, and it is difficult to remove the nickel from the complex effectively. Also, the reactivity of hydroxylamine is rather low. These defects together make the method uneconomic.

Miyamoto *et al.*[6] prepared *N*-formylglutamine (VII) by formylation of the γ-methyl ester of L-glutamic acid (I) with formic acid to mask the

$$\underset{\text{I}}{H_3COOCCH_2CH_2\underset{\displaystyle NH_2}{\underset{|}{C}}HCOOH} \xrightarrow{HCOOH} \underset{\text{VI}}{H_3COOCCH_2CH_2\underset{\displaystyle NHCHO}{\underset{|}{C}}HCOOH}$$

$$\xrightarrow{NH_3} \underset{\text{VII}}{H_2NOCCH_2CH_2\underset{\displaystyle NHCHO}{\underset{|}{C}}HCOOH} \xrightarrow{HCl} \underset{\text{VIII}}{H_2NOCCH_2CH_2\underset{\displaystyle NH_2\cdot HCl}{\underset{|}{C}}HCOOH}$$

$$\xrightarrow{\text{Amberlite IR-4B}} \underset{\text{IV}}{H_2NOCCH_2CH_2\underset{\displaystyle NH_2}{\underset{|}{C}}HCOOH}$$

α-amino group, followed by amidation of the intermediate (VI) with ammonia. The *N*-formylglutamine (VII) was deformylated with hydrochloric acid to give L-glutamine hydrochloride (VIII), which was then treated with an ion-exchange resin to obtain L-glutamine (IV). Although this method consists of a series of reactions of simple operation, the number of steps is sufficient to reduce the overall yield to only 29%.

Schalk *et al.*[7] and Tanaka[8] synthesized L-glutamine (IV) from the γ-methyl ester of L-glutamie acid (I) by treatment of the ester with carbon disulfide and ammonia, to amidize the ester and at the same time mask the α-amino group, followed by acid hydrolysis of the intermediate ammonium salt (IX).

$$\underset{\text{I}}{H_3COOCCH_2CH_2\underset{|\atop NH_2}{C}HCOOH} \xrightarrow{CS_2,\ NH_3} \underset{\text{IX}}{H_2NOCCH_2CH_2\underset{|\atop NHCS_2NH_4}{C}HCOONH_4}$$

$$\xrightarrow{CH_3COOH} \underset{\text{IV}}{H_2NOCCH_2CH_2\underset{|\atop NH_2}{C}HCOOH}$$

The details are as follows. L-Glutamic acid is dissolved in methanol containing sulfuric acid. On leaving the mixture to stand, γ-methyl ester is formed. This ester can easily be obtained in a yield of 85% or higher, although the reaction is of course affected by the amounts of methanol and sulfuric acid used, the temperature, and standing time. The solution is saturated with ammonia after addition of carbon disulfide, kept at room temperature, and then amidated. It is desirable to add two moles of carbon disulfide per mole of L-glutamic acid to obtain a stable yield. This may be due to the fact that an excess of ammonium dithiocarbamate formed from carbon disulfide and ammonia can affect the amidation process. High pressure is not required for amidation, and the reaction is accomplished within 40–48 h at 10–30°C. Ammonium sulfate is crystallized out from the reaction mixture and removed; methanol and ammonia are recovered by condensation of the solution. When 4 N acetic acid is added to the residual solution and the mixture heated at 60°C for 30 min, glutamine is formed from the intermediate ammonium salt. Carbon disulfide can be recovered as excess ammonium dithiocarbamate is simultaneously cleaved. After the reaction mixture has been concentrated, methanol is added and the solution cooled. L-Glutamine then crystallizes out.

The characteristics and advantages of this method are its high yield (70–80%), practical simplicity of operation, and completely continuous process that involves no intermediate separation step.

4.10.2 Resolution of DL-glutamine

Resolution of DL-glutamine is not practiced commercially since L-glutamine is at present prepared by fermentation or synthesized from L-glutamic acid.

References

1. H. B. Vickery, *Biochem. Prepn.*, **1**, 44 (1949).
2. S. Akabori and K. Narita, *Proc. Acad. Sci.* (*Japan*), **29**, 264 (1953).
3. R. W. H. Chang, *U.S. Patent* No. 3084189 (1963).
4. J. A. Roper and H. McIlwain, *Biochem. J.*, **42**, 488 (1948).
5. R. M. Gipson, F. H. Pettit, C. G. Skinner and W. Shive, *J. Org. Chem.*, **28**, 1425 (1963).
6. M. Miyamoto, Y. Kawamatsu, M. Shinohara and Y. Ueno, *Yakugaku Zasshi* (Japanese), **81**, 439 (1961).
7. W. Schalk, P. Wirth, G. E. M. Dannenberg and V. Schmied-Kowarzik, *German Patent* No. 1080113 (1960).
8. M. Tanaka, *Japanese Patent* No. 40–6206 (1965).

4.11 Glycine

Glycine was first found by Braconnot in 1820 in an acid hydrolyzate of gelatin. Its composition was reported by Horsford in 1846 and its structure was elucidated by Cahours, who synthesized glycine from monochloroacetic acid and alcoholic ammonia in 1858.

Since glycine has no asymmetric carbon atom, it has been produced mostly by chemical synthesis. Commercially, it is prepared by amination of monochloroacetic acid, by the Strecker method, or by the Bucherer method. Although a number of other techniques using cyanoacetic acid, cyanocarboxylic acid and 2-aminoethanol as starting material are known, these are of less practical use than the methods just mentioned.

4.11.1 Synthesis from monochloroacetic acid

This method was first proposed by Cahours who obtained glycine (III) in 10–15% yield from monochloroacetic acid (I) and ammonia (II).

$$\underset{\text{I}}{ClCH_2COOH} + \underset{\text{II}}{NH_3} \longrightarrow \underset{\text{III}}{H_2NCH_2COOH}$$

Although this process has the advantage that it does not use hydrogen cyanide, it has the defect that the effective yield of glycine is reduced by the accompanying, subsidiary formation of secondary and tertiary amine analogues. To repress the formation of these subsidiary products, various investigations[1-3] have been carried out, and some improvement has been attained by controlling the quantitative ratio of ammonia to monochloroacetic acid, adding ammonium carbonate to the reaction system, and using aminating reagents other than ammonia. For example, glycine was obtained in 84.5% yield from monochloroacetic acid by careful treatment with 60 molar equivalents of ammonia water at 50°C for 4 h.[1] The method of addition of ammonium carbonate to the system has also improved the yields. When a reaction mixture of monochloroacetic acid:ammonia:carbon dioxide (1:12:5) was incubated at 60°C for 4 h, glycine was obtained in 80.5% yield.[1]

Hillman *et al.*[4] synthesized glycine ethyl ester by refluxing monochloroacetic acid with hexamethylenetetramine in dioxane, and treating the resultant addition product with alcohol containing hydrochloric acid. Guinot[5] obtained glycine in 92% yield by treating a small amount of hexamethylenetetramine with monochloroacetic acid and ammonia water at 75–80°C and pH 6. Williams[6] prepared glycine in 92% yield by reacting 1 mole of monochloroacetic acid with 1.5 moles of formaldehyde and 3 moles of ammonia water at 30°C for 4h.

4.11.2 Synthesis from formalin

A. The Strecker method

In this method, aminoacetonitrile (V) is first prepared from formaldehyde (IV), hydrogen cyanide and ammonia. Hydrolysis of the product (V) then gives glycine.

$$\underset{\text{IV}}{CH_2O} + HCN + NH_3 \longrightarrow \underset{\text{V}}{H_2NCH_2CN} \longrightarrow \underset{\text{III}}{H_2NCH_2COOH}$$

The method of synthesis of aminoacetonitrile (V) has been much improved in the last 25 yr, and this substance is now produced on a large scale by a continuous process. Previously, Cocker and Lapworth[7] prepared aminoacetonitrile by treating 1 mole of aldehyde with 2 moles of hydrogen cyanide and 2 moles of ammonia. However, its synthesis from glycolonitrile (VI) and ammonia was later achieved by Gluud and Klempt,[8] and by Greshman and Schweitzer,[9] who carried out the reaction at a comparatively high temperature for a short time and obtained aminoacetonitrile (V) in about 85% yield.

$$\underset{\text{VI}}{HOCH_2CN} + NH_3 \longrightarrow \underset{\text{V}}{H_2NCH_2CN}$$

There are two methods applicable to continuous reaction equipment. Ajinomoto Co., Inc.[10] synthesized glycine in 87% yield by treating formalin, which was continuously absorbing crude hydrogen cyanide (prepared from methane and ammonia), with ammonia at 120°C for 2 min, and then hydrolyzing the product with alkali. Sanders and Herbert[11] treated formalin with hydrogen cyanide and ammonia at 100°C for 3.6 min in a tube reactor and obtained aminoacetonitrile in 83% yield. This was then hydrolyzed with sulfuric acid,[7,9] hydroxides of alkaline earth metals,[8,12] or sodium hydroxide,[13] etc. For example, glycine was synthesized in 86.7% yield on hydrolysis of the aminoacetonitrile with 5% sodium hydroxide followed by condensation of the solution, from which sodium ion was removed using a weakly acidic cation-exchange resin.

B. The Bucherer method

In this method, glycine is prepared by hydrolyzing hydantoin (VII) with a solution of an alkaline metal base.

$$\begin{array}{c} H_2C-C=O \\ | \quad\; | \\ HN \;\; NH \\ \backslash\,/ \\ C \\ \| \\ O \end{array}$$

VII

Since it was difficult to synthesize hydantoin from formalin,[14] i.e. compared to the preparation of its derivatives from other aldehydes or ketones, an efficient method for glycine synthesis using hydantoin as an intermediate was not developed until recently. Gresham and Schweitzer[15] synthesized hydantion (VII) from aminoacetonitrile (V). White *et al.*[16] obtained hydantoin solution by incubating a solution of ammonium hydrogen carbonate and sodium cyanide, to which paraformaldehyde had been added. The mixture was stirred at room temperature until solution was complete, and heated at 80–85°C for 3h. The dissolved hydantoin has hydrolyzed at 170°C for 30 min with 30% sodium hydroxide solution, and glycine was obtained in 83.2% yield by treatment with a cation-exchange resin. When formalin is substituted for paraformaldehyde, it is necessary to add it slowly, dropwise, at a lower temperature. The reaction conditions and process are then more complicated.

A modified method[17] has been developed which accomplishes the hydrolysis directly, i.e. without the addition of alkali hydroxide, by heat-treatment of a solution of ammonium carbonate and sodium cyanide to which formalin is added simultaneously at room temperature. After the reaction is completed by removing the carbon dioxide evolved during the reaction at 170°C for 2h, the reaction mixture is treated with a weakly acidic cation-exchange resin. Glycine is then obtained in 87.5% yield. In the above methods, the hydantoin yield was only about 40%.[16] Thus, without the process of hydantoin isolation, the method of direct hydrolysis gives a satisfactory overall yield.

Treatment of glycolonitrile (VI) with ammonium carbonate at 80°C for 2 h in an autoclave, followed by treatment with an acid, gives hydantoin (VII) in about 80% yield.[18] Also, another method, similar to the modified one described above in that the whole reaction proceeds at once, has recently been reported.[19] According to this method, glycolonitrile (VI) is treated with one equivalent of ammonium carbonate and 4.5 moles of 8% ammonia water in an autoclave at 160°C for 6 h, and glycine is then obtained directly in 85% yield. This technique is advantageous in that the reaction mixture does not include any sodium ion.

References

1. N. D. Cheronis and K. H. Spitzmueller, *J. Org. Chem.*, **6**, 349 (1941).
2. M. S. Dunn, A. W. Butler and E. H. Frieden, *J. Phys. Chem.*, **45**, 1123 (1941).
3. J. P. Greenstein and M. Winitz, *Chemistry of the Amino Acids*, vol. 3, Wiley, 1961.
4. G. Hillman and A. Hillman, *Z. Physiol. Chem.*, **283**, 71 (1948).
5. M. H. M. Guinot, *French Patent* No. 1237327 (1960).
6. R. E. Williams, *U.S. Patent* No. 3190914 (1965).
7. W. Cocker and A. Lapworth, *J. Chem. Soc.*, **1931**, 1391.
8. W. Gluud and W. Klempt, *Ber. Ges. Kohlentech.*, **5**, 314 (1950).
9. W. F. Gresham and C. E. Schweitzer, *U.S. Patent* No. 2520312 (1950).
10. Ajinomoto Co., Inc., *British Patent* No. 908735 (1962).
11. K. W. Sanders and D. W. Herbert, *U.S. Patent* No. 3167582 (1965).
12. C. E. Schweitzer, *U.S. Patent* No. 2388189 (1945).
13. M. Tokime and Y. Mishima, *Japanese Patent* No. 29–8677 (1954).
14. H. R. Henze and R. J. Speer, *J. Am. Chem. Soc.*, **64**, 522 (1942).
15. W. F. Gresham and C. E. Schweitzer, *U.S. Patent* No. 2402134 (1946).
16. H. C. White and D. V. Wysong, *U.S. Patent* No. 2663712 (1953).
17. T. Shirai and T. Sakurai, *Japanese Patent* No. 36–21315 (1961).
18. T. Shirai, T. Sakurai and Y. Kawamura, *Japanese Patent* No. 39–24807 (1964).
19. Röhm and Haas G.m.b.H., *British Patent* No. 1138599 (1969).

4.12 Histidine

Histidine was discovered independently in 1896 by Kossel in sturine (a protamine of salmon sperm) and by Hedin in casein hydrolyzate. Concerning the chemical structure of histidine, the presence of the imidazole nucleus was suggested by Pauli in 1904 from the coloring reaction of this amino acid with diazobenzene-sulfonic acid. Shortly after, Knoop and Windaus found that treatment of histidine with nitrous acid to remove the α-amino group resulted in the formation of imidazole propionic acid. In 1911, Pyman finally confirmed the chemical structure of histidine by synthesis of the DL-amino acid via diaminoacetone from citric acid.

Nowadays, L-histidine is produced mainly by the isolation method, using hydrolyzate of blood powder of animals. In spite of the extensive studies on both fermentative and enzymatic methods to obtain histidine, no industrial procedure for histidine production has been found. Furthermore, chemical synthesis of histidine is relatively difficult since it has an imidazole nucleus, and a practically efficient method to synthesize histidine has not yet been established. It is therefore strongly hoped that an advantageous method for histidine production, either by synthesis or by fermentation without dependence on any natural protein, will be developed.

The fundamental problem in histidine synthesis is how to prepare the imidazole nucleus economically. Here, the main procedures for synthesis of the imidazole nucleus are first outlined, and then histidine synthesis from the imidazole nucleus is described.

4.12.1 Synthetic methods for hydroxymethylimidazole

The synthesis of hydroxymethylimidazole from fructose was first performed by Parrod[1,2] in 1930. Later investigations[3,4] to improve the reaction elicited a simplification of the operation and a good yield. An outline of this

CH_2OH–$C{=}O$–$(CHOH)_3$–CH_2OH (I) $\xrightarrow[\text{aeration}]{HCHO,\ NH_3,\ CuCO_3\cdot Cu(OH)_2}$ II

CH_2OH–$C{=}O$–CH_2OH (III) $\xrightarrow[\text{aeration}]{HCHO,\ NH_3,\ (CH_3COO)_2Cu}$ II

II: $HC{=}C\text{-}CH_2OH$ / $HN\ \ N$ / C / H (imidazole ring)

reaction is given in the diagram.

Agitation of a mixture of fructose (I), formaldehyde, ammonia and basic cupric carbonate under aeration with heating and subsequent addition of picric acid resulted in the precipitation of hydroxymethylimidazole picrate. By recrystallization and subsequent hydrolysis of the picrate, the free imidazole compound (II) was obtained in a yield of 50–70% based on fructose. When sucrose was used in place of fructose, the yield was rather low.[5]

Using a mixture of dihydroxyacetone, formaldehyde, ammonia and cupric acetate, hydroxymethylimidazole was obtained at a yield of 71% in the same manner as described above.[6] This procedure is expected to be one of the advantageous methods if dihydroxyacetone can be supplied at low cost.

4.12.2 Synthetic methods for histidine

A. Synthesis from chloromethylimidazole

In 1911 Pyman[7] synthesized hydroxymethylimidazole (II) from citric acid. This imidazole derivative was converted to chloromethylimidazole (IV) by the action of phosphorus pentachloride. Condensation of the latter imidazole with ethyl chloromalonate, followed by hydrolysis and decarboxylation with hydrochloric acid, led to the formation of α-chloro-β-imidazole propionic acid (VII). Amination of this compound (VII) with ammonia gave histidine (VIII) in a yield of about 4% based on citric acid or 17.8% based on hydroxymethylimidazole (II).

$$\underset{\text{II}}{\text{HC=C-CH}_2\text{OH (HN–CH=N ring)}} \xrightarrow{\text{PCl}_5} \underset{\text{IV}}{\text{HC=C-CH}_2\text{Cl (HN–CH=N ring)}} + \underset{\text{V}}{\text{ClCH(COOC}_2\text{H}_5)_2} \xrightarrow{\text{C}_2\text{H}_5\text{ONa}} \underset{\text{VI}}{\text{HC=C-CH}_2\text{CCl(COOC}_2\text{H}_5)_2 \text{ (HN–CH=N ring)}}$$

$$\xrightarrow[\text{91\%}]{\text{HCl}} \underset{\text{VII}}{\text{HC=C-CH}_2\text{CHClCOOH (HN–CH=N ring)}} \xrightarrow[\text{38\%}]{\text{NH}_3} \underset{\text{VIII}}{\text{HC=C-CH}_2\text{CH(NH}_2\text{)COOH (HN–CH=N ring)}}$$

With the progress in developing a general method for amino acid synthesis using ethyl acetamidomalonate (IX) or ethyl acetamidocyanoacetate (X), it has become easy to synthesize histidine experimentally by this method.

Condensation of chloromethylimidazole (IV) with acetamidomalonate (IX) or ethyl acetamidocyanoacetate (X) gave the condensation product (XI). This was converted directly to histidine by acid hydrolysis. Based on

(IV), the yield was 30.2% when ethyl acetamidomalonate was used,[8] whereas it was 40% when (IV) was condensed with ethyl acetamidocyanoacetate (X).[9]

$$\text{HC=C-CH}_2\text{Cl (imidazole ring: HN, N, CH) (IV)} + \text{R-HCCOOC}_2\text{H}_5\text{(NHCOCH}_3\text{)} \xrightarrow{\text{C}_2\text{H}_5\text{ONa}} \text{HC=C-CH}_2\text{C(R)(NHCOCH}_3\text{)COOC}_2\text{H}_5 \text{ (XI)}$$

R = $COOC_2H_5$ IX
R = CN X

$$\xrightarrow{\text{H}^+} \text{HC=C-CH}_2\text{CH(NH}_2\text{)COOH (VIII)}$$

B. Method using imidazolealdehyde

Histidine can also be synthesized using imidazolealdehyde as starting material, where this aldehyde is obtained by oxidation of hydroxymethylimidazole. For example, by the hydantoin method of Wheeler and Hoffman, imidazolealdehyde (XII) was condensed with hydantoin (XIII) in the presence of acetic anhydride-sodium acetate. Reduction of the condensation product (XIV) by sodium amalgam and subsequent alkaline hydrolysis led to histidine in a yield of 36% based on (XII).[10]

$$\text{HC=C-CH}_2\text{OH (II)} \xrightarrow{\text{HNO}_3} \text{HC=C-CHO (XII)} + \text{H}_2\text{C-C=O (HN, NH, C=O) (XIII)}$$

$$\xrightarrow{(\text{CH}_3\text{CO})_2\text{O},\ \text{CH}_3\text{COONa}} \text{HC=C-CH=C-C=O (CH}_3\text{CO-N, N; CH}_3\text{CO-N, NH, C=O) (XIV)} \xrightarrow{\text{Na/Hg}} \text{HC=C-CH}_2\text{-CH-C=O (HN, N; HN, NH, C=O)}$$

$$\xrightarrow{\text{OH}^-} \text{HC=C-CH}_2\text{CH(NH}_2\text{)COOH (VIII)}$$

Imidazolealdehyde (XII) was also condensed with hippuric acid (XV) by the application of Erlenmeyer's azlactone method to form the azlactone (XVI). This azlactone was then treated with sodium carbonate, and the resulting benzoyldehydrohistidine (XVII) was reduced with sodium amalgam to *N*-benzoylhistidine (XVIII). Hydrolysis of (XVIII) with hydrochloric acid gave histidine in a yield of about 20% based on the aldehyde (XII).[11]

```
HC=C-CHO  +  H2CCOOH      (CH3CO)2O, CH3COONa      HC=C-CH=C-C=O
 |   |          |         ------------------->      |   |   |  |
HN   N        NHCO-C6H5                    CH3CO-N   N    N  O
  \ //                                          \ //      \\/
   C                                             C         C
   H      XII     XV                       76%   H         |
                                                   XVI    C6H5

  Na2CO3                                Na/Hg
---------> HC=C-CH=CCOOH            ---------> HC=C-CH2CHCOOH
           |   |     |                         |   |      |
           HN  N    NHCO-C6H5                  HN  N     NHCO-C6H5
             \ //                                \ //
        88%   C                            42%    C
              H      XVII                         H      XVIII

---------> HC=C-CH2CHCOOH
   HCl     |   |      |
           HN  N     NH2
             \ //
              C
              H      VIII
```

There has also been a report on the short step synthesis of histidine.[12] Namely, condensation of the aldehyde (XII) with 2-mercaptothiazol-5-one (XIX) and subsequent treatment of the condensation product (XX) with a mixture of red phosphorus and hydrogen iodide, resulted in the formation of histidine in a good yield.

```
HC=C-CHO     H2C-C=O                       HC=C-CH=C-C=O
 |   |        |   |                         |   |   |  |
HN   N    +   HN  S    ---------------->    HN  N   HN  S
  \ //          \ /       CH3COOH             \ //    \ /
   C             C                             C       C
   H             ||                            H       ||
       XII       S     XIX                 80%    XX   S

  HI, red P
-----------> HC=C-CH2CHCOOH
  CH3COOH    |   |      |
             HN  N     NH2
               \ //
                C
          64%   H      VIII
```

4.12.3 Resolution of DL-histidine

Optical resolution of DL-histidine has not yet been conducted industrially since a suitable synthetic method for DL-histidine is itself still awaited. However, if it does become possible to produce DL-histidine at low cost, its optical resolution will no doubt be carried out practically. Of the reported methods for resolution of DL-histidine, only those which are relatively advantageous are discussed here.

Optical resolution of DL-histidine is feasible by direct resolution or by a selective crystallization which selectively crystallizes either of the two enantiomorphs by seeding of an optically active isomer into a supersaturated solution of DL-amino acid. Duschinsky[13)] added 340 g of L-histidine hydrochloride to 660 g of DL-histidine. He obtained 440 g of L-histidine hydrochloride and recovered 46 g of D-histidine hydrochloride and 540 g of DL-histidine. He also reported details of both the resolution and racemization of histidine.[14)] The enzymatic procedure for resolution of DL-histidine depends upon enzymatic asymmetric hydrolysis of the *N*-acyl or amide derivative of the racemic amino acid.[15)] For instance, acetyl-DL-histidine has been asymmetrically hydrolyzed in this way by hog kidney aminoacylase. The liberated free histidine was separated from unhydrolyzed acetyl-D-histidine using the difference in solubility or by employing an ion-exchange resin.[16)] Optical resolution of histidine has also been carried out by asymmetric hydrolysis of DL-histidine amide•HCl using hog kidney amidase.[17)] The yields for both enzymatic methods were about 80%.

References

1. P. Girard and J. Parrod, *Compt. Rend.*, **190**, 328 (1930).
2. J. Parrod, *Bull. Soc. Chim. France*, (4) **51**, 1424 (1932).
3. R. Weidenhagen, R. Herrmann and H. Wegner, *Chem. Ber.*, **70**, 570 (1937).
4. W. J. Darby, H. B. Lewis and J. R. Totter, *J. Am. Chem. Soc.*, **64**, 463 (1942); *Org. Syn. Coll.*, vol. 3, p. 460, 1955.
5. N. F. Albertson and S. Archer, *J. Am. Chem. Soc.*, **67**, 308 (1945).
6. R. Weidenhagen and R. Herrman, *Chem. Ber.*, **68**, 1953 (1935).
7. F. L. Pyman, *J. Chem. Soc.*, **99**, 668, 1386 (1911).
8. N. F. Albertson and S. Archer, *J. Am. Chem. Soc.*, **67**, 308 (1945).
9. N. F. Albertson and B. F. Tullar, *ibid.*, **67**, 502 (1945).
10. V. Deulofeu and A. E. Mitta, *J. Org. Chem.*, **14**, 915 (1949).
11. F. L. Pyman, *J. Chem. Soc.*, **109**, 186 (1916).
12. A. C. Davis and A. L. Levy, *ibid.*, **1949**, 2179.
13. R. Duschinsky, *Chem. Ind.*, **53**, 10 (1934).

14. R. Duschinsky, *Plaquette jubilaire Emile Christophe Barell. Basel*, p. 375, 1936.
15. S. M. Birnbaum and J. P. Greenstein, *Arch. Biochem. Biophys.*, **39**, 108 (1952).
16. C. G. Baker and H. A. Sober, *J. Am. Chem. Soc.*, **75**, 4058 (1953).
17. L. Levintow, V. E. Price and J. P. Greenstein, *J. Biol. Chem.*, **184**, 55 (1950).

4.13 Hydroxyproline

In 1902, Fischer isolated L(—)-hydroxyproline from gelatin hydrolyzate and proved that it is an hydroxy-derivative of proline, since it yields proline on reduction with hydrogen iodide and red phosphorus. From the fact that it occurs only in certain types of protein such as gelatin and collagen, and since it is considered to be a secondary product of L(—)-proline, formed by enzymic oxidation, it is often differentiated from amino acids which are common protein constituents.

Synthesis of this amino acid is comparatively difficult because it has two asymmetric carbon atoms. A resolution process is essential to obtain natural L(—)-hydroxyproline by chemical synthesis since DL-hydroxyproline and DL-allohydroxyproline are all produced. Fortunately, gelatin contains comparatively large quantities of L(—)-hydroxyproline, and the extraction and isolation of L-hydroxyproline and proline from gelatin hydrolyzate is the simplest way to obtain the natural L-isomers. L(—)-Hydroxyproline tastes refreshingly sweet, but it is not a common constituent of protein. Demand for L-hydroxyproline is negligible except for its use as a chemical reagent, and therefore, practical methods for its synthesis have not been developed.

4.13.1 Synthesis from epihalohydrin

Leuchs[1] first synthesized hydroxyproline from epichlorohydrin in an attempt to elucidate the position of the hydroxyl group. The product is a mixture of hydroxyproline and allohydroxyproline (1:3), but they can be separated as a result of the difference in solubilities of the copper salts in water. On hydrolysis, hydroxyproline is obtained as the salt of low solubility (blue), and allohydroxyproline as the salt of high solubility (violet). However, as a matter of course, it is difficult to purify a highly soluble salt.

Thereafter, several papers on the synthesis of hydroxyproline have been presented. They almost all involve α,γ-disubstituted compounds of

$$ClCH_2\text{-}CH(O)CH_2 + CH_2(COOC_2H_5)_2 \xrightarrow{NaOC_2H_5} ClCH_2\text{-}CH\text{-}CH_2\text{-}CH(COOC_2H_5)\text{-}CO\text{-}O \text{ (ring)}$$

$$\xrightarrow[\text{ii) acid hydrolysis}]{\text{i) } Br_2\ (Cl_2;\ SO_2Cl_2)} RCH_2\text{-}CH\text{-}CH_2\text{-}CH(R')\text{-}CO\text{-}O \text{ (ring)} \xrightarrow{NH_4OH} HO\text{-}CH\text{-}CH_2\text{-}CH(COOH)\text{-}NH\text{-}CH_2 \text{ (ring)}$$

R=Cl, R′=Br(Cl)

I

γ-valerolactone (I), as seen in Leuchs method. The resulting ratio of hydroxyproline to allohydroxyproline is at most 1:1 and the overall yield is 54% or lower.

The results of such syntheses via α,γ-disubstituted γ-valerolactone are summarized in Table 4–2.

TABLE 4–2

Synthesis of hydroxyproline/allohydroxyproline via α,γ-disubstituted γ-valerolactone

Raw materials for synthesis		(I) R	(I) R′	Hydroxyproline: allohydroxyproline formation rate	Ref.
$ClCH_2CH-CH_2$ (Br) (epoxide, —O—)	$CH_2(COOC_2H_5)_2$	Cl	Br(Cl)	1 : 3	(1)
		NH_2	Br	allo form	(2)
	$CH_3COCH_2COC\,C_2H_5$	Cl	NH_2	1 : 1	(3, 4)
	$CH(COOC_2H_5)_2$	Br	NH_2		(5)
	| $NHCHO$				(5)
$C_6H_4(CO)_2NCH_2CH-CH_2$ (epoxide, —O—)	$CH_2(COOC_2H_5)_2$	NH_2	Cl(Br)	1 : 1	(6)
$CH_2\ CHCH_2Br$	$CH_2(COOC_2H_5)_2$	Cl(Br)	Cl(Br)	1 : 1	(6)
	$CH(COOC_2H_5)_2$ | $NHCOCH_3$	Br(Cl)	NH_2	1 : 1	(7–9)
	$NCCHCOOC_2H_5$ | $NHCOCH_3$	Br(Cl)	NH_2	1 : 1	(7)

4.13.2 Synthesis from allybbromide

The product of condensation of acetamidodiethylmalonate with allyl bromide is brominated at a double bond to give the dibromide which is then boiled in water for a long period or hydrolyzed with acid. Hydroxyproline is obtained by direct treatment of the resulting product with barium hydroxide without isolation of the intermediate (I).

In our experience,[8] a mixture of diastereomers of (I) (R=Br; R′=NH_2HBr) is obtained in good yield by treating the dibromide, either isolated or not, with hydrogen bromide. These diastereomers can be separated as an insoluble salt (*cis* isomer) and a soluble salt (*trans* isomer) on the basis of their solubility in water. On heating each isomer with a solution of 10% sodium hydrogen carbonate, pure allohydroxyproline is obtained from the *cis* isomer, and crude hydroxyproline from the *trans* isomer. The yield from acetamidomalonate is about 55%. As hydroxyproline is easily purified from its insoluble copper salt, this method is convenient for preparing pure hydroxyproline and allohydroxyproline simultaneously.

$CH_2{=}CHCH_2Br$ + $CH(COOC_2H_5)_2$ (NHCOCH$_3$) ⟶ $CH_2{=}CHCH_2C(COOC_2H_5)_2$ (NHCOCH$_3$)

$\xrightarrow{Br_2/\text{ ether}}$ $CH_2(Br)CH(Br)CH_2C(COOC_2H_5)_2$ (NHCOCH$_3$) + $CH_2(Br)$-CH(O—CO)-CH_2-C($COOC_2H_5$)(NHCOCH$_3$)

(from $CH_2{=}CHCH_2C(COOC_2H_5)_2NHCOCH_3$: i) $Br_2/CHCl_3$ ii) HBr-H_2O, 100°C)

$\xrightarrow{HBr\text{-}H_2O,\ 100^{\circ}C}$ $BrCH_2$-CH(O—CO)-CH_2-CH-NH_2·HBr

$\xrightarrow{NaHCO_3,\ 90^{\circ}C}$ HO-CH-CH_2 / H_2C CH-COOH / N / H (hydroxyproline)

4.13.3 Other synthetic methods

Kuhn *et al.*[10] reported an interesting method via γ-ketoproline, but the yield was low.

Another example which yielded a mixture of optically active hydroxyproline and allohydroxyproline from L-histidine via lactones of erythro- and threo-γ-hydroxy-L-ornithine [(I): R=R′=NH_2] has been reported.[11]

As hydroxyproline is an uncommon amino acid, the number of papers published on its synthesis is less than for other amino acids which are com-

$$\begin{matrix} COOC_2H_5 \\ | \\ CH_2 \\ | \\ NHCOOC_2H_5 \end{matrix} + \begin{matrix} HCCOOC_2H_5 \\ \| \\ HCCOOC_2H_5 \end{matrix} \xrightarrow{Na/C_6H_6} \text{OC—CH-COOC}_2\text{H}_5,\ \text{H}_2\text{C}\ \text{CH-COOC}_2\text{H}_5,\ \text{N-COOC}_2\text{H}_5 \text{ (ring)}$$

$$\xrightarrow[100°C]{HCl} \text{OC—CH}_2,\ \text{H}_2\text{C}\ \text{CH-COOC}_2\text{H}_5,\ \text{N-COOC}_2\text{H}_5 \text{ (ring)}$$

i) $H_2/PtO_2/CH_3OH$; ii) $Ba(OH)_2$ → allohydroxyproline

i) $Al(OC_3H_7)_3$; ii) $Ba(OH)_2$ → hydroxyproline + allohydroxyproline

mon constituents of proteins. However, development of a stereoselective method for synthesizing predominantly a natural racemic mixture is expected.

4.13.4 Resolution of isomers of hydroxyproline

The existence of four isomers of hydroxyproline was proved in early times by Leuchs *et al.*,[12] who resolved the isomers of DL-phenylhydantoic acid derivatives with quinine (L-isomer forms the more insoluble salt). The 3,5-dinitrobenzyl derivatives were resolved with threo-1-*p*-nitrophenyl-2-aminopropan-1,3-diol,[13] and a natural isomer was separated from DL-hydroxyproline by D-amino acid oxidase from hog kidney.[14]

L-hydroxyproline —C_4-inversion[15–17]→ L-allohydroxyproline

L-hydroxyproline —epimerization at C_2[16] (1:2)→ D-allohydroxyproline

L-allohydroxyproline —(1:3)→ D-hydroxyproline

D-allohydroxyproline —C_4-inversion[16]→ D-hydroxyproline

Knowledge about interconversions between stereoisomers is important, especially in a compound which has several stereoisomers. In this case, C_4-inversion via *O*-tosyl derivatives and epimerization at the C_2 asymmetric center have been reported, as shown in the diagram. Based on this, DL-allo isomer can be converted to DL-isomer of the natural type.

References

1. H. Leuchs, *Chem. Ber.*, **38**, 1937 (1905); H. Leuchs, M. Giua and J. F. Brewster, *ibid.*, **45**, 1960 (1912).
2. W. Traube, R. Johow and W. Tepohl, *ibid.*, **56**, 1861 (1923).
3. H. McIlwain and G. M. Richardson, *Biochem. J.*, **33**, 44 (1939).
4. V. V. Feofilaktov and A. Onishchenko, *J. Gen. Chem. U.S.S.R., Engl. Transl.*, **9**, 331 (1939); *Compt. Rend. Acad. Sci. U.R.S.S.*, **20**, 133 (1938).
5. J. Capková-Jirku, J. V. Kostir and M. Vondracek, *Chem. Listy*, **44**, 19 (1950).
6. R. Gaudry and C. Godin, *J. Am. Chem. Soc.*, **76**, 139 (1954).
7. R. Gaudry, L. Berlinguet, A. Langis and G. Paris, *Can. J. Chem.*, **34**, 502 (1956).
8. Y. K. Lee and T. Kaneko, *Bull. Chem. Soc. Japan*, **46**, 2924 (1973).
9. T. Wieland and U. Wintermeyer, *Chem. Ber.*, **90**, 1721 (1957).
10. R. Kuhn and G. Oswald, *ibid.*, **89**, 1423 (1956).
11. B. Witkop and J. Beiler, *J. Am. Chem. Soc.*, **78**, 2882 (1956).
12. H. Leuchs and J. F. Brewster, *Chem. Ber.*, **46**, 986 (1913); H. Leuchs and K. Bormann, **52**, 2086 (1919).
13. L. Velluz, G. Amiard and R. Heymès, *Bull. Soc. Chim. France*, **21**, 1015 (1954).
14. J. R. Parikh, J. P. Greenstein, M. Minitz and S. M. Birnbaum, *J. Am. Chem. Soc.*, **80**, 953 (1958).
15. A. Neuberger, *J. Chem. Soc.*, **1945**, 431.
16. D. S. Robinson and J. P. Greenstein, *J. Biol. Chem.*, **195**, 383 (1952).
17. A. A. Patchett and B. Witkop, *J. Am. Chem. Soc.*, **79**, 185 (1957).

4.14 Isoleucine

In 1901, Fischer identified a substance with different optical rotation and solubility from leucine in crude leucine preparations of protein hydrolyzates. In 1904, Ehrlich isolated the same substance from beet-sugar molasses, and later also derived it from fibrin, wheat gluten and egg albumin. He assigned the name "isoleucine" to the substance, and showed that the properties of this amino acid were different from those of leucine even though both compounds had the same bulk chemical composition. In 1907, he synthesized isoleucine epimers from *d*-valeraldehyde. Four stereoisomers of isoleucine were later synthesized by Abderhalden and Zeisset in 1931. Since isoleucine has two asymmetric carbon atoms, there are four stereoisomers, namely L-, D-isoleucine and L-, D-alloisoleucine. Of these, L-isoleucine is nutritionally available but the other three isomers are not.

For the production of isoleucine, the isolation method, fermentation, and chemical synthesis are all practicable. However, even though isoleucine is widely distributed in proteins, its content is relatively low and purification of this amino acid is difficult. Nevertheless, isolation of an isoleucine-rich fraction is feasible, and this preparation may be employed when contamination with other amino acids such as leucine does not interfere with the purpose of use.

So far as fermentation is concerned, a method accumulating only L-isoleucine, by the addition of D-threonine to the medium as precursor, is practiced. Indeed, the production of isoleucine is expected to be carried out largely by such fermentation methods, in spite of problems such as limitations in the fermentation materials and the formation of other amino acids that are difficult to separate.

Besides the above methods, chemical synthesis is also carried out. However, an epimeric mixture of DL-isoleucine and DL-alloisoleucine is formed by the ordinary synthetic techniques, and so it is necessary to separate L-isoleucine from the mixture. Moreover, racemization of the remaining isomers after the isolation of L-isoleucine is difficult. For these reasons, the chemical synthesis of isoleucine is less favorable than fermentation, a situation which contrasts rather with that of amino acids having one asymmetric carbon.

Although there is some difficulty in obtaining optically pure *d*-valeraldehyde, when the aldehyde is readily available, a mixture of L-isoleucine and D-alloisoleucine can easily be prepared by synthesis. Then, after separating L-isoleucine from the mixture by optical resolution, epimerization of the residual D-alloisoleucine leads again to a mixture of L-isoleucine and D-alloisoleucine. This method is thus advantageous for the production of L-isoleucine.

In the future, it appears that the fermentative method will be developed further. However, in order to improve production by synthesis, formation of the allo form of isoleucine must be kept to a minimum. At the present, no especially efficient synthetic method is known. Thus, in order to compete with fermentation, the development of an asymmetric synthetic technique which specifically forms L-isoleucine is needed.

Many commercial preparations sold under the name "DL-isoleucine" are contaminated with DL-alloisoleucine, or occasionally a mixture of L-isoleucine and D-alloisoleucine. In some published reports on the synthesis of isoleucine, although the formation of a mixture of isoleucine isomers is described, the ratio of isoleucine to alloisoleucine is omitted. Therefore, it is necessary to note the L- or DL-isoleucine content of the products.

4.14.1 The Strecker method

As in the case of other amino acids, isoleucine has been synthesized for a long time from valeraldehyde by the Strecker method. For instance, treatment of valeraldehyde (2-methylbutylaldehyde) (I) with hydrogen cyanide in the presence of pyridine as catalyst gave cyanohydrin (II). Treatment of this with ammonia under pressure led to aminonitrile (III). Subsequent hydrolysis with hydrochloric acid resulted in isoleucine (IV) in an overall yield of 64%.[1)]

$$\underset{\text{I}}{(CH_3CH_2)(CH_3)CHCHO} \xrightarrow[\text{pyridine}]{HCN} \underset{\text{II}}{(CH_3CH_2)(CH_3)CHCH(OH)CN} \ (98\%) \xrightarrow{NH_3} \underset{\text{III}}{(CH_3CH_2)(CH_3)CHCH(NH_2)CN} \ (87.4\%)$$

$$\text{II} \xrightarrow[\text{HCOOH}]{HCONH_2} (CH_3CH_2)(CH_3)CHCH(NHCHO)CN \ (82\%) \longrightarrow \text{IV}$$

$$\text{III} \xrightarrow{HCl} \text{IV} \ (75\%)$$

$$\underset{\text{V}}{(CH_3CH)(CH_3)=CCHO} \xrightarrow[\text{Pt}]{H_2} \text{I}$$

$$\text{I} \xrightarrow{NaCN,\ (NH_4)_2CO_3} \underset{\text{VI}}{(CH_3CH_2)(CH_3)CHCH\text{–}C(=O)\text{–}NH\text{–}C(=O)\text{–}NH} \ (75\%) \xrightarrow{OH^-} \underset{\text{IV}}{(CH_3CH_2)(CH_3)CHCH(NH_2)COOH} \ (81.3\%)$$

It was reported that the DL-isoleucine content of this crude product was 68%, as calculated from the L-isoleucine content determined by the microbioassay method.[2)] Recently, it has been reported that heating of the cyanohydrin (II) with formic acid and formamide at 100–120°C gives *N*-

formyl-α-aminonitrile in a good yield (82%).[3] Hydrolysis of this nitrile yielded isoleucine.

Valeraldehyde (I) obtained by reduction of tiglaldehyde (V) has been used as starting material. Subsequent treatment of the aldehyde with sodium cyanide and ammonium carbonate according to the Bucherer method gave 5-*sec*-butylhydantoin (VI). Hydrolysis of this with barium hydroxide led to isoleucine (IV) in an overall yield of 61%.[4] The DL-isoleucine content of the crude product obtained by this method was calculated as 66% by the microbiassay method.

On the other hand, when *d*-valeraldehyde prepared by the oxidation of optically active *d*-amylalcohol (2-methyl-1-butanol) isolated from fusel oil is employed for the Strecker reaction as first performed by Ehrlich,[5] an epimeric mixture of L-isoleucine and D-alloisoleucine can be prepared. However, Ehrlich has reported that oxidation of the alcohol with chromic acid salt results in a poor yield of only about 25%, the overall yield of isoleucine from the alcohol being only 8%.

The above oxidation process has subsequently been improved to give a better result.[6] Namely, vapour-phase oxidation of *d*-amylalcohol in the presence of a silver catalyst gave *d*-valeraldehyde in a yield of 73%. This was converted to the hydantoin derivative by the Bucherer-Berg method, and subsequent hydrolysis of the hydantoin with sodium hydroxide led to an equimolar mixture of L-isoleucine and D-alloisoleucine in a yield of 74% or 54% based on the starting alcohol. This mixture was treated with formic acid and the resulting formyl derivative recrystallized from methylethylketone to give the insoluble formyl-L-isoleucine. Acid hydrolysis of the latter yielded L-isoleucine.

After separation of the formyl-L-isoleucine, the main ingredient in the mother liquor was formyl-D-alloisoleucine. Heating this with alkali again gave a mixture of L-isoleucine and D-alloisoleucine, from which L-isoleucine could be obtained by repeating the resolution operation described above. The main problem with these methods is how to prepare or obtain optically pure *d*-valeraldehyde, since the *d*-amylalcohol produced from fusel oil by fractional distillation is often contaminated with isoamylalcohol.

4.14.2 Synthesis via halogeno acid

A method based on the amination of α-halogeno acid has long been used for isoleucine synthesis. This method has been improved and remains even now as one of the most advantageous routes for isoleucine synthesis.

Treatment of diethyl malonate (VIII) with *sec*-butylbromide (VII) led to diethyl *sec*-butylmalonate (IX). Hydrolysis of this, followed by bromination and decarboxylation, yielded α-bromo-β-methylvaleric acid (X). The bromo acid was then aminated with aqueous ammonia to give isoleucine (IV) in an overall yield of 27%.[7] The DL-isoleucine content of the product formed by this method was calculated as 89% by the microbioassay method. Such a high content of DL-isoleucine may be attributable to product formation in low yield at the crystallization step.

$$\underset{\text{VII}}{\begin{matrix}CH_3CH_2\diagdown\\ CH_3\diagup\end{matrix}CHBr} + \underset{\text{VIII}}{\overset{COOC_2H_5}{\overset{|}{CH_2COOC_2H_5}}} \xrightarrow[83\sim84\%]{C_2H_5ONa} \underset{\text{IX}}{\begin{matrix}CH_3CH_2\diagdown\\ CH_3\diagup\end{matrix}\overset{COOC_2H_5}{\overset{|}{CHCHCOOC_2H_5}}}$$

$$\xrightarrow{\text{i) KOH, ii) HCl}} \left[\begin{matrix}CH_3CH_2\diagdown\\ CH_3\diagup\end{matrix}\overset{COOH}{\overset{|}{CHCHCOOH}}\right] \xrightarrow{Br_2} \left[\begin{matrix}CH_3CH_2\diagdown\\ CH_3\diagup\end{matrix}\underset{Br}{\underset{|}{\overset{COOH}{\overset{|}{CHCCOOH}}}}\right]$$

$$\xrightarrow[\text{heat}]{} \underset{\text{X}}{\begin{matrix}CH_3CH_2\diagdown\\ CH_3\diagup\end{matrix}\underset{Br}{\underset{|}{CHCHCOOH}}} \xrightarrow[49\%]{NH_3} \underset{\text{IV}}{\begin{matrix}CH_3CH_2\diagdown\\ CH_3\diagup\end{matrix}\underset{NH_2}{\underset{|}{CHCHCOOH}}}$$

66.7% based on IX

In order to obtain α-halogeno acid, a method using the less expensive starting material, methylethylketone, is available. Thus, the Reformatsky reaction between methylethylketone (XI) and ethyl bromoacetate (XII) was carried out to yield ethyl β-hydroxyl-β-methylvalerate (XIII). Without isolating the latter from the reaction mixture, dehydration by the action of

$$\underset{\text{XI}}{\begin{matrix}CH_3CH_2\diagdown\\ CH_3\diagup\end{matrix}C{=}O} + \underset{\text{XII}}{BrCH_2COOC_2H_5} \xrightarrow[I_2]{Zn,Cu} \underset{\text{XIII}}{\left[\begin{matrix}CH_3CH_2\diagdown\\ CH_3\diagup\end{matrix}\underset{OH}{\underset{|}{C}}CH_2COOC_2H_5\right]}$$

$$\xrightarrow[65\%]{ZnCl_2,(CH_3CO)_2O} \underset{\text{XIV}}{\begin{matrix}CH_3CH_2\diagdown\\ CH_3\diagup\end{matrix}C{=}CHCOOC_2H_5} \xrightarrow[89\%]{KOH} \begin{matrix}CH_3CH_2\diagdown\\ CH_3\diagup\end{matrix}C{=}CHCOOH$$

$$\xrightarrow[\text{Raney nickel}]{H_2} \begin{matrix}CH_3CH_2\diagdown\\ CH_3\diagup\end{matrix}CHCH_2COOH \xrightarrow[75\%]{Br_2} \underset{\text{X}}{\begin{matrix}CH_3CH_2\diagdown\\ CH_3\diagup\end{matrix}\underset{Br}{\underset{|}{CHCHCOOH}}}$$

92%

$$\xrightarrow[47.5\%]{NH_3} \underset{\text{IV}}{\begin{matrix}CH_3CH_2\diagdown\\ CH_3\diagup\end{matrix}\underset{NH_2}{\underset{|}{CHCHCOOH}}}$$

zinc chloride and acetic anhydride was carried out to give ethyl β-methyl-β-ethylacrylate (XIV). Hydrolysis and subsequent reduction of the product, followed by bromination, led to α-bromo acid (X) in an overall yield of 40%. Further amination of this acid gave isoleucine (IV) in an overall yield of 19%. The isoleucine content of the product was 71% as determined by chromatography of its dinitrophenyl (DNP) derivative.[8]

In addition to the above methods, a route starting from trichloroethylene (as described in the section on valine) is also applicable to isoleucine synthesis. That is to say, treatment of *sec*-butylchloride (XV) with trichloroethylene (XVI) in the presence of aluminium chloride resulted in the formation of the tetrachloro derivative (XVII). Treatment with sulfuric acid then yielded the α-chloro acid (XVIII). After prolonged heating of this acid in a sealed tube, amination occurred, and isoleucine (IV) was obtained in an overall yield of 41%. Although the isoleucine content of the product proved to be less than 35% by the microbioassay method, this technique is suitable for the purpose of obtaining alloisoleucine.[9]

$$\underset{\text{XV}}{(CH_3CH_2)(CH_3)CHCl} + \underset{\text{XVI}}{(H)(Cl)C{=}CCl_2} \xrightarrow[74\%]{AlCl_3} \underset{\text{XVII}}{(CH_3CH_2)(CH_3)CHCH(Cl)CCl_3}$$

$$\xrightarrow[67\%]{H_2SO_4} \underset{\text{XVIII}}{(CH_3CH_2)(CH_3)CHCH(Cl)COOH} \xrightarrow[83\%]{NH_3} \underset{\text{IV}}{(CH_3CH_2)(CH_3)CHCH(NH_2)COOH}$$

4.14.3 Other miscellaneous methods

Starting from methylethylketone, isoleucine can be synthesized by the azlactone method, as follows.[2] Methylethylketone (XI) was condensed with hippuric acid (XIX) to form azlactone (XX). Alkaline hydrolysis of the latter gave α-benzamide-β-ethyl-β-methylacrylic acid (XXI), and subsequent reduction and acid hydrolysis led to isoleucine (IV) in an overall yield of 33%. The DL-isoleucine content was 78%, as estimated by the microbioassay method, and recrystallization of the product improved the figure to 88%.

Simultaneous hydrolysis and reduction of azlactone by treatment with red phosphorus and hydriodic acid led to isoleucine in an overall yield of 20%. In this case, the isoleucine content of the product was estimated to be 50% by the microbioassay method.

$$\underset{\text{XI}}{(CH_3CH_2)(CH_3)C{=}O} + \underset{\text{XIX}}{CH_2(NHCO{-}C_6H_5)COOH} \xrightarrow{CH_3COONa,\ (CH_3CO)_2O} \underset{\text{XX}}{(CH_3CH_2)(CH_3)C{=}C{-}C{=}O\ (\text{oxazolone ring: N, O, C–}C_6H_5)} \xrightarrow{NaOH}$$

$$\underset{\text{XXI},\ 89\%}{(CH_3CH_2)(CH_3)C{=}C(NHCO{-}C_6H_5)COOH}$$

$$\text{XX} \xrightarrow[42\%]{HI,\ \text{red P},\ CH_3COOH} \text{IV}$$

$$\xrightarrow[\text{Raney nickel}]{H_2} \underset{90\%}{(CH_3CH_2)(CH_3)CHCH(NHCO{-}C_6H_5)COOH} \xrightarrow[89\%]{HCl} \underset{\text{IV}}{(CH_3CH_2)(CH_3)CHCH(NH_2)COOH}$$

Apart from these routes, amination by direct treatment of substituted malonic acid ester with chloramine has been reported.[10)] Diethylmalonate (VII) was reacted with *sec*-butylbromide (VIII) to obtain diethyl butylmalonate (IX). Treatment of the latter ester with chloramine in the presence of morpholine then gave diethyl α-amino-*sec*-butylmalonate (XXII) in a good yield (83%). This method appears to be of interest inasmuch as hydrolysis of the substituted aminomalonic acid ester readily yields isoleucine, although this reaction was not actually described in the above report (ref. 10).

$$\underset{\text{IX}}{(CH_3CH_2)(CH_3)CHCH(COOC_2H_5)COOC_2H_5} \xrightarrow[\text{morpholine}]{NH_2Cl} \underset{\text{XXII},\ 83\%}{(CH_3CH_2)(CH_3)CHC(NH_2)(COOC_2H_5)COOC_2H_5}$$

4.14.4 Separation of isoleucine isomers

As described above, ordinary chemical synthesis leads to the formation of a mixture of DL-isoleucine and DL-alloisoleucine. For the purpose of isolating a desired isoleucine isomer, two kinds of methods are available. In the first, DL-isoleucine and DL-alloisoleucine are first separated by their difference in solubility, and subsequent, appropriate optical resolution of each racemate gives the four isomers. For example, Abderhalden[11)] obtained the less soluble DL-isoleucine from an aqueous ethanol solution of a mixture of DL-isoleucine and DL-alloisoleucine. Esterification of the mother liquor, subsequent fractional distillation and hydrolysis, led to the separation of

DL-alloisoleucine. Each separated racemate was converted to its formyl derivative and then subjected to chemical resolution as the brucine salt, to give all four isomers. Huffman and Ingersol[12] acetylated a synthetically obtained mixture by the Schotten-Baumann method, and separated both isomers by repeated crystallization of the acetylated product from aqueous ethanol, whereby the less soluble acetyl-DL-alloisoleucine precipitated. Each racemate was optically resolved as the quinine salt and *l*-α-fenchylamine salt to give the four stereoisomers.

The second method is based on optical resolution of either the racemate of DL-isoleucine or DL-alloisoleucine, epimerization of each of the L- and D-isomers obtained, and resolution of the resulting mixture of epimers to yield each stereoisomer. For example, Greenstein *et al.*[13,14] treated a mixture of DL-isoleucine and DL-alloisoleucine with excess acetic anhydride in acetic acid to give an equimolar mixture of the four acetyl isomers. By this treatment, the barely soluble pure acetyl-DL-alloisoleucine was separated. This was resolved enzymatically by the action of hog kidney aminoacylase to yield L-alloisoleucine and acetyl-D-alloisoleucine. Each diastereomer was epimerized in acetic anhydride–acetic acid solution to the respective epimeric mixture, i.e. to a mixture of acetyl-L-alloisoleucine and acetyl-D-isoleucine, and a mixture of acetyl-D-alloisoleucine and acetyl-L-isoleucine. These mixtures were then resolved again with aminoacylase to yield the four stereoisomers of isoleucine.

Since aminoacylase from hog kidney can hydrolyze both acetyl-L-isoleucine and acetyl-L-alloisoleucine at almost the same rate, treatment of the mixture of acetyl-DL-isoleucine and acetyl-DL-alloisoleucine with this aminoacylase resulted in a mixture of L-isoleucine and L-alloisoleucine. On the other hand, mould aminoacylase can rapidly hydrolyze acetyl-L-isoleucine but it hydrolyzes acetyl-L-alloisoleucine only at about 7% of the rate for acetyl-L-isoleucine.

A synthesized product of high isoleucine content was therefore acetylated without epimerization. Subsequent treatment of the resulting mixture with aminoacylase from *Aspergillus oryzae* resulted in the rapid hydrolysis of acetyl-L-isoleucine, and almost pure L-isoleucine was obtained by this simple procedure.[15] The above method is thus advantageous, although reuse of the residual mixture of isomers for the production of L-isoleucine is difficult.

When DL-isoleucine containing no allo form is used as a starting material, acetylation without any accompanying epimerization, and subsequent asymmetric hydrolysis by aminoacylase gives L-isoleucine. Also, optical resolution of the ammonium salt of acetyl-DL-isoleucine can be performed by a preferential crystallization procedure.[16] In both cases, the acetyl-D-isoleucine cannot be reused for the production of L-isoleucine since epimerization yields a mixture of D-isoleucine and acetyl-L-alloisoleucine.

The material that is most suitable for L-isoleucine production is a mixture of L-isoleucine and D-alloisoleucine. In this case, the mixture may be acetylated and then resolved by aminoacylase.[13,14] Also, as described under the section on the Strecker method, the formyl derivative of the mixture can be separated by its difference in solubility.[6] The residual D-alloisoleucine may be used repeatedly for resolution due to the fact that epimerization gives a mixture of D-alloisoleucine and L-isoleucine.

References

1. W. F. Gresham and C. E. Schweitzer, *U.S. Patent* No. 2520312 (1950).
2. F. P. Doyle, D. O. Holland, W. Morflitt, J. H. C. Nayler and C. M. O'Connor, *J. Chem. Soc.*, **1955**, 1719.
3. F. Becke and P. Pässer, *Ann. Chem.*, **735**, 27 (1970).
4. J. E. Lirak and E. C. Britton, *U.S. Patent* No. 2553055 (1951).
5. F. Ehrlich, *Chem. Ber.*, **40**, 2538 (1907).
6. Dow Chemical Co., *British Patent* No. 704983 (1954).
7. C. S. Marvel, *Org. Syn. Coll.*, vol. 3, 495, 1955.
8. K. Mori, S. Kondo and H. Kumagae, *Nippon Kagaku Kaishi* (Japanese), **78**, 1174 (1957).
9. H. Kawashima, T. Takami and Y. Yamada, *ibid.*, **86**, 1190 (1965).
10. M. Horiike, J. Oda, Y. Inoue and M. Ohno, *Agr. Biol. Chem.*, **33**, 292 (1969).
11. E. Abderhalden and W. Zeisset, *Z. Physiol. Chem.*, **195**, 121 (1931).
12. W. A. Huffman and A. W. Ingersoll, *J. Am. Chem. Soc.*, **73**, 3366 (1951).
13. J. P. Greenstein, L. Levintow, C. G. Baker and J. White, *J. Biol. Chem.*, **188**, 647 (1951).
14. J. P. Greenstein, S. M. Birnbaum and L. Levintow, *Biochem. Prepn.*, **3**, 84 (1953).
15. I. Chibata, S. Yamada and H. Ito, *Japanese Patent* No. 37–12261 (1962).
16. M. Shibasaki, *Japanese Patent* No. 40–1576 (1965).

4.15 Leucine

The history of the discovery of leucine goes back to 1819 when Proust isolated a crude white substance from fermented milk curd. In the following year, Braconnot, the discoverer of glycine, isolated the same substance in a crystalline state from muscle protein hydrolyzate, and designated it "leucine" from the Greek work for white (*leukos*), based on the bright white color of the crystals. The chemical structure of leucine was confirmed later. In 1893, Schulze *et al.* synthesized leucine via the Strecker method, starting from isovaleraldehyde, and confirmed that the synthetic product was identical with that obtained by racemization of natural leucine. In 1904, Fischer carried out optical resolution of synthetic leucine, as the brucine salt of its formyl derivative, and obtained the optically active isomer.

Leucine production can be carried out by the isolation method, by chemical synthesis or by fermentation. However, most leucine is at present produced by the isolation method since it is relatively easily obtained from protein hydrolyzates. In particular, mixtures of leucine and isoleucine can readily be isolated, and such mixtures are utilized when a combination of both amino acids is permissible, For the purpose of obtaining pure, optically active leucine which is not contaminated with other amino acids, optical resolution of synthetic leucine is sometimes carried out.

The production of leucine by fermentation has also been studied and developed, and if the demand for leucine increases in the future, either chemical synthesis or fermentation is considered suitable to replace the isolation method. A large number of reports on the chemical synthesis of leucine have appeared, but only those procedures interesting from the standpoint of industrial production are treated here.

4.15.1 The Strecker method

Synthesis of leucine by the Strecker method, starting from isovaleraldehyde, has been employed for more than 100 yr and has been considerably refined. Even now, this synthetic method is one of the most advantageous techniques, so long as highly pure isovaleraldehyde is readily available.

For example, treatment of isovaleraldehyde (I) with liquid hydrogen cyanide in the presence of pyridine catalyst gave cyanohydrin (II). Subsequent reaction with ammonia under high pressure yielded aminonitrile (III), and hydrolysis of this with hydrochloric acid gave leucine (IV) in an overall yield of 74%.[1)] Also, by the Bucherer method, isovaleraldehyde (I) was converted to the hydantoin derivative (V) and autoclaving of this with

$$(CH_3)_2CHCH_2CHO \text{ (I)} \xrightarrow[\text{pyridine}]{HCN} (CH_3)_2CHCH_2CH(OH)CN \text{ (II)} \xrightarrow{NH_3} (CH_3)_2CHCH_2CH(NH_2)CN \text{ (III)} \xrightarrow{HCl} (CH_3)_2CHCH_2CH(NH_2)COOH \text{ (IV)}$$

$$(CH_3)_2CHCH_2CHO \text{ (I)} \xrightarrow{NaCN,\ (NH_4)_2CO_3} (CH_3)_2CHCH_2\text{-}CH\text{-}C{=}O \ (\text{HN–C(=O)–NH ring}) \text{ (V)} \xrightarrow{NaOH} (CH_3)_2CHCH_2CH(NH_2)COOH \text{ (IV)}$$

sodium hydroxide resulted in the formation of leucine (IV) in a yield of 84%.[2)]

4.15.2 Synthesis via halogeno acid

Leucine synthesis via α-halogeno acid has long been employed by Fischer and other workers. The method has been considerably improved and is now one of the most advantageous methods for leucine synthesis.

$$(CH_3)_2CHCH_2CH_2COOH \text{ (VI)} \xrightarrow[PCl_3]{Br_2} \underset{63\sim66\%}{(CH_3)_2CHCH_2CH(Br)COOH} \text{ (VII)} \xrightarrow{NH_3} \underset{43\sim45\%}{(CH_3)_2CHCH_2CH(NH_2)COOH} \text{ (IV)}$$

For example, Marvel[3)] added bromine to isocaproic acid (VI) in the presence of phosphorus trichloride (with heating) to obtain α-bromoisocaproic acid (VII). Then, treatment of this halogeno acid with ammonia yielded L-leucine (IV) in an overall yield of 28%. Also, the amination step can be carried out at a yield of 71% by the method using hexamethylenetetramine described in the section on valine (4.26).[4)]

4.15.3 Synthesis from acetamidomalonic acid ester or acetamidocyanoacetic acid ester

Leucine can be synthesized by alkylation of ethyl acetamidomalonate and subsequent hydrolysis of the alkylated product. Namely, condensation of diethylacetamidomalonate (VIII) with isobutylbromide (IX) in the presence of sodium ethoxide gave the condensation product (X). Alkaline hydrolysis of this, followed by decarboxylation, led to *N*-acetyl-DL-leucine in an overall yield of 29%.[5)] The acetyl derivative is convenient for optical resolution.

$$
\begin{array}{l}
\underset{\text{VIII}}{\mathrm{HC(COOC_2H_5)_2NHCOCH_3}} \xrightarrow[\mathrm{C_2H_5ONa}]{(\mathrm{CH_3})_2\mathrm{CHCH_2Br}\ (\mathrm{IX})} \underset{\text{X},\ 46\%}{(\mathrm{CH_3})_2\mathrm{CHCH_2C(COOC_2H_5)_2NHCOCH_3}} \xrightarrow{-\mathrm{OH}} \underset{\text{XI},\ 60\%}{(\mathrm{CH_3})_2\mathrm{CHCH_2CH(NHCOCH_3)COOH}} \xrightarrow{\mathrm{H^+}} \underset{\text{IV},\ 87\%}{(\mathrm{CH_3})_2\mathrm{CHCH_2CH(NH_2)COOH}} \\
\text{X} \xrightarrow{\mathrm{HBr}} \text{IV} \\
\text{VIII} \xrightarrow[\mathrm{C_2H_5ONa}]{\mathrm{CH_2{=}C(CH_3)CH_2Cl}\ (\text{XII})} \underset{\text{XIII},\ 79\%}{\mathrm{CH_2{=}C(CH_3)CH_2C(COOC_2H_5)_2NHCOCH_3}} \xrightarrow[\text{Raney nickel}]{\mathrm{H_2}} \text{X}\ (\text{quantitative})
\end{array}
$$

Furthermore, use of the cheap and readily available methylalkylchloride (XII), in place of isobutylbromide, facilitated the process of the reaction. By reduction of the resulting unsaturated compound (XIII) with Raney nickel, the same condensation product (X) as that formed with isobutylbromide was obtained in good yield. Hydrolysis of this by heating with hydrobromic acid led to leucine (IV) in an overall yield of 69%.[6)]

Alkylation of ethyl acetamidocyanoacetate (XIV) with isobutylbromide (IX), instead of diethylacetamidomalonate, and subsequent hydrolysis of the condensation product (XV) with hydrochloric acid, led to leucine in an overall yield of 51%.[7)]

$$
\underset{\text{IX}}{(\mathrm{CH_3})_2\mathrm{CHCH_2Br}} + \underset{\text{XIV}}{\mathrm{NC{-}CH(COOC_2H_5)NHCOCH_3}} \xrightarrow{\mathrm{C_2H_5ONa}} \underset{\text{XV},\ 65\%}{(\mathrm{CH_3})_2\mathrm{CHCH_2C(CN)(COOC_2H_5)NHCOCH_3}} \xrightarrow{\mathrm{HCl}} \underset{\text{IV},\ 79\%}{(\mathrm{CH_3})_2\mathrm{CHCH_2CH(NH_2)COOH}}
$$

4.15.4 Other miscellaneous methods

Methods for leucine synthesis using diketene, which seems promising as an industrial starting material, have been reported. Reaction of diketene (XVI) with ammonia gave acetoacetic amide (XVII) in almost quantitative yield. Condensation of this with isobutylbromide (IX) in the presence of sodium ethoxide yielded 2-isobutylacetoacetic amide (XVIII). Treatment of (XVIII) with a solution of potassium hypobromic acid caused the simultaneous occurrence of a Hofmann reaction and haloform reaction, and

$$\underset{\text{XVI}}{CH_2{=}C\text{-}CH_2\ (O\text{-}C{=}O)} \xrightarrow{NH_3} \underset{\text{XVII}}{CH_3COCH_2CONH_2} + \underset{\text{IX}}{(CH_3)_2CHCH_2Br}$$

$$\xrightarrow{C_2H_5ONa} \underset{41\%\quad \text{XVIII}}{(CH_3)_2CHCH_2CH(CONH_2)COCH_3} \xrightarrow{KOBr} \underset{50\%\quad \text{IV}}{(CH_3)_2CHCH_2CH(NH_2)COOH}$$

leucine was obtained in an overall yield of 20.5% based on acetoacetic amide.[8]

In addition, leucine synthesis via imino ester is of some interest, although certain operational problems arise because the process involves many reactions that require completely anhydrous conditions.[9] Namely, treatment of isopentylcyanide (XIX) with hydrochloride in absolute methanol led to iminoester hydrochloride (XX). Reaction of this with sodium hypochlorite gave *N*-chloroiminoester (XXI). Subsequent treatment of (XXI) with potassium butoxide resulted in a conversion reaction, and hydrolysis with hydrochloride gave leucine via the two intermediates (XXII, XXIII). Following this short reaction process, leucine (IV) was obtained in an overall yield of 52–59%. However, by changing the conditions of hydrolysis, leucine ethylester hydrochloride (XXIV) may also first be formed.

$$\underset{\text{XIX}}{(CH_3)_2CHCH_2CH_2CN} \xrightarrow[HCl]{CH_3OH} \underset{90\%\quad \text{XX}}{(CH_3)_2CHCH_2CH_2C({=}NH_2Cl)OCH_3} \xrightarrow{HOCl}$$

$$\underset{\text{XXI}}{(CH_3)_2CHCH_2CH_2C({=}NCl)OCH_3} \xrightarrow{tert\text{-}C_4H_9OH} \underset{\text{XXII}}{\left[(CH_3)_2CHCH_2CH{-}C(OCH_3)\ (\text{ring with }N)\right]} \xrightarrow{tert\text{-}C_4H_9OH}$$

$$\underset{\text{XXIII}}{\left[(CH_3)_2CHCH_2CH{-}C(OCH_3)(O\text{-}tert\text{-}C_4H_9)\ (\text{ring with }NH)\right]} \xrightarrow{HCl} \underset{65\%\text{ based on XX}\quad \text{XXIV}}{(CH_3)_2CHCH_2CH(NH_2\cdot HCl)COOCH_3}$$

$$\longrightarrow \underset{58\sim66\%\text{ based on XX}\quad \text{IV}}{(CH_3)_2CHCH_2CH(NH_2)COOH}$$

4.15.5 Resolution of DL-leucine

A number of methods for the optical resolution of DL-leucine have been reported. Those techniques which seem relatively useful for mass production of the amino acid are presented, here.

In the case of enzymatic resolution, acetyl-DL-leucine can be asymmetrically hydrolyzed by treatment with aminoacylase obtained from hog kidney[10] or moulds.[11] Optical resolution is then possible in good yield by separating the resultant products, L-leucine and acetyl-D-leucine. Such asymmetric hydrolysis has also been achieved with insolubilized mould aminoacylase.[12]

In the case of physicochemical resolution, preferential crystallization of either one of the optically active isomers from a supersaturated solution of acetyl-DL-leucine is known.[13]

References

1. W. F. Gresham and C. E. Schweitzer, *U.S. Patent* No. 2520312 (1950).
2. H. C. White, *U.S. Patent* No. 2557920 (1951).
3. C. S. Marvel, *Org. Syn. Coll.*, vol. 3, p. 523, 1955.
4. G. Hillmann and A. Hillmann, *Z. Physiol. Chem.*, **283**, 71 (1948).
5. H. R. Snyder, J. F. Shekleton and C. D. Lewis, *J. Am. Chem. Soc.*, **67**, 310 (1945).
6. N. F. Albertson and S. Archer, *ibid.*, **67**, 308 (1945).
7. N. F. Albertson and B. F. Tuller, *ibid.*, **67**, 502 (1945).
8. M. Yamato and K. Oshima, *Yakugaku Zasshi* (Japanese), **85**, 943 (1965).
9. H. E. Baumgarten, J. E. Dirks, J. M. Peterson and R. L. Zey, *J. Org. Chem.*, **31**, 3708 (1966).
10. S. M. Birnbaum, L. Levintow, R. B. Kingsley and J. P. Greenstein, *J. Biol. Chem.*, **194**, 455 (1952).
11. K. Michi and H. Tsuda, *Bull. Agr. Chem. Soc. Japan*, **21**, 235 (1957).
12. T. Tosa, T. Mori, N. Fuse and I. Chibata, *Enzymologia*, **32**, 153 (1967).
13. M. Shibasaki, *Japan Patent Appl.* No. 40–1576 (1965).

4.16 Lysine

In 1889, Drechsel isolated a basic substance (in fact containing lysine) and named it "lysatin". Shortly after, it was demonstrated that lysatin was impure, and contaminated with arginine. In 1891, a basic amino acid was isolated in a pure state from lysatin and designated as "lysine".

The chemical structure of lysine was presumed to be diaminocaproic acid based on the fact that putrefaction of lysine yielded cadaverine (pentamethylene diamine). In 1902, the final structure of this amino acid was confirmed by Fischer and Weigert, who showed that the chemical properties of lysine synthesized from γ-cyanophenylmalonic ester were coincident with those of the racemized natural amino acid. Although L-lysine can be produced by isolation from protein hydrolyzates and chemical synthesis, it is nowadays produced mainly by fermentation. In the case of chemical synthesis, extensive research on methods for L-lysine production (combined with appropriate optical resolution) has been carried out. With the recent progress in petroleum chemistry, and in other branches of the chemical industry, there is every indication that materials suitable for chemical synthesis may be supplied economically in large quantities. Therefore, in the future, mass production of L-lysine by chemical synthesis will undoubtedly be industrialized and developed, and so compete with the fermentation method.

4.16.1 Synthesis from caprolactam

The possibility for synthesis of lysine from caprolactam as starting material has been known for a long time. Recently, a large number of such procedures have been developed because caprolactam is now produced in large quantities as material for use in the manufacture of synthetic fiber.

The synthetic methods of lysine from caprolactam can be divided into two types. One consists of cleavage of the lactam ring, conversion to the acyl-derivative of ε-aminocaproic acid, halogenation and subsequent amination. In the other method, the lactam is halogenated directly.

A. Halogenation of ε-aminocaproic acid

The synthesis of lysine from caprolactam via α-bromo acid, as outlined below, was reported by Eck and Marvel[1)] some 30 yr ago. It is one of the fundamental procedures for lysine synthesis.

Beckmann rearrangement of hydroxyiminocyclohexanone (I), as mediated by hot sulfuric acid, led to the formation of caprolactam (II). (A number of different procedures have been published for the synthesis of the

$$\underset{\text{I}}{\text{cyclo-}(CH_2)_5C{=}NOH} \xrightarrow{H_2SO_4} \underset{\text{II}}{\text{cyclo-}CH_2{-}CH_2{-}C{=}O\ \ CH_2\ \ CH_2{-}CH_2{-}NH} \xrightarrow{\text{i) } H_2SO_4 \text{ ii) } NaOH,\ C_6H_5COCl}$$

$$\underset{\text{65\sim72\% based on I} \quad \text{III}}{C_6H_5CONH(CH_2)_5COOH} \xrightarrow[\text{red P}]{Br_2} \underset{64\sim89\% \quad \text{IV}}{C_6H_5CONH(CH_2)_4CH(Br)COOH}$$

$$\xrightarrow{NH_4OH} \underset{70\sim81\% \quad \text{V}}{C_6H_5CONH(CH_2)_4CH(NH_2)COOH} \xrightarrow{HCl} \underset{76\sim85\% \quad \text{VI}}{NH_2(CH_2)_4CH(NH_2\cdot 2HCl)COOH}$$

$$\xrightarrow[\text{alcohol}]{\text{pyridine}} \underset{91\sim94\% \quad \text{VII}}{NH_2(CH_2)_4CH(NH_2\cdot HCl)COOH}$$

caprolactam, although these are not dealt with here.) Acid hydrolysis of the caprolactam, and subsequent benzoylation of the hydrolyzed product, gave ε-benzoylamino-*n*-caproic acid (III). This was brominated, and treatment of the bromo acid (IV) so obtained with aqueous ammonia resulted in the formation of ε-*N*-benzoyllysine (V). Hydrolysis of the latter with hydrochloric acid, and further treatment with pyridine in alcohol, gave lysine monohydrochloride (VII) in an overall yield of 30%.

As an improved method for bromination, the ε-benzoylaminocaproic acid (III) was treated with red phosphorus and bromine in the presence of a small amount of water, and the α-bromo acid (IV) was obtained in a good yield (95%).[2)]

For the amination step of this α-bromo acid, autoclaving with ammonia, ammonium carbonate and cuprous carbonate gave ε-benzoyllysine in a very good yield (98–99%).[3)] A method via α-chloro acid, in place of the α-bromo acid, has also been reported.[4,5)] Namely, ε-benzoylaminocaproic acid (III), obtained by alkaline hydrolysis of caprolactam (II) and benzoylation, was

$$\underset{\text{II}}{\text{cyclo-}CH_2{-}CH_2{-}C{=}O\ \ CH_2\ \ CH_2{-}CH_2{-}NH} \xrightarrow{\text{i) } NaOH \text{ ii) } C_6H_5COCl} \underset{93.5\%^{4)},\ 96.4\%^{5)} \quad \text{III}}{C_6H_5CONH(CH_2)_5COOH}$$

$$\text{(cyclic)} \xrightarrow[\text{or } SO_2Cl_2,\ 91.1\%^{5)}]{SO_2Cl_2,\ I_2,\ 96\sim97.5\%^{4)}} C_6H_5\text{-}CONH(CH_2)_4\underset{\underset{Cl}{|}}{C}HCOOH \quad \text{VIII}$$

$$\xrightarrow[\text{or } NH_4OH,\ (NH_4)_2CO_3,\ KI,\ 83.7\%^{5)}]{NH_4OH,\ 60\%^{4)}} C_6H_5\text{-}CONH(CH_2)_4\underset{\underset{NH_2}{|}}{C}HCOOH\ (\text{V}) \xrightarrow[93.3\%^{5)}]{HCl} NH_2(CH_2)_4\underset{\underset{NH_2\cdot HCl}{|}}{C}HCOOH\ (\text{VII})$$

converted to the α-chloro acid (VIII) by treatment with sulfonylchloride in the presence of iodine. Amination of this α-chloro acid (VIII) gave benzoyl-lysine (V), which was hydrolyzed with hydrochloric acid to lysine mono-hydrochloride (VII) by the Eck and Marvel procedure described above.

Subsequently, the individual steps in this synthetic method were investigated further. Thus, without using iodine in the chlorination step, the crude chloro compound (VIII) obtained was purified from acetone, and subsequent amination was carried out by treatment with aqueous ammonia and ammonium carbonate in the presence of potassium iodide. This resulted in a better yield at the amination step. By this procedure, lysine was obtained in a good overall yield (68.6% based on caprolactam).[5]

B. Synthesis via halogenocaprolactam

In this method, caprolactam is subjected directly to halogenation, without being hydrolyzed. The acylation process of the ε-amino group for protection, which is required in the above method, can therefore be omitted. As shown in the following scheme, Brenner *et al.* found a route for lysine synthesis via halogenocaprolactam obtained by the direct chlorination of caprolactam.[6]

Caprolactam (II) was treated successively with phosphorus oxychloride, phosphorus pentachloride and sulfuryl chloride, whereupon α,α-dichloro-caprolactam (IX) was obtained. Hydrogenation of this dihalocaprolactam with Raney nickel in the presence of triethanolamine gave α-monochloro-caprolactam (X), which was then converted to α-azidocaprolactam (XI)

$$\underset{\text{II}}{\text{caprolactam: } CH_2(CH_2\text{-}CH_2\text{-}C{=}O)(CH_2\text{-}CH_2\text{-}NH)} \xrightarrow[75\sim80\%]{\text{i) } POCl_3\ \text{ii) } PCl_5\ \text{iii) } SO_2Cl_2} \underset{\text{IX}}{CH_2(CH_2\text{-}CCl_2\text{-}C{=}O)(CH_2\text{-}CH_2\text{-}NH)}$$

$\xrightarrow[\text{Raney nickel, } N(C_2H_5O)_3]{H_2}$ 91% — X: α-chlorocaprolactam (CH_2–$CH(Cl)$–$C{=}O$ / CH_2 / CH_2–CH_2–NH) $\xrightarrow{NaN_3}$ 78~85% — XI: α-azidocaprolactam (CH_2–$CH(N_3)$–$C{=}O$ / CH_2 / CH_2–CH_2–NH)

$\xrightarrow[\text{Raney nickel}]{H_2}$ 94% — XII: α-aminocaprolactam (CH_2–$CH(NH_2)$–$C{=}O$ / CH_2 / CH_2–CH_2–NH) $\xrightarrow{HCl}$ 91~94% — VII: $NH_2(CH_2)_4CH(NH_2\cdot HCl)COOH$

by treatment with sodium azide. Reduction of the product (XI) with Raney nickel gave α-aminocaprolactam (XII), and subsequent acid hydrolysis yielded lysine monohydrochloride (VII) in an average overall yield of 50% based on caprolactam. Brenner *et al.* also carried out optical resolution of DL-α-aminocaprolactam using L-pyrrolidonecarboxylic acid, and so obtained L-lysine. This technique is described below under section 4.16.4.

Since Brenner's original work, synthesis of lysine via halogenocaprolactam has been studied extensively. For example, caprolactam was converted to dichlorocaprolactam (IX) in good yield (91%) either by treatment with phosphorus pentachloride and chlorine[7] or by treatment with phosphorus oxychloride and phosphorus pentachloride.[8] Reduction of the dichlorocaprolactam (IX) with a palladium–carbon catalyst gave monochlorocaprolactam (X) in a yield of 88–89%. However, conversion of the monochloro derivative (X) into aminocaprolactam (XII), by direct amination with ammonia and subsequent hydrolysis with hydrochloric acid, resulted in a poor yield of 33%. Another route via α-bromocaprolactam (XIV), in place of the chloro intermediate, has also been reported.[7,8]

II: caprolactam (CH_2–CH_2–$C{=}O$ / CH_2 / CH_2–CH_2–NH) $\xrightarrow[PCl_5]{Br_2}$ 83.8%[7] — XIII: α,α-dibromocaprolactam (CH_2–CBr_2–$C{=}O$ / CH_2 / CH_2–CH_2–NH) $\xrightarrow[\text{Pd-C, } CH_3COOH,\ CH_3COONa]{H_2}$ 94.8%[7] — XIV: α-bromocaprolactam (CH_2–$CH(Br)$–$C{=}O$ / CH_2 / CH_2–CH_2–NH)

II $\xrightarrow{Br_2,\ PBr_3}$ XIV 67%[8]

$\xrightarrow{\text{liq. ammonia}}$ 61.5%[8] — XII: (CH_2–$CH(NH_2)$–$C{=}O$ / CH_2 / CH_2–CH_2–NH) $\xrightarrow{HCl}$ 44% based on XIV[7] — VII: $NH_2(CH_2)_4CH(NH_2\cdot HCl)COOH$

In the case of the dichloro derivative described above, dibromocaprolactam (XIII) could be converted to monobromocaprolactam (XIV) by reduction. It was also possible to obtain the monobromo derivative (XIV) directly from caprolactam.[8] Amination of (XIV) and acid hydrolysis of the aminated product yielded lysine.

Acid hydrolysis of the α-chloro- or α-bromo-caprolactams (X, XIV) described above, led to ε-amino-α-halogenocaproic acid (XV) in a yield of 95%, and further treatment with acetic anhydride gave ε-acetoamide-α-halogenocaproic acid (XVI). Amination of (XVI) with excess liquid ammonia, followed by hydrolysis with hydrochloric acid, yielded lysine.[7]

$$\underset{\text{X (X=Cl), XIV (X=Br)}}{\text{caprolactam ring: } CH_2-CH(X)-C{=}O,\ CH_2,\ CH_2-CH_2-NH} \xrightarrow[95\%]{HX} \underset{\text{XV}}{NH_2(CH_2)_4CH(X)COOH} \xrightarrow[CH_3COOH]{(CH_3CO)_2O,\ CH_3COONa}$$

$$\underset{\text{XVI}}{CH_3CONH(CH_2)_4CH(X)COOH}\ (95\%) \xrightarrow{NH_3} \left[CH_3CONH(CH_2)_4CH(NH_2)COOH\right]$$

$$\xrightarrow{HCl} \underset{\text{VII}}{NH_2(CH_2)_4CH(NH_2 \cdot HCl)COOH}$$

69% based on XIV
57% based on X

The yields via the bromo and chloro intermediates were 69% and 57% respectively. Thus, although this route has more steps than that involving the amination of α-halogenocaprolactam and subsequent hydrolysis, the yield is higher than that in the latter case.

Besides these techniques, there is also an intermediary method between the above methods, i.e. between that of hydrolysis of caprolactam, acylation and halogenation and that of direct halogenation of lactam.[9] The principle of this intermediary method is based on the fact that *N*-benzoylcaprolactam obtained by benzoylation of caprolactam (II), can be monochlorinated by sulfurylchloride. Ring cleavage of the *N*-benzoyl-α-chlorocaprolactam (XVIII) with sodium hydroxide gave ε-benzoylamino-α-chlorocaproic acid (VIII), and further amination and hydrolysis led to lysine in a good overall yield (70% based on caprolactam). Treatment of the *N*-benzoyl-α-chlorocaprolactam (XVIII) with concentrated sulfuric acid at room temperature gave α-chlorocaprolactam (X), and ammonolysis with aqueous ammonia followed by hydrolysis led to lysine in an overall yield of 60% based on caprolactam.

$$\underset{\text{II}}{\text{caprolactam}} \xrightarrow[\text{dimethylalanine}]{C_6H_5COCl} \underset{\text{XVII}}{(CH_2)_5\text{-lactam-}N\text{-}COC_6H_5} \; (97.3\%) \xrightarrow{SO_2Cl_2} \underset{\text{XVIII}}{\alpha\text{-Cl-}N\text{-benzoylcaprolactam}} \; (89.3\%)$$

$$\text{XVIII} \xrightarrow{NaOH} \underset{\text{VIII}}{C_6H_5CONH(CH_2)_4CH(Cl)COOH} \; (94.5\%) \xrightarrow{NH_3} \underset{\text{V}}{C_6H_5CONH(CH_2)_4CH(NH_2)COOH} \; (91\%) \xrightarrow{HCl} \text{VII} \; (94.5\%)$$

$$\text{XVIII} \xrightarrow{H_2SO_4} \underset{\text{X}}{\alpha\text{-chlorocaprolactam}} \; (95.3\%) \xrightarrow{NH_4OH} \left[\underset{\text{XII}}{\alpha\text{-aminocaprolactam}}\right] \xrightarrow{HCl} \underset{\text{VII}}{NH_2(CH_2)_4CH(NH_2 \cdot HCl)COOH} \; (72\%)$$

For the purpose of introducing amino group into caprolactam, an interesting process via α-nitrocaprolactam has been reported. This was developed by the Dutch State Mines (Stamicarbon N.V.) group.[10,11] Treatment of caprolactam (II) with phosgene gave 2-chloroazacyclo-2,3-heptene-*N*-carbochloride (XIX). This was converted to α-nitrocaprolactam (XX) by treatment with a nitric/sulfuric acid mixture. Reduction of the α-nitro derivative (XX) with Raney nickel yielded α-aminocaprolactam (XII) in a high yield. Optical resolution of the DL-α-aminocaprolactam was carried out using L-pyrrolidonecarboxylic acid. The L-α-aminocaprolactam obtained was hydrolyzed with alkali, and subsequent addition of hydrochloric acid gave L-lysine monohydrochloride (VII). The D-isomer of α-aminocaprolactam could be racemized to the DL-form, which was then reused for L-lysine production. The flowsheet for this procedure is shown in Fig. 4–4.

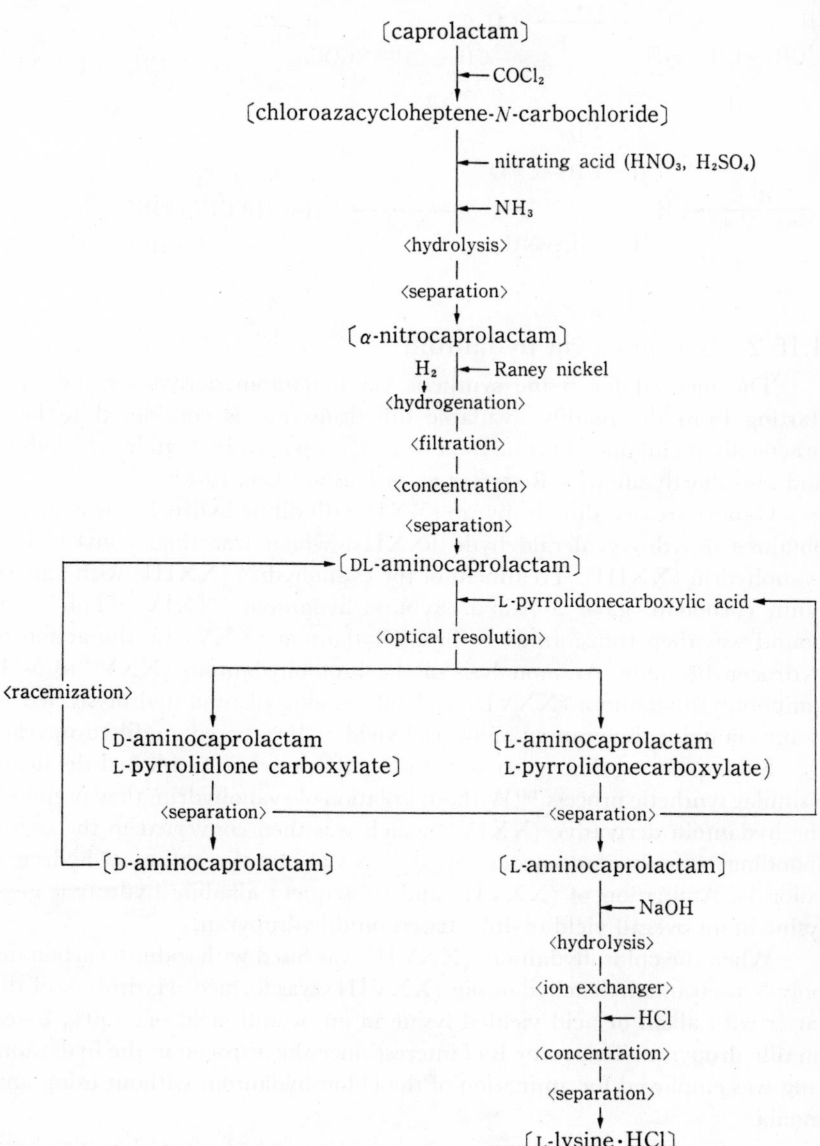

Fig. 4–4. General flowsheet for L-lysine synthesis from caprolactam.

$$\underset{\text{II}}{\begin{array}{l} \text{CH}_2\text{—CH}_2\text{—C=O} \\ \text{CH}_2 \qquad\qquad\quad | \\ \text{CH}_2\text{—CH}_2\text{—NH} \end{array}} \xrightarrow[96.6\%]{\text{COCl}_2} \underset{\text{XIX}}{\begin{array}{l} \text{CH}_2\text{—CH=C—Cl} \\ \text{CH}_2 \qquad\qquad\quad | \\ \text{CH}_2\text{—CH}_2\text{—NCOCl} \end{array}} \xrightarrow[\text{H}_2\text{SO}_4 \;\; 93\%]{\text{HNO}_3} \underset{\text{XX}}{\begin{array}{l} \qquad\quad \text{NO}_2 \\ \text{CH}_2\text{—CH—C=O} \\ \text{CH}_2 \qquad\qquad\quad | \\ \text{CH}_2\text{—CH}_2\text{—NH} \end{array}}$$

$$\xrightarrow[\text{Raney nickel} \;\; 99\%]{\text{H}_2} \underset{\text{XII}}{\begin{array}{l} \qquad\quad \text{NH}_2 \\ \text{CH}_2\text{—CH—C=O} \\ \text{CH}_2 \qquad\qquad\quad | \\ \text{CH}_2\text{—CH}_2\text{—NH} \end{array}} \xrightarrow{\text{NaOH}} \underset{\text{VII}}{\begin{array}{l} \text{NH}_2(\text{CH}_2)_4\text{CHCOOH} \\ \qquad\qquad\quad \text{NH}_2 \cdot \text{HCl} \end{array}}$$

4.16.2 Synthesis via hydantoin

The method for lysine synthesis via hydantoin derivatives, i.e. that starting from the readily available dihydropyran, is considered to be a practically useful one. This method was first reported by Gaudry in 1948,[12] and also shortly after by Rogers *et al.* of DuPont Co., Ltd.[13]

Gaudry treated dihydropyran (XXI) with dilute hydrochloric acid and obtained δ-hydroxyvaleraldehyde (XXII), which was then converted to cyanohydrin (XXIII). Treatment of the cyanohydrin (XXIII) with ammonium carbonate gave 5-(4-hydroxybutyl)hydantoin (XXIV). This compound was then transformed to bromohydantoin (XXV) by the action of hydrogen bromide. Ammonolysis of the bromohydantoin (XXV) to 5-(4-aminobutyl)hydantoin (XXVI), and subsequent alkaline hydrolysis, led to lysine via lysine dipicrate in an overall yield of 35% based on dihydropyran.

Subsequent to the report of Gaudry, Rogers *et al.* published details of a similar synthetic process.[13] Without isolation of cyanohydrin, they prepared the hydantoin derivative (XXIV), which was then converted to the corresponding chlorohydantoin derivative (XXVII) by the action of hydrogen chloride. Amination of (XXVII) and subsequent alkaline hydrolysis gave lysine in an overall yield of 46% based on dihydropyran.

When the chlorohydantoin (XXVII) was fused with sodium carbonate, poly-3,5-tetramethylenehydantoin (XXVIII) was formed. Hydrolysis of the latter with alkali or acid yielded lysine in an overall yield of >50% based on dihydropyran. This route is of interest since the nitrogen in the hydantoin ring was employed for amination of the chlorohydantoin without using ammonia.

Another similar synthetic method starting from furfural has also been reported.[14,15] This route involves conversion of furfural to furylhydantoin, followed by reduction of the furylhydantoin to 5-(4-oxybutyl)hydantoin (XXIV).

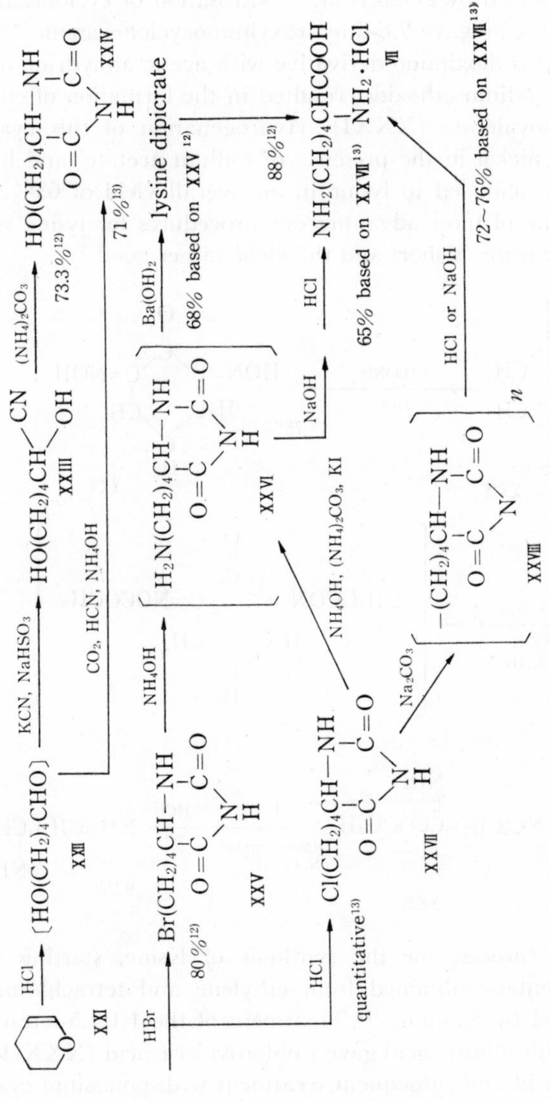

XXI
HCl
(HO(CH₂)₄CHO)
XXII
KCN, NaHSO₃
HO(CH₂)₄CH(CN)OH
XXIII
(NH₄)₂CO₃
73.3%12)
HO(CH₂)₄CH—NH, O=C—C=O, N H
XXIV
CO₂, HCN, NH₄OH
71%13)
HBr
80%12)
Br(CH₂)₄CH—NH, O=C, C=O, N H
XXV
NH₄OH
H₂N(CH₂)₄CH—NH, O=C, C=O, N H
XXVI
Ba(OH)₂
68% based on XXV 12)
lysine dipicrate
88%12)
NaOH
HCl
65% based on XXVII 13)
NH₂(CH₂)₄CHCOOH, NH₂·HCl
VII
HCl
quantitative13)
Cl(CH₂)₄CH—NH, O=C, C=O, N H
XXVII
NH₄OH, (NH₄)₂CO₃, KI
Na₂CO₃
—(CH₂)₄CH—NH, O=C, C=O, N
n
XXVIII
HCl or NaOH
72~76% based on XXVIII 13)

4.16.3 Other miscellaneous methods

Synthesis of lysine from cyclohexanone via dihydroxyimino derivative has been reported by Ferris *et al.*[16] Nitrosation of cyclohexanone (XXIX) with methylnitrite gave 2,6-dihydroxyiminocyclohexanone (XXX). Partial cleavage of this dioximino derivative with acetic anhydride in the presence of ethanolic sodium ethoxide, resulted in the formation of ethyl δ-cyano-α-hydroxyiminovalerate (XXXI). Hydrogenation of this cyano derivative with Raney nickel in the presence of sodium acetate, and hydrolysis with hydrochloric acid, led to lysine in an overall yield of 63%. This method represents one of most advantageous procedures for lysine synthesis, since the synthetic route is short and the yield rather good.

XXIX (cyclohexanone) $\xrightarrow[HCl]{CH_3ONO}$ 75% XXX ($HON{=}C$–$C({=}O)$–$C{=}NOH$ ring with H_2C, CH_2, CH_2)

$\xrightarrow[C_2H_5OH,\ NaOC_2H_5]{(CH_3CO)_2O}$ [$CH_3OCON{=}C$–$C({=}O)$–$C{=}NOCOCH_3$ ring with H_2C, CH_2, CH_2]

$\longrightarrow NC(CH_2)_3C({=}NOH)COOC_2H_5$ (XXXI, 92%) $\xrightarrow[\text{Raney nickel},\ NaOC_2H_5]{H_2}\xrightarrow{HCl} NH_2(CH_2)_4CH(NH_2\cdot HCl)COOH$ (VII, 92%)

Another process for the synthesis of lysine, starting from 1,1,1,5-tetrachloropentane obtained from ethylene and tetrachloromethane, has been reported by Saotome.[17] Treatment of the 1,1,1,5-tetrachloropentane (XXXII) with sulfuric acid gave ε-chlorovaleric acid (XXXIII). Esterification of this acid and subsequent treatment with potassium cyanide, yielded ethyl δ-cyanovalerate (XXXIV). Reduction of the cyano derivative (XXXIV), followed by benzoylation, yielded ε-benzoylaminocaproic acid (III). Finally, as described previously under the method starting from caprolactam,

$$\underset{\text{XXXII}}{Cl(CH_2)_4CCl_3} \xrightarrow[H_2SO_4]{} \underset{\text{XXXIII}}{Cl(CH_2)_4COOH}\ (93\%) \xrightarrow[H_2SO_4]{C_2H_5OH} Cl(CH_2)_4COOC_2H_5\ (100\%)$$

$$\xrightarrow{KCN} \underset{\text{XXXIV}}{NC(CH_2)_4COOC_2H_5}\ (98\%) \xrightarrow{\text{i) } H_2\text{, Raney nickel; ii) } C_6H_5COCl\text{, NaOH}} \underset{\text{III}}{C_6H_5CONH(CH_2)_5COOH}\ (95\%)$$

$$\xrightarrow{SO_2Cl_2,\ I_2} \underset{\text{VIII}}{C_6H_5CONH(CH_2)_4CH(Cl)COOH}\ (95\%)$$

$$\xrightarrow{NH_4OH,\ (NH_4)_2CO_3} \underset{\text{V}}{C_6H_5CONH(CH_2)_4CH(NH_2)COOH}\ (84\%) \xrightarrow{HCl} \underset{\text{VII}}{NH_2(CH_2)_4CH(NH_2 \cdot HCl)COOH}\ (96\%)$$

chlorination of the acid (III), followed by amination, led to DL-lysine in an overall yield of 66% based on the tetrachloropentane (XXXII).

4.16.4 Resolution of DL-lysine

Although a great number of reports have been devoted to the resolution of lysine, only those typical procedures which seem relatively advantageous for industrial application are mentioned here.

Several examples of enzymatic resolution have been given, as follows. Treatment of ε-benzoyl-α-acetyl-DL-lysine (obtained by acetylation of ε-benzoyl-DL-lysine, which is an intermediate in lysine synthesis) with mould aminoacylase resulted in the precipitation of the barely soluble ε-benzoyl-lysine. Hydrolysis of this benzoyl derivative with hydrochloric acid gave L-lysine.[18] Enzymatic resolution of DL-lysine has also been carried out by treatment of diacetyl-DL-lysine with mould aminoacylase.[19]

Asymmetric hydrolysis of ε-benzoyl-DL-lysine by bacterial ε-lysine acylase produced by *Achromobacter pestifer* allowed the formation of L-lysine.[20] By utilizing this method, certain processes required in the above aminoacylase method, such as α-acylation and chemical hydrolysis of the acyl group, could be omitted. For the chemical resolution of lysine, the less soluble L-lysine–L-glutamate could be separated after reacting DL-lysine with L-glutamic acid.[21-23] The D-α-phenoxypropionic acid salt of DL-lysine was prepared and spray-dried. Subsequent methanol treatment gave the less soluble L-lysine D-α-phenoxypropionate.[24]

In addition to these methods, and in the case of the synthesis via aminocaprolactam described in section 4.16.1B, chemical resolution can be achieved at the α-aminocaprolactam stage. Namely, reaction of DL-α-aminocaprolactam with L-pyrrolidonecarboxylic acid gave the less soluble L-α-aminocaprolactam–L-pyrrolidonecarboxylic acid. Subsequent hydrolysis of L-aminocaprolactam yielded L-lysine.[6,25] D-Aminocaprolactam remained in the mother liquor of the reaction mixture, and could be reused after it had been racemized by heat distillation.

Insofar as physicochemical methods are concerned, an effective technique based on preferential crystallization from an aqueous solution of DL-lysine *p*-aminobenzenesulfonate has been developed.[26] In addition to this, patents for resolution by preferential crystallization using certain salts of lysine, such as the 3,5-dinitrobenzoate,[27] anthraquinone-β-sulfonate, 1-chloronaphthalene-4-sulfonate,[28] have been reported.

References

1. J. C. Eck and C. S. Marvel, *J. Biol. Chem.*, **106**, 387 (1934); *Org. Syn. Coll.*, vol. 2, p. 74, 76, 374, 1943.
2. E. E. Howe and E. W. Pietrusza, *J. Am. Chem. Soc.*, **71**, 2581 (1949).
3. D. C. Sayles and E. F. Degering, *ibid.*, **71**, 3161 (1949).
4. A. Galat, *ibid.*, **69**, 86 (1947).
5. K. Odo and S. Himizu, *Yuki Gosei Kagaku Kaishi* (Japanese), **11**, 386 (1953).
6. M. Brenner and H. R. Rickenbacher, *Helv. Chim. Acta*, **41**, 181 (1958).
7. R. J. Wineman, E. T. Hsu and C. E. Anagnostopoulos, *J. Am. Chem. Soc.*, **80**, 6233 (1958).
8. W. C. Francis, J. R. Thornton, J. C. Werner and T. R. Hopkins, *ibid.*, **80**, 6238 (1958).
9. R. Tull, R. C. O'Neil, E. P. McCarthy, J. J. Pappas and J. M. Chemerda, *J. Org. Chem.*, **29**, 2425 (1964).
10. Stamicarbon N. V., *Japanese Patent* No. 37–7243 (1962); 37–11108 (1962); 38–22859 (1963); 39–15842 (1964); 39–6820 (1964); 44–15764 (1969).
11. J. H. Ottenheim and J. W. Gielkens, *Hydrocarbon Processing*, **47**, 176 (1968).
12. R. Gaudry, *Can. J. Res.*, **26B**, 287 (1948).
13. A. O. Rogers, R. D. Emmick, L. W. Tyran, L. B. Phillips, A. A. Levine and N. D. Scott, *J. Am. Chem. Soc.*, **71**, 1837 (1949).
14. D. D. Coffman and M. W. Fariow, *U.S. Patent* No. 2728777 (1955).
15. H. Conroy, *U.S. Patent* No. 2786848 (1957); 2786849 (1957).
16. A. F. Ferris, G. S. Johnson, F. E. Gould and H. Stange, *J. Org. Chem.*, **25**, 1302 (1960).
17. K. Saotome, *Kogyo Kagaku Zasshi* (Japanese), **65**, 1059, 1061 (1962).
18. I. Chibata, S. Yamada and S. Yamada, *Bull. Agr. Chem. Soc. Japan*, **20**, 174 (1956); **21**, 296 (1957).

19. K. Michi and H. Tsuda, *ibid.*, **22**, 283 (1958).
20. I. Chibata, T. Tosa and T. Ishikawa, *Arch. Biochem. Biophys.*, **104**, 231 (1964).
21. R. D. Emmick, *U.S. Patent* No. 2556907 (1951).
22. A. O. Rogers, *U.S. Patent* No. 2657230 (1953).
23. M. C. Baker, *U.S. Patent* No. 2942024 (1960).
24. Stamicarbon N. V., *Japanese Patent* No. 41–21222 (1956).
25. J. E. Nelemans, A. H. Pecasse, W. Pesh and U. Veristrijden, *U.S. Patent* No. 3105067 (1963).
26. I. Chibata, S. Yamada and M. Yamamoto, *Japanese Patent* No. 45–29963 (1970).
27. T. Uzuki and N. Sato, *Japanese Patent* No. 43–3601 (1968).
28. T. Uzuki and N. Sato, *Japanese Patent* No. 45–852 (1970).

4.17 Methionine

In 1921, Mueller found a sulfur-containing compound essential for bacterial cell growth in protein hydrolyzates, and reported in 1922 that it is a sulfur-containing amino acid. In 1928, Barger and Coyne succeeded in synthesizing this amino acid, proving their assumption that its structure was α-amino-γ-methylthiolbutyric acid, and naming it "methionine".

At present, methionine is produced commercially on a large scale by chemical synthesis, although it is second to glutamic acid in total production. From the work of Jackson and Rose, it has also been shown that L-methionine and D-methionine are of equal nutritive value. Therefore, the complicated process of resolution of the two isomers can be omitted, and DL-methionine is supplied directly for practical use. Indeed, this is one reason why the chemical synthesis of methionine has been so easily industrialized.

Although methionine is a constituent of almost all known proteins, its content is generally so low as to prevent direct isolation from proteins in large quantities at low cost. Also, a technique for the fermentative production of methionine has not yet been developed.

The method applied at present for the industrial production of methionine utilizes acrolein as starting material. By this convenient technique, DL-methionine is synthesized from acrolein and methylmercaptan through 3 or 4 steps. Another method that is expected to be of practical importance employs γ-butyrolactone; however, it is still necessary to develop means for supplying the γ-butyrolactone at low cost. Several other methods for methionine synthesis have been reported, although none are at present applicable for production on an industrial scale.

4.17.1 Synthesis from acrolein

Catch *et al.*[1] synthesized DL-methionine (V) from acrolein (I) in an overall yield of 29%. Acrolein (I) was reacted with methylmercaptan and the resulting β-methylthiopropionaldehyde (II) treated with liquid hydrogen cyanide (a modification of the Strecker reaction). Methionine (V) was obtained on hydrolysis of the aminonitrile (IV) derived from (II).

$$\underset{\text{I}}{CH_2{=}CHCHO} \xrightarrow{CH_3SH} \underset{\text{II}}{CH_3SCH_2CH_2CHO} \xrightarrow{HCN} \underset{\text{III}}{CH_3SCH_2CH_2\underset{\displaystyle OH}{\underset{|}{C}}HCN}$$

$$\xrightarrow{NH_3} \underset{\text{IV}}{CH_3SCH_2CH_2\underset{\displaystyle NH_2}{\underset{|}{C}}HCN} \xrightarrow{H_2O} \underset{\text{V}}{CH_3SCH_2CH_2\underset{\displaystyle NH_2}{\underset{|}{C}}HCOOH}$$

Prior to this work, Kaneko and Mitsui[2)] had treated acrolein (I) with methylmercaptan in the presence of mercury methylmercaptide as catalyst, and so prepared β-methylthiopropionaldehyde (II) in 85.5% yield.

Later, Pierson, Giella and Tishler[3)] found that methylmercaptan could be easily added to acrolein in the presence of copper methylmercaptide. They converted the resulting β-methylthiopropionaldehyde (II) to 5-(β-methylthioethyl)hydantoin (VI) by the Bucherer reaction, and hydrolysis of this gave DL-methionine (V) in an overall yield of 50%.

$$\underset{\text{II}}{CH_3SCH_2CH_2CHO} \xrightarrow{(NH_4)_2CO_3,\ NaCN} \underset{\text{VI}}{CH_3SCH_2CH_2\underset{\substack{|\\NH}}{CH}-\underset{\substack{|\\NH}}{CO}} \ (\text{NH, NH bridged by CO}) \xrightarrow{OH^-} \underset{\text{V}}{CH_3SCH_2CH_2\underset{\substack{|\\NH_2}}{C}HCOOH}$$

Gaudry and Nadeau[4)] also obtained DL-methionine in 60% overall yield by the above procedure. On the hand, the same workers[4)] prepared α-hydroxy-γ-chlorobutyronitrile (IX) from acrolein (I) via acetal (VII) and the aldehyde (VIII). The cyanohydrin (IX) was converted to 5-(β-chloroethyl)hydantoin (X) with ammonium carbonate. Treatment with sodium methylmercaptide yielded 5-(β-methylthioethyl)hydantoin (VI), which was then hydrolyzed to give DL-methionine (V) in 35% yield based on acetal (VII).

$$\underset{\text{I}}{CH_2=CHCHO} \longrightarrow \underset{\text{VII}}{ClCH_2CH_2CH(OC_2H_5)_2} \xrightarrow{HCl} \underset{\text{VIII}}{ClCH_2CH_2CHO}$$

$$\xrightarrow{NaCN} \underset{\text{IX}}{ClCH_2CH_2\underset{\substack{|\\OH}}{C}HCN} \xrightarrow{(NH_4)_2CO_3} \underset{\text{X}}{ClCH_2CH_2\underset{\substack{|\\NH}}{CH}-\underset{\substack{|\\NH}}{CO}} \ (\text{NH, NH bridged by CO})$$

$$\xrightarrow{CH_3SNa} \underset{\text{VI}}{CH_3SCH_2CH_2\underset{\substack{|\\NH}}{CH}-\underset{\substack{|\\NH}}{CO}} \ (\text{NH, NH bridged by CO}) \xrightarrow{H_2O} \underset{\text{V}}{CH_3SCH_2CH_2\underset{\substack{|\\NH_2}}{C}HCOOH}$$

Many patents are devoted to DL-methionine synthesis. However, the methods at present of practical use in industry are based on the principle of DL-methionine synthesis via a hydantoin intermediate, after the direct addition of methylmercaptan to acrolein. With these techniques, the most important question is how to synthesize the acrolein and methylmercaptan economically. Acrolein can be prepared from (1) glycerine, (2) acetaldehyde and formalin, (3) propylene, or (4) acetylene and formaldehyde. Methylmercaptan can be synthesized from (1) thiourea, (2) hydrogen sulfide and sodium hydroxide, or (3) methanol and hydrogen sulfide.

A flowsheet for the synthetic method which is at present practiced commercially is given in Fig. 4–5.[5] Acrolein is prepared from propylene by catalytic oxidation, and methylmercaptan is prepared from methanol and hydrogen sulfide by dehydration in the gaseous phase. β-Methylthiopropionaldehyde, an addition product of the reaction between acrolein and methylmercaptan, is treated with hydrogen cyanide and ammonium carbonate. The resulting hydantoin is then hydrolyzed to give DL-methionine.

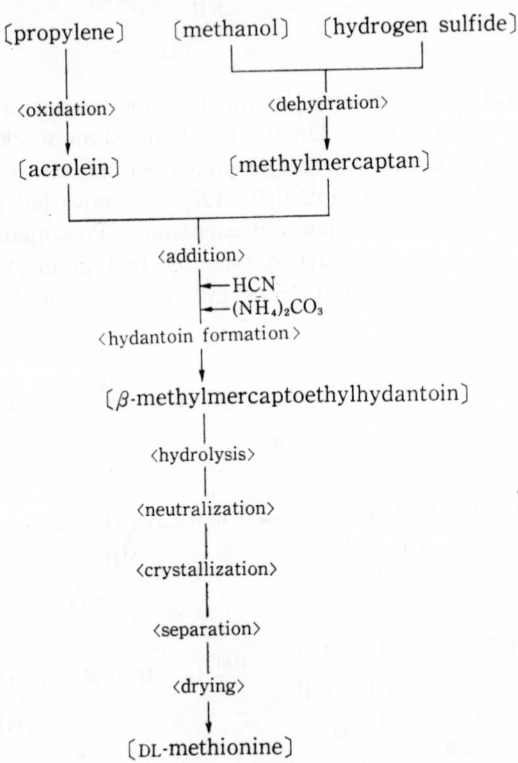

Fig. 4–5. General flowsheet for DL-methionine synthesis.

4.17.2 Synthesis from γ-butyrolactone

Livak *et al.*[6] brominated γ-butyrolactone (XI) and then prepared α-amino-γ-hydroxybutyric acid (XIII) by amination of the resulting bromo compound (XII). The acid (XIII) was converted to its *N*-carbamoyl derivative (XIV) with potassium cyanate, and 5-(β-bromoethyl)hydantoin (XV) was obtained by cyclization. The hydantoin (XV) was then treated with sodium methylmercaptide, and the resulting 5-(β-methylthioethyl)-hydantoin (VI) hydrolyzed with alkali to give DL-methionine (V) in an overall yield of 17% from γ-butyrolactone.

$$\underset{\text{XI}}{\overset{CH_2CH_2CH_2}{\underset{O\text{———}CO}{|\quad\quad|}}} \xrightarrow{PBr_3} \underset{\text{XII}}{\overset{CH_2CH_2CHBr}{\underset{O\text{———}CO}{|\quad\quad|}}} \xrightarrow{NH_3} \underset{\text{XIII}}{HOCH_2CH_2\underset{NH_2}{\underset{|}{C}}HCOOH}$$

$$\xrightarrow{KCNO} \underset{\text{XIV}}{HOCH_2CH_2\underset{NHCONH_2}{\underset{|}{C}}HCOOH} \xrightarrow{HBr} \underset{\text{XV}}{BrCH_2CH_2CH\text{—}CO \;(\text{NH—CO—NH ring})}$$

$$\xrightarrow{CH_3SNa} \underset{\text{VI}}{CH_3SCH_2CH_2CH\text{—}CO \;(\text{NH—CO—NH ring})} \xrightarrow[OH^-]{} \underset{\text{V}}{CH_3SCH_2CH_2\underset{NH_2}{\underset{|}{C}}HCOOH}$$

Plieninger[7] brominated γ-butyrolactone (XI) in order to prepare α,γ-dibromobutyric acid (XVI), which was then cyclized by heating. The α-bromo-γ-butyrolactone (XII) formed was converted with liquid ammonia to α-amino-γ-butyrolactone (XVII), which was then treated with sodium methylmercaptide. DL-Methionine (V) was so obtained in an overall yield of 40%.

$$\underset{\text{XI}}{\overset{CH_2CH_2CH_2}{\underset{O\text{———}CO}{|\quad\quad|}}} \xrightarrow[P]{Br_2} \underset{\text{XVI}}{\underset{Br\quad\quad\quad Br}{CH_2CH_2CHCOOH}} \xrightarrow{heat} \underset{\text{XII}}{\overset{CH_2CH_2CHBr}{\underset{O\text{———}CO}{|\quad\quad|}}}$$

$$\xrightarrow{NH_3} \underset{\text{XVII}}{\overset{CH_2CH_2CHNH_2}{\underset{O\text{———}CO}{|\quad\quad|}}} \xrightarrow{CH_3SNa} \underset{\text{V}}{CH_3SCH_2CH_2\underset{NH_2}{\underset{|}{C}}HCOOH}$$

Snyder *et al.*[8] derived α-hydroximino-γ-butyrolactone (XIX) from α-aceto-γ-butyrolactone (XVIII) by treatment with ethyl nitrite in methanol. Reduction of the product gave α-amino-γ-butyrolactone (XVII), which was converted to 3,6-bis(β-hydroxyethyl)-2,5-diketopiperazine (XX) by heating.

Reaction with thionyl chloride then gave 3,6-bis(β-chloroethyl)2,5-diketopiperazine (XXI), which was treated with sodium methylmercaptide. The resulting 3,6-bis(β-methylthiolethyl)-2,5-diketopiperazine (XXII) was hydrolyzed, and DL-methionine (V) obtained in an overall yield of 26%.

$$\underset{\text{XVIII}}{\underset{O\text{———}CO}{CH_2CH_2CHCOCH_3}} \xrightarrow[CH_3OH]{C_2H_5NO_2} \underset{\text{XIX}}{\underset{O\text{———}CO}{CH_2CH_2C{=}NOH}} \xrightarrow[Pd]{H_2} \underset{\text{XVII}}{\underset{O\text{———}CO}{CH_2CH_2CHNH_2}}$$

$$\xrightarrow[heat]{} \underset{\text{XX}}{HOCH_2CH_2CH\langle{}^{CONH}_{NHCO}\rangle CHCH_2CH_2OH} \xrightarrow{SOCl_2} \underset{\text{XXI}}{ClCH_2CH_2CH\langle{}^{CONH}_{NHCO}\rangle CHCH_2CH_2Cl}$$

$$\xrightarrow{CH_3SNa} \underset{\text{XXII}}{CH_3SCH_2CH_2CH\langle{}^{CONH}_{NHCO}\rangle CHCH_2CH_2SCH_3} \xrightarrow{H^+} \underset{\text{V}}{\underset{NH_2}{CH_3SCH_2CH_2CHCOOH}}$$

Snyder and Cannon[9] converted 3,6-bis(β-chloroethyl)-2,5-diketopiperazine (XXI) to the thiuronium derivative (XXIII) by reaction with

$$\underset{\text{XXI}}{ClCH_2CH_2CH\langle{}^{CONH}_{NHCO}\rangle CHCH_2CH_2Cl}$$

$$\xrightarrow{NH_2CSNH_2} \underset{\text{XXIII}}{HCl\cdot NH_2\overset{NH}{\overset{\|}{C}}SCH_2CH_2CH\langle{}^{CONH}_{NHCO}\rangle CHCH_2CH_2S\overset{NH}{\overset{\|}{C}}NH_2\cdot HCl}$$

$$\xrightarrow[NaOH]{(CH_3)_2SO_4} \underset{\text{XXII}}{CH_3SCH_2CH_2CH\langle{}^{CONH}_{NHCO}\rangle CHCH_2CH_2SCH_3}$$

$$\xrightarrow{H_2O} \underset{\text{V}}{\underset{NH_2}{CH_3SCH_2CH_2CHCOOH}}$$

thiourea. After treatment with dimethyl sulfate and sodium hydroxide, they obtained DL-methionine (V) in 66% overall yield by hydrolysis of the resulting 3,6-bis(β-methylthiol)-2,5-diketopiperazine (XXII).

Hill and Robson[10] prepared *N*-benzoylmethionine ethyl ester (XXVI) by the reaction of sodium methylmercaptide with ethyl α-benzamido-γ-chlorobutyrate (XXV) derived from α-benzamido-γ-butyrolactone (XXIV) by treatment with hydrogen chloride in ethanol. On saponification and hydrolysis of the ester (XXVI), DL-methionine (V) was obtained in 31% yield.

$$\underset{\text{XXIV}}{\begin{array}{l}CH_2CH_2CHNHCOC_6H_5\\ |\qquad\quad\ \ |\\ O\text{———}CO\end{array}} \xrightarrow[C_2H_5OH]{HCl} \underset{\text{XXV}}{\begin{array}{l}ClCH_2CH_2CHCOOC_2H_5\\ \qquad\qquad\ \ |\\ \qquad\quad\ NHCOC_6H_5\end{array}}$$

$$\xrightarrow{CH_3SNa} \underset{\text{XXVI}}{\begin{array}{l}CH_3SCH_2CH_2CHCOOC_2H_5\\ \qquad\qquad\qquad\ |\\ \qquad\qquad\ NHCOC_6H_5\end{array}} \xrightarrow[NaOH]{} \underset{\text{XXVII}}{\begin{array}{l}CH_3SCH_2CH_2CHCOOH\\ \qquad\qquad\qquad\ |\\ \qquad\qquad\ NHCOC_6H_5\end{array}}$$

$$\xrightarrow{H_2O} \underset{\text{V}}{\begin{array}{l}CH_3SCH_2CH_2CHCOOH\\ \qquad\qquad\qquad\ |\\ \qquad\qquad\quad NH_2\end{array}}$$

In another process, the α-benzamido-γ-chloroethylbutyrate (XXV) was treated with sodium benzylmercaptide instead of sodium methylmercaptide, to cause reductive release of the benzyl group. The product was treated with methyl iodide, resulting in substitution of the benzyl group with methyl group, and DL-methionine was then obtained.[11]

4.17.3 Miscellaneous methods of synthesis

Barger and Coyne[12] synthesized DL-methionine in the 1920's in an attempt to elucidate its structure. β-Methylthiopropionaldehyde (II) was prepared via β-methylthiopropionaldehyde diethyl acetal (XXVIII), which was derived from β-chloropropionaldehyde diethyl acetal (VII) by treatment with sodium mercaptide. On hydrolysis of the aminonitrile intermediate (XXIX) formed from the aldehyde (II) by the Strecker reaction, DL-methionine (V) was obtained in 6% overall yield. Holland and Nayler[13] later modified this method and successfully improved the overall yield.

$$ClCH_2CH_2CH(OC_2H_5)_2 \xrightarrow{CH_3SNa} CH_3SCH_2CH_2CH(OC_2H_5)_2 \xrightarrow{HCl} CH_3SCH_2CH_2CHO$$

VII XXVIII II

$$\xrightarrow{HCN,\ NH_3} CH_3SCH_2CH_2CH(NH_2)CN \xrightarrow{H_2O} CH_3SCH_2CH_2CH(NH_2)COOH$$

XXIX V

Windus and Marvel[14] prepared 2-methylthioethylmalonic acid (XXXIII) by saponification of the ethyl ester (XXXII), which was itself a condensation product from 2-methylthiochloroethyl (XXX) and ethyl malonate (XXXI). Amination of the bromo derivative (XXXIV) obtained by bromination of the acid (XXXIII) yielded the α-aminomalonate derivative (XXXV), and heat treatment of the resulting α-aminomalonate derivative (XXXV) gave DL-methionine (V) in 24% yield from (XXXIII). Emerson *et al.*[15] later modified this method to give an improved yield.

$$CH_3SCH_2CH_2Cl + CH_2(COOC_2H_5)_2 \xrightarrow[C_2H_5ONa]{} CH_3SCH_2CH_2CH(COOC_2H_5)_2$$

XXX XXXI XXXII

$$\xrightarrow[\text{i) } OH^- \text{ ii) } H^+]{} CH_3SCH_2CH_2CH(COOH)_2 \xrightarrow{Br_2} CH_3SCH_2CH_2C(Br)(COOH)_2$$

XXXIII XXXIV

$$\xrightarrow{NH_3} CH_3SCH_2CH_2C(NH_2)(COOH)_2 \xrightarrow[\text{heat}]{} CH_3SCH_2CH_2CH(NH_2)COOH$$

XXXV V

Barger and Weichselbaum[16] treated 2-methylthiochloroethyl (XXX) with the sodium salt of ethyl phthalimidomalonate (XXXVI), saponified the condensation product (XXXVII), and then simultaneously hydrolyzed and decarboxylated the resulting tricarboxylic acid (XXXVIII) to obtain DL-methionine (V) in 58% overall yield. Booth *et al.* have also reported a similar result.[17]

$$CH_3SCH_2CH_2Cl + NaC(COOC_2H_5)_2(N(CO)_2C_6H_4) \longrightarrow CH_3SCH_2CH_2C(COOC_2H_5)_2(N(CO)_2C_6H_4)$$

XXX XXXVI XXXVII

$$\xrightarrow[\text{ii) } H^+]{\text{i) } OH^-} CH_3SCH_2CH_2C(COOH)_2(N(COOH)CO\text{-}C_6H_5) \xrightarrow{H_2O} CH_3SCH_2CH_2CH(NH_2)COOH$$

XXXVIII V

Albertson and Tullar[18] obtained DL-methionine in 48% overall yield by hydrolyzing the condensation product (XL) formed from the reaction of 2-methylthiochloroethyl (XXX) and ethylacetamidocyanoacetate (XXXIX) in the presence of sodium ethoxide.

$$CH_3SCH_2CH_2Cl + HC(COOC_2H_5)(NHCOCH_3)CN \xrightarrow{C_2H_5ONa} CH_3SCH_2CH_2C(COOC_2H_5)(NHCOCH_3)CN$$

XXX XXXIX XL

$$\longrightarrow CH_3SCH_2CH_2CH(NH_2)COOH$$

V

Goldsmith and Tishler,[19] using ethylacetamidomalonate instead of ethylacetamidocyanoacetate, obtained DL-methionine in 60.5% overall yield.

Riemschneider and Kluge[20] treated 2-methylthiochloroethyl (XXX) with ethyl acetoacetate (XLI), reacted the resulting ethyl β-methylthioethylacetoacetate (XLII) with hydrogen azide, and then saponified the product

$$CH_3SCH_2CH_2Cl + CH_2(COCH_3)COOC_2H_5 \longrightarrow CH_3SCH_2CH_2CH(COCH_3)COOC_2H_5$$

XXX XLI XLII

$$\xrightarrow{HN_3} CH_3\overset{+}{S}(\rightarrow \overset{-}{N}H)CH_2CH_2CH(NHCOCH_3)COOC_2H_5 \xrightarrow{OH^-} CH_3S(\rightarrow O)CH_2CH_2CH(NH_2)COOH$$

XLIII XLIV

$$\xrightarrow{CaSO_3} CH_3SCH_2CH_2CH(NH_2)COOH$$

V

(XLIII). When the methionine sulfoxide (XLIV) so obtained was reduced with calcium sulfite, DL-methionine (V) was formed in 70% yield from (XLII).

Feofilaktov and Ivanova[21] converted ethyl β-methylthioethylaceto-acetate (XLII) to the phenylhydrazone (XLV). DL-methionine (V) was then obtained in 35% overall yield by saponification (to give the free acid (XLVI)) and subsequent reduction.

$$\underset{\text{XLII}}{\underset{|\atop COCH_3}{CH_3SCH_2CH_2CHCOOC_2H_5}} \xrightarrow{C_6H_5N_2Cl} \underset{\text{XLV}}{\underset{\|\atop NNHC_6H_5}{CH_3SCH_2CH_2CCOOC_2H_5}}$$

$$\xrightarrow[OH^-]{} \underset{\text{XLVI}}{\underset{\|\atop NNHC_6H_5}{CH_3SCH_2CH_2CCOOH}} \xrightarrow[Zn,\ CH_3COOH]{} \underset{\text{V}}{\underset{|\atop NH_2}{CH_3SCH_2CH_2CHCOOH}}$$

Gagnon *et al.*[22] converted the condensation product (XLVIII) of 2-methylthiochloroethyl (XXX) and ethyl cyanoacetate (XLVII) to the azide (L) via the hydrazide (XLIX). The urethane (LI) then obtained by the Curtius reaction was hydrolyzed to give DL-methionine (V).

$$\underset{\text{XXX}}{CH_3SCH_2CH_2Cl} + \underset{\text{XLVII}}{\underset{|\atop CN}{CH_2COOC_2H_5}} \xrightarrow[C_2H_5ONa]{} \underset{\text{XLVIII}}{\underset{|\atop CN}{CH_3SCH_2CH_2CHCOOC_2H_5}}$$

$$\xrightarrow{NH_2NH_2} \underset{\text{XLIX}}{\underset{|\atop CN}{CH_3SCH_2CH_2CHCONHNH_2}} \xrightarrow{HNO_2} \underset{\text{L}}{\underset{|\atop CN}{CH_3SCH_2CH_2CHCON_3}}$$

$$\xrightarrow{C_2H_5OH} \underset{\text{LI}}{\underset{|\atop CN}{CH_3SCH_2CH_2CHNHCOOC_2H_5}} \xrightarrow[HCl]{} \underset{\text{V}}{\underset{|\atop NH_2}{CH_3SCH_2CH_2CHCOOH}}$$

Billimoria and Cook[23] prepared 2-thio-4-(2-methylthioethylidene)-thiazolid-5-one (LIII) as the condensation product of β-methylthiopropion-aldehyde (II) and 2-thiothiazol-5-one (LII). Simultaneous reduction and hydrolysis of (LIII) by treatment with red phosphorus and hydrogen iodide then yielded DL-methionine (V).

$$CH_3SCH_2CH_2CHO + \underset{\text{LII}}{\begin{array}{c}CH_2-CO\\ |\quad\ \ |\\ NH\quad S\\ \diagdown\ \diagup\\ CS\end{array}} \longrightarrow \underset{\text{LIII}}{CH_3SCH_2CH_2CH{=}\begin{array}{c}C-CO\\ |\quad\ |\\ NH\ S\\ \diagdown\diagup\\ CS\end{array}}$$

II

$$\xrightarrow[\text{red P}]{\text{HI}} \underset{\text{V}}{CH_3SCH_2CH_2\underset{NH_2}{\underset{|}{C}}HCOOH}$$

4.17.3 Resolution of DL-methionine

When methionine is to be used for the enrichment of food or animal feed, optical resolution of synthesized DL-methionine is not necessary. However, since the optically active isomer is required specifically for medical purposes, resolution is practiced commercially. There have been many reports on resolution methods. They include chemical, enzymatic and physical techniques.

Chemical resolution is based on the difference in solubility between diastereomers obtained from DL-methionine derivatives and optically active acids or bases, e.g. the brucine salt of formylmethionine,[24,25] α-bromo-*d*-camphor-π-sulfonate of methionine or methionine methyl ester,[26,27] D-tartrate of methionine amide,[28] and the optically active salt of pyrrolidonecarboxylic acid.[29]

Among the enzymatic methods, asymmetric hydrolysis of acylated methionine by aminoacylase,[30-35] especially fungal aminoacylase,[31-33] is advantageous for practical use. A continuous resolution technique using a column of insolubilized fungal aminoacylase has been developed and practiced commercially.[36] In addition to these methods, there are several other techniques such as asymmetric hydrolysis of methionine isopropyl ester by esterase[37,38] or of methionine amide by amidase,[39] and synthesis of asymmetric anilide of *N*-substituted methionine using papain.[40]

Physical resolution is based on preferential crystallization. That is to say, although it is not possible to crystallize out preferentially either of the isomers from a solution of free DL-methionine, preferential crystallization has been reported after conversion of the DL-methionine to a simple derivative such as the ammonium salt of *N*-acetylmethionine.[41]

References

1. J. R. Catch, A. H. Cook, A. R. Graham and I. Heilbron, *J. Chem. Soc.*, **1947**, 1609.
2. T. Kaneko and S. Mitsui, *Nippon Kagaku Kaishi* (Japanese), **59**, 1382 (1938).
3. E. Pierson, M. Giella and M. Tishler, *J. Am. Chem. Soc.*, **70**, 1450 (1948).
4. R. Gaudry and G. Nadeau, *Can. J. Res.*, **26B**, 226 (1948).
5. T. Yamaguchi, *Japan Chem. Ind. Assoc. Monthly*, **1967**, no. 3, 163.
6. J. E. Livak, E. C. Britton, J. C. Vanderweele and M. F. Murray, *J. Am. Chem. Soc.*, **67**, 2218 (1945).
7. H. Plieninger, *Chem. Ber.*, **83**, 265 (1950).
8. H. R. Snyder, J. H. Andreen, G. W. Cannon and C. F. Peters, *J. Am. Chem. Soc.*, **64**, 2082 (1942).
9. H. R. Snyder and G. W. Cannon, *ibid.*, **66**, 511 (1944).
10. E. M. Hill and W. Robson, *Biochem. J.*, **30**, 248 (1936).
11. H. Tarver and C. L. A. Schmidt, *J. Biol. Chem.*, **146**, 69 (1942).
12. G. Barger and F. P. Coyne, *Biochem. J.*, **22**, 1417 (1928).
13. D. O. Holland and J. H. C. Nayler, *J. Chem. Soc.*, **1952**, 3403.
14. W. Windus and C. S. Marvel, *J. Am. Chem. Soc.*, **52**, 2575 (1930).
15. O. H. Emerson, P. L. Kirk and C. L. A. Schmidt, *J. Biol. Chem.*, **92**, 449 (1931).
16. G. Barger and T. E. Weichselbaum, *Biochem. J.*, **25**, 997 (1931).
17. E. Booth, V. C. E. Burnop and W. E. Jones, *J. Chem. Soc.*, **1944**, 666.
18. N. F. Albertson and B. F. Tullar, *J. Am. Chem. Soc.*, **67**, 502 (1945).
19. D. Goldsmith and M. Tishler, *ibid.*, **68**, 144 (1946).
20. R. Riemschneider and A. Kluge, *Monatsh. Chem.*, **84**, 522 (1953).
21. V. V. Feofilaktov and T. N. Ivanova, *Zh. Obshch. Khim.*, **21**, 1684 (1951).
22. P. E. Gagnon, K. Savard, R. Gaudry and E. M. Richardson, *Can. J. Res.*, **25B**, 28 (1947).
23. J. D. Billimoria and A. H. Cook, *J. Chem. Soc.*, **1949**, 2323.
24. R. W. Jackson and R. W. Block, *J. Biol. Chem.*, **122**, 426 (1938).
25. W. Windus and C. S. Marvel, *J. Am. Chem. Soc.*, **53**, 3490 (1931).
26. G. P. Wheeler and A. W. Ingersoll, *ibid.*, **73**, 4604 (1951).
27. S. Tatsuoka, M. Honjo and T. Kinoshita, *Yakugaku Zasshi* (Japanese), **71**, 718 (1951).
28. S. Tatsuoka and M. Honjo, *ibid.*, **73**, 357 (1953).
29. J. H. Sbarklop, *Japanese Patent* No. 45–32250 (1970).
30. V. E. Price, J. B. Gilbert and J. P. Greenstein, *J. Biol. Chem.*, **179**, 1169 (1949).
31. C. Neuberg and J. Mandl, *Enzymologia*, **14**, 28 (1950).
32. S. Yamada, I. Chibata and S. Yamada, *Yakugaku Zasshi* (Japanese), **75**, 113 (1955).
33. I. Chibata, T. Ishikawa and S. Yamada, *Bull. Agr. Chem. Soc. Japan*, **21**, 304 (1957).
34. S. Kameda, *Japanese Patent* No. 37–11662 (1962).
35. Y. Kimura, *Japanese Patent* No. 47–671 (1972).
36. T. Tosa, T. Mori, N. Fuse and I. Chibata, *Biotechnol. Bioeng.*, **9**, 603 (1967).

37. M. Brenner and V. Kocher, *Helv. Chim. Acta*, **32**, 333 (1949).
38. K. A. J. Wretlind, *Acta Physiol. Scand.*, **20**, 1 (1950).
39. S. Tatsuoka and M. Honjo, *Yakugaku Zasshi* (Japanese), **73**, 355 (1953).
40. C. A. Dekker and J. S. Fruton, *J. Biol. Chem.*, **173**, 471 (1948).
41. M. Shibasaki and T. Fukuro, *Japanese Patent* No. 39–24440 (1964).

4.18 Ornithine

In 1877, Jaffe gave the name "ornithine" to a substance isolated as a mineral acid salt from hydrolyzates of the ornithuric acid (dibenzylornithine) occurring in the excreta of chickens fed with benzoic acid. In 1897, it was proved that a base obtained from alkali hydrolyzates of arginine was also ornithine, and Fischer, synthesizing this amino acid from diethyl phthalimidopropylmalonate, elucidated its structure in 1901.

The occurrence of ornithine in proteins has not been demonstrated in spite of many compositional analyses, and it has been recognized that ornithine is probably not a constituent of proteins. It occurs in the liver as an intermediate of the urea cycle, and plays an important role in the detoxification of ammonia. Arginine is hydrolyzed by arginase to yield urea and ornithine which reacts with carbamyl phosphate to form citrulline.

Ornithine is produced commercially by fermentation. It is also obtainable by alkali hydrolysis of arginine under relatively mild conditions. The methods for ornithine synthesis can be classified into three broad groups, as follows: (1) use of acylamidomalonate derivatives, (2) amination of α-halogeno acid, and (3) use of a route via hydantoin. Of these, method (3) is anticipated to be of practical importance for commercial production if the demand for DL-ornithine increases greatly.

4.18.1 Synthesis from acylamidomalonate derivatives

This method is advantageous for production on a small scale, and uses derivatives of acylamidomalonic ester such as acetylamidomalonic ester.[1-5] For example, acrylonitrile (I) is condensed with ethyl acetamidomalonate (II) in the presence of an alkali catalyst, and the product obtained is converted to aminopiperidone (III) by reductive cyclization, Ornithine (IV) is then obtained on hydrolysis.

$$\underset{\text{I}}{NCCH{=}CH_2} + \underset{\text{II}}{\underset{|}{CH(COOC_2H_5)_2}\atop NHCOCH_3} \xrightarrow{NaOC_2H_5} \underset{|}{NCCH_2CH_2C(COOC_2H_5)_2}\atop NHCOCH_3$$

$$\xrightarrow[\text{Raney nickel}]{H_2} \underset{\text{III}}{\text{3-acetamido-3-carbethoxy-2-piperidone: } CH_2CH_2CH_2\text{–}NH\text{–}C{=}O\text{–}C(NHCOCH_3)(COOC_2H_5)} \xrightarrow{\quad} \underset{\text{IV}}{NH_2CH_2CH_2CH_2\underset{|}{C}HCOOH \atop NH_2} \quad 80\%$$

4.18.2 Synthesis by amination of α-halogeno acid

In 1901, Fischer[6] re-emphasized the structure of ornithine by synthesizing it via α-bromo-δ-phthalimidovaleric acid.

$$\mathrm{Br(CH_2)_3Br} + \mathrm{CH_2(COOC_2H_5)_2} + \mathrm{C_6H_4(CO)_2NK} \longrightarrow \mathrm{C_6H_4(CO)_2N(CH_2)_3CH(COOC_2H_5)_2}$$

$$\xrightarrow{Br_2} \mathrm{C_6H_4(CO)_2N(CH_2)_3C(Br)(COOC_2H_5)_2} \xrightarrow{HBr} \mathrm{C_6H_4(CO)_2N(CH_2)_3CH(Br)COOH}$$

$$\xrightarrow{NH_3} \mathrm{C_6H_4(CO)_2N(CH_2)_3CH(NH_2)COOH} \xrightarrow{HCl} \underset{\text{IV}}{\mathrm{NH_2CH_2CH_2CH_2CH(NH_2)COOH}}$$

Various methods have since been examined for the synthesis of α-halogeno acids.[7-9] For example, cyclopentanone (V) was converted to δ-benzamidovaleric acid (VI), of which the α-position was then brominated. The bromo acid (VII) was aminated with ammonia and the masking group of the δ-amino group removed.[7]

$$\underset{\text{V}}{\text{cyclopentanone}} \longrightarrow \text{cyclopentanone oxime (C=NOH)} \longrightarrow \text{2-piperidone (CH}_2\text{)}_4\text{C=O, NH} \longrightarrow \mathrm{NH_2(CH_2)_4COOH}$$

$$\longrightarrow \underset{\text{VI}}{\mathrm{C_6H_5CONH(CH_2)_4COOH}} \xrightarrow{Br_2} \underset{\text{VII}}{\mathrm{C_6H_5CONH(CH_2)_3CH(Br)COOH}}$$

$$\xrightarrow[\text{ii) } H^+]{\text{i) } NH_3} \underset{\text{IV}}{\mathrm{NH_2CH_2CH_2CH_2CH(NH_2)COOH}}$$

4.18.3 Synthesis via hydantoin

β-Cyanopropionaldehyde dimethylacetal (VIII), an intermediate in glutamic acid synthesis, is reduced and acetylated to form the acetal (IX) from which γ-acetylamidobutyraldehyde (X) is prepared. Ornithine (IV) is obtained after the aldehyde (X) has been treated by the Bucherer method for hydantoin synthesis.[10)]

$$\underset{\text{VIII}}{NCCH_2CH_2CH(OCH_3)_2} \xrightarrow[\text{Raney nickel}]{H_2} NH_2(CH_2)_3CH(OCH_3)_2$$

$$\xrightarrow{(CH_3CO)_2O} \underset{\text{IX}}{CH_3CONH(CH_2)_3CH(OCH_3)_2} \xrightarrow[H^+]{} \underset{\text{X}}{CH_3CONH(CH_2)_3CHO}$$

$$\xrightarrow[HCN]{NH_4OH,\ CO_2} CH_3CONH(CH_2)_3CH\text{—}CO\ (\text{NH—CO—NH ring}) \xrightarrow{NaOH} \underset{\text{IV}}{NH_2CH_2CH_2CH_2CH(NH_2)COOH}$$

5-Cyanoethylhydantoin (XI) may be prepared by hydantoination of β-cyanopropionaldehyde formed by hydroformylation of acrylonitrile, an intermediate in glutamic acid synthesis. After reduction of the cyano group of the hydantoin (XI), ornithine (IV) is obtained in good yield on hydrolysis of the resulting intermediate product. To prevent subsidiary formation of

$$NCCH_2CH_2CHO \xrightarrow{NH_4OH,\ CO_2,\ HCN} \underset{\text{XI}}{NC(CH_2)_2CH\text{—}CO\ (\text{NH—CO—NH ring})} \xrightarrow[\text{Raney nickel, } NH_4OH]{H_2}$$

$$\left[\begin{array}{l} \underset{\text{XII}}{NH_2(CH_2)_3CH\text{—}CO\ (\text{NH—CO—NH ring})} \\ + \\ \underset{\text{XIII}}{\overline{NH(CH_2)_3CHCO}\ (NHCONH_2)} \end{array}\right] \xrightarrow[90\%]{NaOH} \underset{\text{IV}}{NH_2CH_2CH_2CH_2CH(NH_2)COOH}$$

secondary amine, the cyano group of the hydantoin is reduced in ammonia water at 70–100°C in the presence of Raney nickel catalyst. The product is a mixture of 5-aminopropylhydantoin (XII) and 2-ureido-5-oxopiperidine (XIII), both of which are converted to ornithine on hydrolysis.[11)]

$$CH_2{=}CHCHO \xrightarrow[\text{alkali catalyst}]{HCN} \underset{\text{XIV}}{NCCH_2CH_2CH(OH)CN} \xrightarrow{NH_4OH,\ CO_2} \underset{\text{XI}}{NC(CH_2)_2CH\text{—}CO\ (\text{NH—CO—NH ring})}$$

$$\longrightarrow \underset{\text{IV}}{NH_2CH_2CH_2CH_2CH(NH_2)COOH}$$

Another method also exists for the synthesis of 5-cyanoethylhydantoin (XI). Two moles of hydrogen cyanide are added to one mole of acrolein, and the resulting α-hydroxyglutarodinitrile (XIV) is converted to 5-cyanoethylhydantoin (XI) by treatment with ammonia and carbon dioxide.[12]

It has also been reported that 5-aminopropylhydantoin (XII) can be synthesized by the amination of 5-γ-bromopropylhydantoin (XVIII) derived from butenediol (XV) via γ-hydroxybutyraldehyde (XVI) and 5-γ-hydroxypropylhydantoin (XVII).[13]

$$CH \equiv CH + 2HCHO \longrightarrow HOCH_2C{\equiv}CCH_2OH \longrightarrow \underset{XV}{HOCH_2CH{=}CHCH_2OH}$$

$$\longrightarrow \begin{array}{c} CH{=}CH \\ | \quad\;\; | \\ CH_2 \;\; CH_2 \\ \diagdown \;\; \diagup \\ O \end{array} \xrightarrow{NaOH} \begin{array}{c} CH_2{-}CH \\ | \qquad \| \\ CH_2 \quad CH \\ \diagdown \;\; \diagup \\ O \end{array} \xrightarrow{H^+} \underset{XVI}{HO(CH_2)_3CHO}$$

$$\xrightarrow{NH_4OH,\ HCN,\ CO_2} \underset{XVII}{\begin{array}{c} HO(CH_2)_3CH{-}CO \\ | \qquad | \\ NH \quad NH \\ \diagdown \;\; \diagup \\ CO \end{array}} \xrightarrow{HBr} \underset{XVIII}{\begin{array}{c} Br(CH_2)_3CH{-}CO \\ | \qquad | \\ NH \quad NH \\ \diagdown \;\; \diagup \\ CO \end{array}}$$

$$\xrightarrow{NH_3} \underset{XII}{\begin{array}{c} NH_2(CH_2)_3CH{-}CO \\ | \qquad | \\ NH \quad NH \\ \diagdown \;\; \diagup \\ CO \end{array}} \longrightarrow \underset{IV}{\begin{array}{c} NH_2CH_2CH_2CH_2CHCOOH \\ | \\ NH_2 \end{array}}$$

4.18.4 Resolution of DL-ornithine

Chemically synthesized DL-orinithine can be resolved by an enzymatic method. For example, α,δ-dichloroacetyl-DL-ornithine is hydrolyzed asymmetrically by aminoacylase.[14] Also, after DL-ornithine has been treated with L-glutamic acid in water, L-ornithine–L-glutamate separates out due to the difference in solubilities of the diastereomers formed. L-Ornithine is then obtained on hydrolysis.[15] In addition, when a solution of isolated D-ornithine is heated at 120–150°C, DL-β-amino-α-piperidone is prepared by cyclization. Subsequent hydrolysis gives DL-ornithine.[16] However, since the L-ornithine used for medical purposes is produced by fermentation, resolution of synthesized DL-ornithine has not yet been practiced on a commercial scale.

References

1. J. Kato, H. Ishihara and S. Kanao, *Japanese Patent* No. 29–1578 (1954).
2. N. F. Albertson and S. Archer, *J. Am. Chem. Soc.*, **67**, 2043 (1945).
3. D. T. Warner and O. A. Moe, *ibid.*, **70**, 2765 (1948).
4. R. Gaudry, *Can. J. Chem.*, **31**, 1060 (1953).
5. S. Akabori, Y. Izumi and T. Okuda, *Nippon Kagaku Kaishi* (Japanese), **77**, 490 (1956).
6. E. Fischer, *Chem Ber.*, **34**, 454 (1901).
7. L. F. Schniepp and C. S. Marvel, *J. Am. Chem Soc.*, **57**, 1557 (1935).
8. J. C. Eck and C. S. Marvel, *J. Biol. Chem.*, **106**, 387 (1934).
9. K. Saotome, *Kogyo Kagaku Zasshi* (Japanese), **65**, 1061 (1962).
10. Union Carbide Co., *Canadian Patent* No. 635216 (1962).
11. Y. Komachiya and S. Sakurai, *Japanese Patent* No. 40–6207 (1965).
12. Shell Oil Co., *French Patent* No. 1360750 (1964).
13. R. Gaudry, *Can. J. Chem.*, **29**, 544 (1951).
14. L. Levintow and J. P. Greenstein, *J. Biol. Chem.*, **188**, 643 (1951).
15. K. Togo, M. Hara and K. Ohno, *Japanese Patent* No. 40–11606 (1965).
16. M. Hara, K. Togo and Y. Komachiya, *Japanese Patent* No. 40–18085 (1965).

4.19 Phenylalanine

In 1879, Schulze and Barbieri published a preliminary report on the presence of a new amino acid in lupine seedling extract. Two years later, they reported the detailed results, estimating the chemical structure of the amino acid as phenylaminopropionic acid. Erlenmeyer and Lipp synthesized phenyl-α-aminopropionic acid during the period 1882–83, and named it "phenylalanine". Further, in 1885, Schulze confirmed that the amino acid extracted from lupine seedlings in fact coincided with synthesized phenylalanine. Later, he also revealed the broad occurrence of this amino acid in proteins.

Since the phenylalanine content of proteins is generally low, isolation from proteins is not an an efficient method for its industrial production. Although methods for enzymatic and fermentative production are also being investigated, most phenylalanine is today obtained by chemical synthesis. As a future method for phenylalanine production, however, fermentation is expected to develop together with chemical synthesis. Nevertheless, since phenylalanine is considered to be one of the amino acids that are relatively difficult to produce on a large scale by the fermentative method, chemical synthesis is expected to continue to play the more significant role in phenylalanine production.

Concerning phenylalanine synthesis, there is at present no established industrial method, even though a large number of studies on methods using benzylchloride or benzaldehyde have been reported. Typical procedures that have been investigated are described below.

4.19.1 The Strecker method

This is the method used for the first synthesis of phenylalanine by

C_6H_5–CH_2CHO (I) $\xrightarrow{HCN}$ C_6H_5–$CH_2CH(OH)CN$ (II) $\xrightarrow{NH_3}$ C_6H_5–$CH_2CH(NH_2)CN$ (III) $\xrightarrow{HCl}$ C_6H_5–$CH_2CH(NH_2)COOH$ (IV)

I $\xrightarrow{NaHSO_3}$ C_6H_5–$CH_2CH(OH)SO_3Na$ $\xrightarrow{KCN, (NH_4)_2CO_3}$ C_6H_5–CH_2–CH–CO(NH–CO–NH) (V) $\xrightarrow{OH^-}$ IV

Erlenmeyer and Lipp.[1] Namely, starting from phenylacetaldehyde (I), phenylalanine (IV) was synthesized via cyanohydrin (II) and aminonitrile (III) in a yield of 72% based on cyanohydrin.

By the Bucherer method, phenylacetoaldehyde (I) was converted to benzylhydantoin (V), and subsequent alkaline hydrolysis of this gave phenylalanine (IV) in an overall yield of 40%.[2,3] Both this method, and the Strecker method, appear to be suitable techniques for industrial production, if phenylacetaldehyde can be supplied economically as the starting material.

4.19.2 Synthesis via α-halogeno acid

Synthesis of phenylalanine via α-bromo acid was performed by Fischer a long time ago, but the original method has now been much improved.

Marvel[4] carried out condensation of benzylchloride (VI) with malonic acid ester (VII). The benzylmalonic acid ester (VIII) so obtained was hydrolyzed and then brominated to form the α-bromo acid (IX). Amination of this led to phenylalanine (IV) in an overall yield of 34%.

$$C_6H_5\text{-}CH_2Cl\ (\text{VI}) + CH_2(COOC_2H_5)COOC_2H_5\ (\text{VII}) \xrightarrow{C_2H_5ONa} C_6H_5\text{-}CH_2CH(COOC_2H_5)COOC_2H_5\ (\text{VIII})\quad 51\sim57\%$$

$$\xrightarrow{\text{i) KOH, ii) } Br_2} C_6H_5\text{-}CH_2CH(Br)COOH\ (\text{IX}) \xrightarrow{NH_4OH} C_6H_5\text{-}CH_2CH(NH_2)COOH\ (\text{IV})\quad 62.4\%\text{ based on VIII}$$

In place of benzylmalonic acid ester, ethyl benzylacetoacetate (X) has been brominated in an alkaline solution.[5] Subsequent treatment with liquid ammonia resulted in both deacetylation and amination, and further acid hydrolysis led to phenylalanine (IV) in a yield of 54%.

$$C_6H_5\text{-}CH_2CH(COCH_3)COOC_2H_5\ (\text{X}) \xrightarrow[NaOH]{Br_2} \left[C_6H_5\text{-}CH_2C(COCH_3)(Br)COOH\right]$$

$$\xrightarrow{\text{liq. ammonia}} \left[C_6H_5\text{-}CH_2CH(NH_2)COONH_4\right] \xrightarrow{HCl} C_6H_5\text{-}CH_2CH(NH_2)COOH\ (\text{IV})$$

A method via α-chloro acid as the α-halogeno acid is also known. For example, benzenediazonium chloride (XI), which could be obtained from aniline by the Meerwein addition reaction, was condensed with methyl acrylate (XII) or acrylonitrile (XIII). The condensation product ((XIV) or (XV)) was converted to the α-chloro acid (XVI) by hydrolysis, and subsequent amination of this gave phenylalanine in an overall yield of 48–50% based on aniline.[6)]

$$C_6H_5NH_2 \longrightarrow \underset{\text{XI}}{C_6H_5N{=}NCl} + \underset{\text{XII }(R = COOCH_3),\ \text{XIII }(R = CN)}{CH_2{=}CHR} \xrightarrow[55\sim65\%]{CuCl_2} \underset{\text{XIV }(R = COOCH_3),\ \text{XV }(R = CN)}{C_6H_5CH_2CH(Cl)R}$$

$$\xrightarrow[85\sim95\%]{HCOOH\text{-}HCl} \underset{\text{XVI}}{C_6H_5CH_2CH(Cl)COOH} \xrightarrow[80\sim95\%]{\text{liq. ammonia}} \underset{\text{IV}}{C_6H_5CH_2CH(NH_2)COOH}$$

4.19.3 Synthesis from acetamidomalonic acid ester or acetamidocyanoacetic acid ester

Acylaminomalonic acid ester, which is widely employed as a convenient intermediate for the synthesis of other amino acids, can also be used in the case of phenylalanine synthesis. Although several acyl groups such as formyl, acetyl and benzoyl have been employed, ethyl acetamidomalonate (XVII) is most widely used for this reaction. Thus, the ester (XVII) was reacted with benzylchloride (VI). The condensation product (XVII) was treated with hydrobromic acid for simultaneous hydrolysis and decarboxylation, which resulted in the formation of phenylalanine (IV) in an overall yield of 60%.[7)] On the other hand, acetyl derivatives of phenylalanine are commonly employed for optical resolution. *N*-Acetylphenylalanine (XIX) was

$$\underset{\text{VI}}{C_6H_5CH_2Cl} + \underset{\text{XVII }(R = COOC_2H_5),\ \text{XX }(R = CN)}{HC(R)(NHCOCH_3)COOC_2H_5} \xrightarrow{C_2H_5ONa} \underset{\text{XVIII }(R = COOC_2H_5),\ \text{XXI }(R = CN)}{C_6H_5CH_2C(R)(NHCOCH_3)COOC_2H_5}$$

$$\text{XVIII} \xrightarrow[\text{83\% based on XVIII}]{\text{i) NaOH, ii) HCl}} \underset{\text{XIX}}{C_6H_5CH_2CH(NHCOCH_3)COOH}$$

$$\text{XVIII, XXI} \xrightarrow{H^+ \text{ or } OH^-} \underset{\text{IV}}{C_6H_5CH_2CH(NH_2)COOH}$$

obtained in an overall yield of 68% by hydrolysis of the above condensation product (XVII), first with sodium hydroxide and then with hydrochloric acid.[8]

Condensation of ethyl acetamidocyanoacetate (XX) with benzylchloride (VI), and subsequent alkaline hydrolysis of the condensation product (XXI), led to phenylalanine (IV) in an overall yield of 62%.[9]

4.19.4 Synthesis via hydroxyimino acid

Conversion of hydroxyimino acid to amino acid by reduction can be applied to the synthesis of phenylalanine. Treatment of ethyl benzylacetoacetate (X) with butylnitrite gave hydroxyimino ester (XXII). After conversion of the ester to hydroxyimino acid (XXIII), reduction of this acid with a palladium–carbon catalyst led to phenylalanine (IV) in 80% yield.[10]

$C_6H_5-CH_2CH(COCH_3)COOC_2H_5$ (X) $\xrightarrow[H_2SO_4]{C_4H_9ONO}$ $C_6H_5-CH_2C(=NOH)COOC_2H_5$ (XXII)

$C_6H_5-CH_2CH(COOC_2H_5)COOC_2H_5$ (VIII) $\xrightarrow[C_2H_5ONa]{C_2H_5ONO}$ $C_6H_5-CH_2C(=NOH)COOC_2H_5$ (XXII), 92%

XXII $\xrightarrow{NaOH}$ $C_6H_5-CH_2C(=NOH)COOH$, 89% based on X

XXII $\xrightarrow{H_2,\ \text{Raney nickel}}$ $C_6H_5-CH_2CH(NH_2)COOC_2H_5$ (XXIV), 53%

$C_6H_5-CH_2C(=NOH)COOH$ $\xrightarrow{H_2,\ Pd\text{-}C,\ HCl}$ $C_6H_5-CH_2CH(NH_2)COOH$ (IV), 89%

In a similar manner, treatment of ethyl benzylmalonate (VIII) with ethyl nitrite gave the hydroxyimino ester (XXII), Subsequent Raney-nickel catalyzed reduction resulted in the formation of phenylalanine ethylester (XXIV) in a yield of 49%.[11]

Besides these techniques, starting from the readily available benzaldehyde (XXV) and acetone, hydroxyiminoketone (XXVII) has been prepared via benzylacetone (XXVI) as shown in the following scheme. Oxidation of the ketone (XXVII) with sodium hypobromite then led to hydroxyimino acid (XXIII), and subsequent reduction gave phenylalanine (IV) in an overall yield of 48%.[12]

In a similar manner, treatment of benzylacetone (XXVI), first with methyl nitrite and subsequently with ethyl sulfuric acid, led to ethoxyiminoketone (XXVIII). Phenylalanine (IV) was then synthesized from this ketone via ethoxyimino acid (XXIX) in an overall yield of 50%.[13]

$$C_6H_5CHO + CH_3COCH_3 \xrightarrow{NaOH} C_6H_5CH{=}CHCOCH_3$$

XXV — 82%

$$\xrightarrow[\text{Raney nickel}]{H_2} C_6H_5CH_2CH_2COCH_3 \xrightarrow[HCl]{C_4H_9ONO} C_6H_5CH_2C(=NOH)COCH_3$$

93% XXVI — 88% XXVII

XXVII $\xrightarrow[NaOH]{NaOBr}$ $C_6H_5CH_2C(=NOH)COOH$ (XXIII), 87%

XXVI $\xrightarrow[HCl]{CH_3ONO}$ $\xrightarrow[NaOH]{(C_2H_5)_2SO_4}$ $C_6H_5CH_2C(=NOC_2H_5)COCH_3$ (XXVIII), 77%

XXVIII $\xrightarrow{NaOCl}$ $C_6H_5CH_2C(=NOC_2H_5)COOH$ (XXIX), 70%

XXIX $\xrightarrow[Pd\text{-}C]{H_2}$ $C_6H_5CH_2CH(NH_2)COOH$ (IV), 93%

XXIII $\xrightarrow[\text{Raney nickel}]{H_2}$ IV

4.19.5 Synthesis via condensation with benzaldehyde

Numerous reports have been published detailing methods for phenylalanine synthesis involving the condensation of benzaldehyde with a compound having active methylene group and amino group. For example, the

$$C_6H_5CHO + H_2C(NHCOR)COOH$$

XXV; XXX ($R = C_6H_5$), XXXII ($R = CH_3$)

$\xrightarrow[CH_3COONa]{(CH_3CO)_2O}$ $C_6H_5CH{=}C{-}C{=}O$ (azlactone: N=C(R)–O ring)

62~64% XXXI ($R = C_6H_5$)
74~77% XXXIII ($R = CH_3$)

XXXIII $\longrightarrow$ $C_6H_5CH{=}C(NHCOCH_3)COOH$ (XXXIV), 80~90%

XXXI $\xrightarrow{\text{red P, HI}}$ $C_6H_5CH_2CH(NH_2)COOH$ (IV); $(CH_3CO)_2O$, 63.6%~67% based on XXXI

XXXIII $\xrightarrow[\text{Raney nickel}]{H_2}$ XIX, 98%

XXXIV $\longrightarrow$ $C_6H_5CH_2CH(NHCOCH_3)COOH$ (XIX)

XIX $\xrightarrow{HCl}$ IV, 85~86% based on XXXIV

Erlenmeyer method via azlactone has long been used for phenylalanine synthesis, and various improvements to this reaction have been made. For example, condensation of benzaldehyde (XXV) with hippuric acid (XXX) gave the azlactone (XXXI). Treatment of this with a mixture of red phosphorus, hydriodic acid and acetic anhydride, followed by reduction and hydrolysis, led to phenylalanine (IV) in an overall yield of 41%.[14)]

When acetylglycine (XXXII) was used in place of hippuric acid, the azlactone (XXXIII) formed was converted to acetoamido cinnamic acid (XXXIV), and reduction with platinum oxide catalyst then yielded acetylphenylalanine (XIX). Further hydrolysis of the acetyl amino acid gave phenylalanine (IV) in an overall yield of 55%[14,15)] Raney-nickel catalyzed reduction of the azlactone (XXXIII) in sodium hydroxide solution also led to acetylphenylalanine (XIX) in a good yield (98%).[16)]

The Wheeler-Hoffman method involving condensation of aldehyde with hydantoin (XXXV) has also been employed for the synthesis of phenylalanine. Namely, the condensation product, benzalhydantoin (XXXVI), was treated with red phosphorus and hydriodic acid. The benzylhydantoin (V) obtained was hydrolyzed with barium hydroxide to give phenylalanine (IV) in an overall yield of 57%.[17)]

$$C_6H_5CHO + \underset{\text{XXXV}}{H_2C(CO)NH\text{-}CO\text{-}NH} \xrightarrow[CH_3COONa]{(CH_3CO)_2O} \underset{\text{XXXVI}}{C_6H_5CH{=}C(CO)NH\text{-}CO\text{-}NH}$$

$$\xrightarrow{\text{red P, HI}} \underset{\text{V}}{C_6H_5CH_2\text{-}CH(CO)NH\text{-}CO\text{-}NH} \xrightarrow{OH^-} \underset{\text{IV}}{C_6H_5CH_2CH(NH_2)COOH}$$

In addition to the above, 2-thio-1-benzoylhydantoin (XXXVII), 2-mercaptothiazol-5-one (XXXVIII), rhodanine (XXXIX), diketopiperazine (XL), etc. have been employed in the condensation reaction with benzaldehyde for phenylalanine synthesis.

In any assessment of the various synthetic methods using benzaldehyde, i.e. as regards their industrial potential, due consideration must be given to the compound condensed with the aldehyde and to the means employed in the reduction step. Also, the kind of process that is most advantageous for phenylalanine production will be affected by the availability of the necessary

XXXVII XXXVIII XXXIX XL

raw materials. Although no method is yet practically established for industrial production, a general flowsheet for DL-phenylalanine and L-phenylalanine production by the azlactone method, starting from benzaldehyde and acetylglycine, is presented in Fig. 4–6.

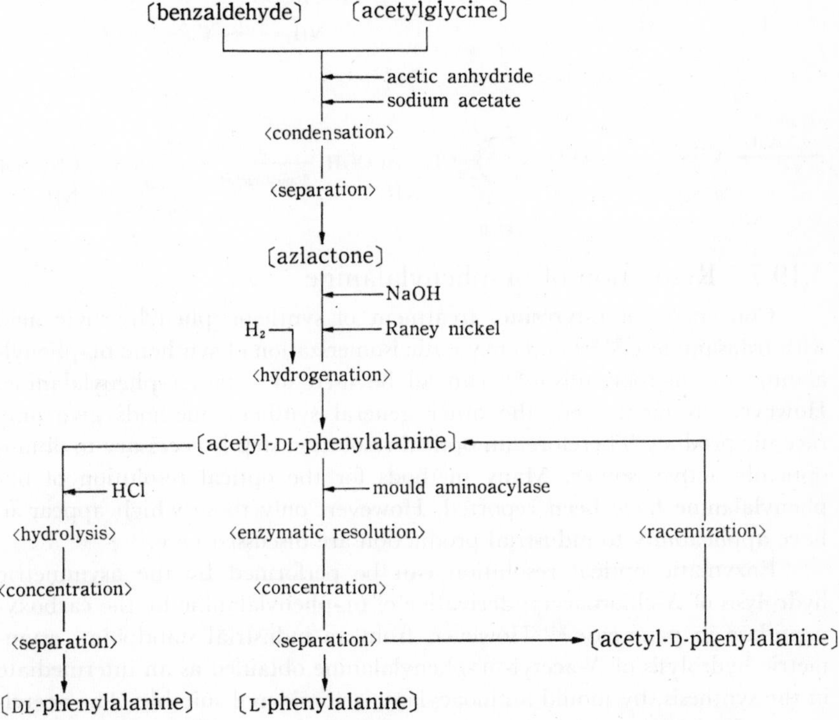

Fig. 4–6. General flowsheet for DL,L-phenylalanine synthesis by the azlactone method.

4.19.6 Conversion of tyrosine

The phenylalanine obtained by customary synthetic methods is racemic. However, a technique for synthetic preparation of the L-isomer of phenylalanine by conversion of L-tyrosine, which is readily isolated from protein hydrolyzates, has been studied. That is to say, tyrosine was converted to phenylalanine by Raney-nickel reduction in an alkaline solution with heating.[18] Moreover, L-phenylalanine has recently been obtained in an overall yield of 40% by tosylation of the copper salt (XLII) of L-tyrosine, followed by reduction of *O*-tosyl-L-tyrosine (XLIII) with Raney nickel.[19]

$$\text{HO-}C_6H_4\text{-}CH_2CH(NH_2)COOH \ (\text{XLI}) \xrightarrow[95\%]{Cu^{2+}} [\text{HO-}C_6H_4\text{-}CH_2CH(NH_2)COO]_2Cu \ (\text{XLII})$$

$$\xrightarrow{CH_3\text{-}C_6H_4\text{-}SO_2Cl} [CH_3\text{-}C_6H_4\text{-}SO_2O\text{-}C_6H_4\text{-}CH_2CH(NH_2)COO]_2Cu$$

$$\xrightarrow[80\sim85\%]{\text{i) HCl; ii) NaOH}} CH_3\text{-}C_6H_4\text{-}SO_2O\text{-}C_6H_4\text{-}CH_2CH(NH_2)COOH \ (\text{XLIII}) \xrightarrow[\text{Raney nickel}]{H_2} C_6H_5\text{-}CH_2CH(NH_2)COOH \ (54.6\%)$$

4.19.7 Resolution of DL-phenylalanine

Conversion of L-tyrosine, treatment of synthetic phenylpyruvic acid with transaminase,[20-22] and enzymatic isomerization of synthetic DL-phenylalanine by microorganisms[23] can all be used to obtain L-phenylalanine. However, as mentioned, the other general synthetic methods give only racemic product. Therefore, an optical resolution step is necessary to obtain optically active isomer. Many methods for the optical resolution of DL-phenylalanine have been reported. However, only those which appear to have applicability to industrial production are discussed here.

Enzymatic optical resolution can be performed by the asymmetric hydrolysis of *N*-chloroacetyl derivative of DL-phenylalanine by the carboxypeptidase in pancrease.[24] However, from an industrial standpoint, asymmetric hydrolysis of *N*-acetyl-DL-phenylalanine obtained as an intermediate in the synthesis, by mould aminoacylase, is considered suitable. The acetyl-D-phenylalanine remaining in the mother liquor can be reused after its racemization. In particular, enzymatic hydrolysis using a column packed with insolubilized aminoacylase is advantageous, since continuous operation is then possible.[25]

Besides the enzymatic resolution methods described above, physicochemical methods based on the preferential crystallization of isomers from a super-saturated solution of acetyl-DL-phenylalanine ammonium salt are also available.[26]

References

1. E. Erlenmeyer and A. Lipp, *Ann. Chem.*, **219**, 179 (1883).
2. H. T. Bucherer and V. A. Liebe, *J. Prakt. Chem.*, **141**, 5 (1934).
3. R. Gaudry, *Can. J. Res.*, **26B**, 773 (1948).
4. C. S. Marvel, *Org. Syn. Coll.*, vol. 3, p. 705, 1955.
5. S. Takizawa, *Sci. Rept. Res. Inst. Tohoku Univ. Ser. A.*, **4**, 311 (1952); *Japanese Patent* No. 26–7516 (1951).
6. R. Filler, L. Gorelic and B. Taqui-Khan, *Proc. Chem. Soc.*, **1962**, 117.
7. N. F. Albertson and S. Archer, *J. Am. Chem. Soc.*, **67**, 308 (1945).
8. H. R. Snyder, J. F. Shekleton and C. D. Lewis, *ibid.*, **67**, 310 (1945).
9. N. F. Albertson and B. F. Tuller, *ibid.*, **69**, 502 (1945).
10. K. E. Hamlin Jr. and W. H. Hartung, *J. Biol. Chem.*, **145**, 349 (1942).
11. J. C. Shivers and C. R. Hauser, *J. Am. Chem. Soc.*, **69**, 1264 (1947).
12. K. Mori, *Nippon Kagaku Kaishi* (Japanese), **29**, 1239 (1958).
13. A. F. Ferris, *J. Org. Chem.*, **24**, 1726 (1959).
14. H. B. Gillespie and H. R. Snyder, *Org. Syn. Coll.*, vol. 2, p. 491, 1943.
15. R. M. Herbst and D. Shemin, *ibid.*, vol. 2, p. 1, 1943.
16. T. Okuda and Y. Fujii, *Bull. Chem. Soc. Japan*, **30**, 698 (1957).
17. H. L. Wheeler and C. Hoffman, *Am. Chem. J.*, **45**, 368 (1911).
18. M. Métayer, *Ann. Chim.*, (12), **4**, 196 (1949).
19. T. Kishi, Y. Kato and M. Tanaka, *Nippon Nogei Kagaku Kaishi* (Japanese), **42**, 238 (1968).
20. S. Sakurai, *J. Biochem.*, **43**, 851 (1956).
21. T. Asai, K. Aida and K. Oishi, *J. Gen. Appl. Microbiol.*, **5**, 150 (1959).
22. A. Kitai, I. Kitamura and N. Miyachi, *Amino Acids* (Japanese), **6**, 16 (1962).
23. I. Chibata, T. Tosa and R. Sano, *Appl. Microbiol.*, **13**, 618 (1965).
24. J. B. Gilbert, V. E. Price and J. P. Greenstein, *J. Biol. Chem.*, **180**, 473 (1949).
25. T. Tosa, T. Mori, N. Fuse and I. Chibata, *Biotech. Bioeng.*, **9**, 603 (1967).
26. M. Shibasaki and T. Fukuro, *Japanese Patent* No. 39–2440 (1964).

4.20 Proline

Willstatter synthesized DL-proline in 1900 by heating α,δ-dibromovaleric acid with ammonia. In 1901, Fischer isolated proline by distillation from esterified hydrolyzate of casein, and gave the amino acid its name. L-Proline is the only amino acid soluble in alcohol, and its solubility in water is also high. For isolation and purification, insoluble salts such as picrate and Reinecke's salt are used.

In living organisms, proline is formed from L-glutamic acid via L-glutamic-γ-semialdehyde. It occurs in large quantities in collagen and may be involved in the function of joints and tendons.

L-Proline is prepared chiefly by extraction from casein hydrolyzate, but recently it has also been produced by fermentation. Commercially practical methods of synthesis have not been developed yet in spite of various studies. Methods of synthesis may be classified into two groups. One involves the reductive cyclization of the γ-carboxyl group of glutamic acid, and the other, cyclization of *n*-valeric acid in which either the α- or δ-position is substituted with an amino group and the remainder with halogen atoms. Many examples of the latter method have been reported.

4.20.1 Synthesis from glutamic acid

As mentioned above, proline is synthesized in living organisms via glutamic-γ-semialdehyde, formed by reduction of the γ-carboxyl group of glutamic acid. In chemical syntheses, the reduction of the γ-carboxyl group of glutamic acid has also been proposed. By this method, L-proline can be prepared from L-glutamic acid without racemization. After treatment of glutamic acid with hydrochloric acid in methanol, the solution is neutralized to obtain γ-methyl glutamate (I). On reduction of the ester (I) with sodium borohydride, proline (II) is obtained in poor yield.[1]

$$\underset{\text{I}}{CH_3OOCCH_2CH_2CH(NH_2)COOH} \xrightarrow{NaBH_4} \left[OHCCH_2CH_2CH(NH_2)COOH \rightarrow \begin{matrix} CH_2-CH_2 \\ | \quad\quad | \\ CH \quad CH-COOH \\ \backslash\backslash \quad / \\ N \end{matrix} \right] \xrightarrow[42\%]{} \underset{\text{II}}{\begin{matrix} CH_2-CH_2 \\ | \quad\quad | \\ CH_2 \quad CH-COOH \\ \backslash \quad / \\ N \\ H \end{matrix}}$$

In another method, *N*-benzyloxycarbonyl-L-glutamine (III) is esterified with methanol and thionyl chloride, and methyl α-benzyloxycarbonylamido-γ-cyanobutyrate (IV) is prepared by dehydration of the γ-amide group. On catalytic reduction in water, the γ-cyano group is converted to aldehyde and the benzyloxycarbonyl group is liberated simultaneously. Finally, L-proline (II) is obtained by reduction.[2)]

$NH_2OCCH_2CH_2CH(NHCOOCH_2C_6H_5)COOH$ (III) $\xrightarrow{SOCl_2,\ CH_3OH}$ $NH_2OCCH_2CH_2CH(NHCOOCH_2C_6H_5)COOCH_3$

$\xrightarrow{CH_3-C_6H_4-SO_2Cl}$ $NCCH_2CH_2CH(NHCOOCH_2C_6H_5)COOCH_3$ (IV) $\xrightarrow[\text{silk-Pd}]{H_2}$ methyl prolinate (ring: CH_2-CH_2, CH_2, $CH-COOCH_3$, NH)

$\longrightarrow$ L-proline (II) (ring: CH_2-CH_2, CH_2, $CH-COOH$, NH) 57%

When methyl pyrrolidonecarboxylate (V), which is easily derived from glutamic acid, is reduced with lithium aluminium hydride, the carbinol (VI) is formed and is oxidized with chromic acid to yield proline (II).[3)]

V (ring: CH_2-CH_2, $O=C$, $CH-COOCH_3$, NH) $\xrightarrow{LiAlH_4}$ VI (ring: CH_2-CH_2, CH_2, $CH-CH_2OH$, NH) $\xrightarrow{H_2SO_4,\ CrO_3}$ II (ring: CH_2-CH_2, CH_2, $CH-COOH$, NH)

15% based on glutamic acid

Synthesis of L-proline by electrolytic reduction of L-pyrrolidonecarboxylic acid (VII) has been reported.[4)] As acquisition of the material is easy and the process is short, this method may be practical if the yield (35%) can be improved.

VII (ring: CH_2-CH_2, $O=C$, $CH-COOH$, NH) $\xrightarrow{\text{electrolytic reduction}}$ II (ring: CH_2-CH_2, CH_2, $CH-COOH$, NH) 35%

4.20.2 Synthesis by cyclization of aminohalogenovaleric acid

After chlorinating the α-amino group of L-ornithine with nitrosyl chloride, D-proline is obtained on treatment with alkali to cause cyclization.[5] L-Proline is yielded on treatment D-ornithine with nitrous acid followed by cyclization.[6]

$$NH_2(CH_2)_3\underset{NH_2}{CHCOOH} \xrightarrow{NOCl} \underset{\text{VIII}}{NH_2(CH_2)_3\underset{Cl}{CHCOOH}} \xrightarrow{69\%} \underset{\text{II}}{\text{proline (}CH_2\text{—}CH_2\text{, }CH_2\text{, }CH\text{—}COOH\text{, }NH\text{ ring)}}$$

There is a method for the synthesis of proline by cyclization with alkali of δ-amino-α-halogenovaleric acid, which can be prepared by various methods. For example, when 3-carboethoxypiperid-2-one (X), which is prepared by reducing the condensation product (IX) of acrylonitrile with diethyl malonate, is treated with thionyl chloride and then hydrolyzed, α-chloro-δ-aminovaleric acid is obtained.[7] Several other methods have also been reported.[8-11]

$$NCCH=CH_2 + CH_2(COOC_2H_5)_2 \xrightarrow[NaOC_2H_5]{} \underset{\text{IX}}{NC(CH_2)_2CH(COOC_2H_5)_2}$$

$$\xrightarrow[\text{Raney nickel}]{H_2} \underset{\text{X}}{\text{(ring: }CH_2\text{, }CH_2\text{, }CH_2\text{, }CH\text{—}COOC_2H_5\text{, }CO\text{, }NH)} \xrightarrow{SOCl_2} \underset{\text{XI}}{\text{(ring: }CH_2\text{, }CH_2\text{, }CH_2\text{, }C(Cl)(COOC_2H_5)\text{, }CO\text{, }NH)}$$

$$\longrightarrow NH_2(CH_2)_3\underset{Cl}{CHCOOH} \xrightarrow{20\%} \underset{\text{II}}{\text{proline (}CH_2\text{—}CH_2\text{, }CH_2\text{, }CH\text{—}COOH\text{, }NH\text{ ring)}}$$

4.20.3 Resolution of DL-proline

There are chemical and enzymatic methods for the resolution of proline. The former is based on the solubility of diastereomers formed from acyl derivatives and optically active bases such as cinchonine.[12,13] The latter method permits isolation of the isomers by means of an ion-exchange resin after L-proline and D-prolinamide are prepared by asymmetric hydrolysis of DL-prolinamide by amidase.[14] However, L-proline has been produced on a large scale by fermentation, and resolution of the product is generally not required.

References

1. S. Kinoshita, M. Tanaka and M. Kishi, *Japanese Patent* No. 37–9232 (1962).
2. T. Itoh, *Bull. Chem. Soc. Japan*, **36**, 25 (1963); S. Akabori, Y. Izumi and T. Itoh, *Japanese Patent* No. 39–6532 (1964).
3. R. Buyle, *Chem. Ind.*, **1956**, 380.
4. M. Viscontini and H. Bühler *Helv. Chim. Acta.*, **49**, 2524 (1966).
5. P. B. Hamilton and P. J. Ortiz, *J. Biol. Chem.*, **184**, 607 (1950); **187**, 733 (1950).
6. Y. Yamada, I. Kumashiro and T. Takenishi, *Japanese Patent* No. 41–8940 (1966).
7. N. F. Albertson and J. L. Fillman, *J. Am. Chem. Soc.*, **71**, 2818 (1949).
8. R. Willstätter, *Chem. Ber.*, **33**, 1160 (1900).
9. H. Plieninger, *ibid.*, **83**, 271 (1950).
10. R. Gaudry, *Can. J. Chem.*, **29**, 544 (1951).
11. K. Hasse and A. Wieland, *Chem. Ber.*, **93**, 1686 (1960).
12. E. Fischer and G. Zemplen, *ibid.*, **42**, 2989 (1909).
13. L. Velluz, G. Amiard and R. Heymès, *Bull. Soc. Chim. France*, **21**, 1015 (1954).
14. V. E. Price, L. Levintow, J. P. Greenstein and R. B. Kingsley, *Arch. Biochem.*, **26**, 92 (1950).

4.21 Sarcosine

Sarcosine (*N*-methylglycine) was first isolated in 1847 by Liebig, who hydrolyzed creatine with hot barium hydroxide. This was in fact the first reported example of an *N*-alkyl amino acid. Sarcosine was synthesized by Vohlhard in 1862, about 15 yr after its finding in nature. He treated ethylchloroacetate with excess methylamine, hydrolyzed the product with barium hydroxide, and so obtained sarcosine. Later, this amino acid was also prepared by hydrolysis of caffeine or creatinine with barium hydroxide. However, since sarcosine has no asymmetric carbon atom in its molecule, as in the case of glycine, it is easily obtainable by chemical synthesis without the resolution step that is generally required in the preparation of other α-amino acids.

Several synthetic methods have been reported, i.e. synthesis from monochloroacetic acid, hydrolysis of sarcosinonitrile (*N*-methylglycinonitrile) with acid or alkali,[1,2] reduction of methylene glycine prepared by condensation of glycine with formalin,[3,4] and hydrolysis of benzylsulfosarcosine.[5] However, two methods are practiced commercially at present for synthesizing sarcosine. One is the amination of α-halogeno acetic acid, and the other the hydrolysis of sarcosinonitrile with acid or alkali (the Strecker method).

4.21.1 Synthesis from α-halogeno acetic acid

The reactions in this method are illustrated in the following equation.

$$\underset{\text{I}}{XCH_2COOR} + \underset{\text{II}}{CH_3NH_2} \longrightarrow CH_3NHCH_2COOR + HX$$

Blicke and Norris[6] obtained sarcosine in 40.5% yield from a solution of monochloroacetic acid (I) (R=H, X=Cl) with a large excess of aqueous methylamine (II). According to a British Patent,[7] sarcosine can be obtained in 75–90% yield by treatment of the acid (I) (R=Na or K) with liquid or aqueous amine at room temperature (or lower) for 1 h. This method is better than that of P.B. Report,[8] where the yield was 60–70% after treatment at 75°C and 2–3 atm for 12h. In any method using α-halogeno acetic acid, a large excess of amine (II) is required for aminating the acid (I). This may be regarded as a demerit since methylamine is more expensive than the ammonia used for glycine synthesis. However, it has recently been reported that the addition of formalin to the reaction system decreases the amount of ammonia required for glycine synthesis,[9] and such an improvement is also expected in the case of sarcosine synthesis.

4.21.2 The Strecker method

This process takes place via the following two steps. First, sarcosinonitrile (V) is prepared from hydrogen cyanide (III), formalin and methylamine (II), or from methylamine (II) and glycolonitrile (IV) which is prepared from hydrogen cyanide and formalin. Then, the aminonitrile (V) is hydrolyzed with acid or alkali to give sarcosine.

$$\text{①}\quad \left.\begin{array}{l} \underset{\text{III}}{HCN} + CH_2O + \underset{\text{II}}{CH_3NH_2} \\ \text{or} \\ \underset{\text{IV}}{HOCH_2CN} + \underset{\text{II}}{CH_3NH_2} \end{array}\right\} \longrightarrow \underset{\text{V}}{CH_3NHCH_2CN}$$

$$\text{②}\quad CH_3NHCH_2CN \xrightarrow[H^+ \text{ or } OH^-]{} CH_3NHCH_2COOH$$

Methods for aminonitrile synthesis have been reported by Biltz and Slotta,[10] Cook and Cox[11] and Wakaomi *et al.*[12] Biltz and Slotta prepared aminonitrile in 95.6% yield by adding sodium cyanide solution to a mixture of the hydrochloride of the amine (II) and formalin. Cook and Cox reported that aminonitrile could be obtained in 93% yield by treating cyanohydrin (IV) with the amine (II). Wakaomi *et al.*, using a tube reactor, continuously treated a solution of cyanohydrin (IV) with a solution of the amine (II) at 100–120°C for 1–7 min and obtained aminonitrile in 80–90% yield. This latter method is of particular interest since the process can be carried out continuously.

Hydrolysis of the aminonitrile (V) with acid or alkali is performed either with aminonitrile solution prepared from hydrogen cyanide, formalin and amine, or with a salt of the aminonitrile isolated.

Gluud *et al.*[13] treated 1 mole of cyanohydrin (IV) with 3 moles of liquid amine (II) at −10 to −20°C, hydrolyzed the aminonitrile in the solution with barium hydroxide after excess amine had been removed, and obtained sarcosine in 75% yield. Schütte[14] synthesized sarcosine in 90% yield by hydrolyzing the sulfate of the aminonitrile (V) with barium hydroxide. Leak and Brakebill,[15] treating a mixture of the amine and formalin with liquid hydrogen cyanide, hydrolyzed the resulting solution with 96% sulfuric acid to obtain sarcosine sulfate. This salt was reacted with calcium hydroxide and then with barium hydroxide, and the crude sarcosine solution treated with benzyl alcohol to remove organic impurities. Sarcosine of high purity was then obtained in an overall yield of 69.3%.

As described above, there are several similarities between the methods for synthesis of sarcosine and of glycine. The major difference between them lies in fact that the Bucherer-Berg method is not employed for sarcosine synthesis.

References

1. G. W. Heimrod, *Chem. Ber.*, **47**, 347 (1914).
2. L. Baumann, *J. Biol. Chem.*, **21**, 563 (1915); W. Eshweiler, *Ann. Chem.*, **279**, 39 (1894).
3. W. Löb, *Biol. Chem. Z.*, **51**, 123 (1912).
4. H. Scheibler and H. Neef, *Chem. Ber.*, **59**, 1504 (1926).
5. T. B. Johnson and J. A. Ambla, *J. Am. Chem. Soc.*, **36**, 372 (1914).
6. F. F. Blicke and D. E. Norris, *ibid.*, **76**, 3213 (1954).
7. Colgate-Palmolive Co., *British Patent* No. 775738 (1957).
8. P. B. Report, 63, 822 (1946).
9. R. E. Williams, *U.S. Patent* No. 3190914 (1963).
10. H. Biltz and K. Slotta, *J. Prakt. Chem.*, **113**, 252 (1926).
11. A. H. Cook and S. F. Cox, *J. Chem. Soc.*, **1949**, 2334.
12. M. Wakaomi, A. Nishi, S. Agatsuma and M. Takeyama, *Japanese Patent* No. 38–15611 (1963).
13. W. Gluud, W. Klempt and A. Kürzinger, *German Patent* No. 656350 (1938).
14. E. Schütte, *Z. Physiol. Chem.*, **279**, 61 (1943).
15. P. H. Leak and E. K. Brakebill, *U.S. Patent* No. 3009954 (1961).

4.22 Serine

Serine was isolated as early as 1865 when Cramer obtained a sweet crystal from the sulfuric acid hydrolyzate of sericine, a silk protein. In 1880, Erlenmeyer presumed its structure to be α-amino-β-hydroxypropionic acid, and in 1902 Fischer and Leuchs confirmed this structure. In 1907, Fischer proved that serine in proteins is optically active, and in 1942 Bergman succeeded in isolating L-serine as its *p*-hydroxyazobenzene-*p*′-sulfonate from an acid hydrolyzate of silk.

After the structure of serine was elucidated, many studies on its synthesis were carried out and practical methods of synthesis have been developed. On the other hand, fermentative production of L-serine has been investigated recently. The production of L-serine by direct fermentation from sugars is not practicable at present, but L-serine can be prepared by culture of D-serine-catabolizing bacteria, which remove D-serine from a medium containing DL-serine, or by culture of L-serine-producing bacteria in a medium containing DL-glyceric acid and nitrogen.

Methods of DL-serine synthesis can be classified into 3 main groups. The first is the synthesis of serine from alkoxyacetaldehyde through preparation of the hydantoin via the cyanohydrin, hydrolysis of the hydantoin to *O*-alkylserine, and dealkylation thereof. The second method involves various condensation reactions. A hydroxymethyl group is introduced into the active methylene groups of hippuric acid, aminomalonate derivatives, or copper glycine. The product is hydrolyzed, after it is reduced in some cases, to give DL-serine. The third method uses vinyl compounds such as acrylic acid ester, acrylonitrile, and vinyl acetate as a starting material. After halogen is added to the double bond of these materials, one of the halogen atoms is replaced with a hydroxyl group and the other with an amino group.

There are also other methods which cannot be included in the above categories. However, methods of the third group are considered to be the most suitable for commercial production of serine.

4.22.1 Synthesis from alkoxyacetaldehyde

Fischer and Leuchs[1)] obtained DL-serine (III) in 9% overall yield by hydrolyzing the aminonitrile (II) which was prepared by treatment of glycolaldehyde (I) with hydrogen cyanide and ammonia. Since glycolalde-

$$\underset{\text{I}}{HOCH_2CHO} \xrightarrow{HCN,\ NH_3} \underset{\text{II}}{HOCH_2\underset{\displaystyle NH_2}{\underset{|}{C}}HCN} \xrightarrow{H_2O} \underset{\text{III}}{HOCH_2\underset{\displaystyle NH_2}{\underset{|}{C}}HCOOH}$$

hyde (I) is easily polymerized, the yield was low.

Leuchs and Geiser[2] attempted to improve the yield using ethoxyacetaldehyde instead of glycolaldehyde. Ethoxyacetaldehyde (VII) can be prepared from the chloroacetal (IV) by ethoxidation to yield an ethoxyacetal derivative (V),[2-6] or it can be directly synthesized by oxidation of ethyleneglycol ethyl ether (VI), which is readily obtainable.[7,8] When the nitrile (VIII), which is prepared by treating ethoxyacetaldehyde (VII) with hydrogen cyanide and ammonia, is hydrolyzed with hydrogen bromide, DL-serine (III) is obtained in 14% yield from the chloroacetal (IV) and in 51% yield from ethyleneglycol ethyl ether (VI).

$$\underset{\text{IV}}{ClCH_2CH(OC_2H_5)_2} \xrightarrow{NaOC_2H_5} \underset{\text{V}}{C_2H_5OCH_2CH(OC_2H_5)_2}$$

$$\underset{\text{VI}}{C_2H_5OCH_2CH_2OH} \xrightarrow{O_2} \underset{\text{VII}}{C_2H_5OCH_2CHO} \xrightarrow{HCN,\ NH_3} \underset{\text{VIII}}{C_2H_5OCH_2CH(NH_2)CN}$$

(V → VII)

$$\xrightarrow[HBr]{} \underset{\text{III}}{HOCH_2CH(NH_2)COOH}$$

Nadeau and Gaudry,[9] treating methoxyacetaldehyde (IX) with hydrogen cyanide and reacting the cyanohydrin (X) so formed with ammonium carbonate, prepared 5-methoxymethylhydantoin (XI) which was then hydrolyzed to give *O*-methylserine (XII). DL-serine (III) was obtained by cleaving the methoxy group of *O*-methylserine (XII). The overall yield of DL-serine from methoxyacetaldehyde was 38%.

$$\underset{\text{IX}}{CH_3OCH_2CHO} \xrightarrow{HCN} \underset{\text{X}}{CH_3OCH_2CH(OH)CN} \xrightarrow{(NH_4)_2CO_3} \underset{\text{XI}}{CH_3OCH_2\text{-}CH\text{-}CO\text{-}NH\text{-}CO\text{-}NH\ (\text{ring})}$$

$$\xrightarrow[Ba(OH)_2]{} \underset{\text{XII}}{CH_3OCH_2CH(NH_2)COOH} \xrightarrow[HBr]{} \underset{\text{III}}{HOCH_2CH(NH_2)COOH}$$

4.22.2 Synthesis by condensation

Erlenmeyer and Stoop[10)] obtained DL-serine (III) in 48% yield from ethyl hippurate (XIV). Ethyl hippurate (XIV) was condensed with ethyl formate (XIII) in the presence of sodium ethoxide to form hydroxymethylene hippuric ester (XV). This ester was then reduced to benzoylserine ethyl ester (XVI) with sodium amalgam, and the resulting ester hydrolyzed with acid to give DL-serine (III).

$$\underset{\text{XIII}}{HCOOC_2H_5} + \underset{\text{XIV}}{\begin{array}{l}CH_2COOC_2H_5\\|\\NHCOC_6H_5\end{array}} \xrightarrow{C_2H_5ONa} \underset{\text{XV}}{\begin{array}{l}HOCH{=}CCOOC_2H_5\\\qquad\;|\\\qquad NHCOC_6H_5\end{array}}$$

$$\xrightarrow{Na/Hg} \underset{\text{XVI}}{\begin{array}{l}HOCH_2CHCOOC_2H_5\\\qquad\;|\\\qquad NHCOC_6H_5\end{array}} \xrightarrow{H^+} \underset{\text{III}}{\begin{array}{l}HOCH_2CHCOOH\\\qquad\;|\\\qquad NH_2\end{array}}$$

Matra[11)] obtained DL-serine in 28% yield from phthalimidomalonic ester (XVIII) which was condensed with monochlorodimethyl ether (XVII) in the presence of sodium ethoxide. The condensation product (XIX) was hydrolyzed with alkali. Maeda, Terumi and Suzuki[12)] have improved the method and doubled the yield.

$$\underset{\text{XVII}}{CH_3OCH_2Cl} + \underset{\text{XVIII}}{\begin{array}{l}COOC_2H_5\\|\\CH{-}N(CO)_2C_6H_4\\|\\COOC_2H_5\end{array}} \xrightarrow{NaOC_2H_5} \underset{\text{XIX}}{\begin{array}{l}\qquad\quad COOC_2H_5\\\qquad\quad|\\CH_3OCH_2CN(CO)_2C_6H_4\\\qquad\quad|\\\qquad\quad COOC_2H_5\end{array}}$$

$$\xrightarrow{H^+} \underset{\text{III}}{\begin{array}{l}HOCH_2CHCOOH\\\qquad\;|\\\qquad NH_2\end{array}}$$

King[13)] treated diethyl acetamidomalonate (XXI) with formaldehyde (XX) in the presence of an alkaline catalyst and saponified the condensation product (XXII), preparing the sodium salt of the malonate derivative (XXIII) which was decarboxylated to give acetyl serine (XXIV) by heating under acidic conditions. DL-Serine was obtained in 65% yield by acid hydrolysis of the acetyl serine (XXIV). This method is useful for the synthesis of DL-serine on an experimental scale.

$$\underset{\text{XX}}{HCHO} + \underset{\text{XXI}}{HC(COOC_2H_5)_2NHCOCH_3} \xrightarrow{OH^-} \left[\underset{\text{XXII}}{HOCH_2C(COOC_2H_5)_2NHCOCH_3}\right] \xrightarrow{NaOH}$$

$$\left[\underset{\text{XXIII}}{HOCH_2C(COONa)_2NHCOCH_3}\right] \xrightarrow{H^+} \underset{\text{XXIV}}{HOCH_2CH(NHCOCH_3)COOH} \xrightarrow{H^+} \underset{\text{III}}{HOCH_2CH(NH_2)COOH}$$

Sato *et al.*[14] succeeded in synthesizing threonine from copper glycinate and acetaldehyde. The principle of this method is also applicable to the synthesis of serine, and DL-serine is obtained in 30% yield by condensing formaldehyde (XX) with copper glycinate (XXV).[15]

$$\underset{\text{XX}}{HCHO} + \underset{\text{XXV}}{Cu(NH_2CH_2COO)_2} \xrightarrow{OH^-}$$

$$\underset{\text{XXVI}}{Cu(NH_2CH(CH_2OH)COO)_2} \xrightarrow{-Cu^{2+}} \underset{\text{III}}{HOCH_2CH(NH_2)COOH}$$

4.22.3 Synthesis from vinyl compounds

Carter *et al.*[16,17] synthesized DL-serine from methyl acrylate. Methyl acrylate (XXVII) was converted to methyl α-acetoxymercury-β-methoxypropionate (XXVIII) which was then treated with potassium bromide. The resulting methyl α-bromomercury-β-methoxypropionate (XXIX) was converted to methyl α-bromo-β-methoxypropionate (XXX) by bromine. The ester (XXX) was saponified and aminated. On demethylation of the resulting β-methoxyserine (XII) with hydrogen bromide, DL-serine (III) was obtained in about 40% yield.

$$\underset{\text{XXVII}}{CH_2=CHCOOCH_3} \xrightarrow[CH_3OH]{(CH_3COO)_2Hg} \underset{\text{XXVIII}}{CH_3OCH_2CH(HgOCOCH_3)COOCH_3} \xrightarrow{KBr}$$

$$CH_3OCH_2\underset{\underset{HgBr}{|}}{C}HCOOCH_3 \xrightarrow{Br_2} CH_3OCH_2\underset{\underset{Br}{|}}{C}HCOOCH_3 \xrightarrow{NaOH} CH_3OCH_2\underset{\underset{Br}{|}}{C}HCOOH$$

XXIX XXX XXXI

$$\xrightarrow{NH_3} CH_3OCH_2\underset{\underset{NH_2}{|}}{C}HCOOH \xrightarrow{HBr} HOCH_2\underset{\underset{NH_2}{|}}{C}HCOOH$$

XII III

Wood and du Vigneaud[18] brominated ethylacrylate (XXXII) to prepare ethyl α,β-dibromopropionate (XXXIII), which was converted to ethyl α-bromo-β-ethoxypropionate (XXXIV) by treatment with sodium ethoxide. DL-Serine (III) was then obtained from the ester (XXXIV) by saponification followed by amination and dealkylation in the same manner as in the method of Carter *et al.* The overall yield was about 40%. The presence of a catalytic amount of mercury acetate improves the yield in the step from ethyl α,β-dibromopropionate (XXXIII) to ethyl α-bromo-β-ethoxypropionate (XXXIV).[19]

$$CH_2{=}CHCOOC_2H_5 \xrightarrow{Br_2} BrCH_2\underset{\underset{Br}{|}}{C}HCOOC_2H_5 \xrightarrow{NaOC_2H_5} C_2H_5OCH_2\underset{\underset{Br}{|}}{C}HCOOC_2H_5$$

XXXII XXXIII XXXIV

$$\xrightarrow[NaOH]{} C_2H_5OCH_2\underset{\underset{Br}{|}}{C}HCOOH \xrightarrow{NH_3} C_2H_5OCH_2\underset{\underset{NH_2}{|}}{C}HCOOH \xrightarrow[HBr]{} HOCH_2\underset{\underset{NH_2}{|}}{C}HCOOH$$

XXXV XXXVI III

Mattocks and Hartung,[20] treating methyl acrylate (XXVII) with hydrobromite, prepared methyl α-hydroxy-β-bromo propionate (XXXVII), which was then aminated to give the *N*-benzylamino acid ester (XXXVIII). The ester (XXXVIII) was saponified and the resulting compound reduced with palladium-hydrogen. DL-Serine (III) was obtained in 32% yield.

$$CH_2{=}CHCOOCH_3 \xrightarrow{HBrO} HOCH_2\underset{\underset{Br}{|}}{C}HCOOCH_3 \xrightarrow{C_6H_5CH_2NH_2} HOCH_2CHCOOCH_3 \atop NHCH_2C_6H_5$$

XXVII XXXVII XXXVIII

$$\xrightarrow[NaOH]{} HOCH_2\underset{\underset{NHCH_2C_6H_5}{|}}{C}HCOOH \xrightarrow[Pd]{H_2} HOCH_2\underset{\underset{NH_2}{|}}{C}HCOOH$$

XXXIX III

Brockmann and Musso,[21] modifying the method of Wood and du Vigneaud,[18] used acrylonitrile (XL) instead of acrylic acid ester as starting material. Acrylonitrile (XL) was converted to α-bromo-β-methoxypropionitrile (XLII), which was hydrolyzed, and the resulting α-bromo-β-methoxypropionic acid (XXXI) was aminated to give DL-serine (III) in 52–59% yield.

$$\underset{\text{XL}}{CH_2{=}CHCN} \xrightarrow{Br_2} \underset{\text{XLI}}{BrCH_2CH(Br)CN} \xrightarrow{NaOCH_3} \underset{\text{XLII}}{CH_3OCH_2CH(Br)CN} \xrightarrow{H^+}$$

$$\underset{\text{XXXI}}{CH_3OCH_2CH(Br)COOH} \xrightarrow{NH_3} \underset{\text{XII}}{CH_3OCH_2CH(NH_2)COOH} \xrightarrow[HBr]{} \underset{\text{III}}{HOCH_2CH(NH_2)COOH}$$

Gundermann and Rose[22] adopted chlorination, instead of bromination, of acrylonitrile and prepared *O*-methylserine in 69% yield in the same manner as with acrylic acid ester. *O*-methylserine was demethylated with hydrochloric acid. White and Wysong,[23] using acrylamide as starting material instead of acrylic acid ester, synthesized DL-serine in 65–70% yield via α-bromo-β-alkoxypropionamide. Wolf *et al.*[24] also synthesized DL-serine in 33% yield from acrylamide via α-chloro-β-methoxypropionamide, adopting chlorination instead of bromination.

Geipel *et al.*[25] synthesized DL-serine from vinyl acetate. 1,1,2-Triacetoxyethane (XLV) which was derived from vinyl acetate (XLIII) via 1-acetoxy-1,2-dibromoethane (XLIV), a brominated derivative of (XLIII), was subjected to the Strecker reaction, and the resulting nitrile intermediate

$$\underset{\text{XLIII}}{CH_2{=}CHOCOCH_3} \xrightarrow{Br_2} \left[\underset{\text{XLIV}}{BrCH_2CH(Br)OCOCH_3}\right] \xrightarrow{CH_3COONa,\ (CH_3CO)_2O}$$

$$\underset{\text{XLV}}{CH_3COOCH_2CH(OCOCH_3)OCOCH_3} \xrightarrow{NH_3,\ KCN,\ NH_4Cl} \left[\underset{\text{XLVI}}{CH_3COOCH_2CH(NH_2)CN}\right]$$

$$\xrightarrow[HCl]{} \underset{\text{III}}{HOCH_2CH(NH_2)COOH}$$

(XLVI) was hydrolyzed to give DL-serine (III). When the reaction was carried out continuously without separation of the intermediate, the overall yield of DL-serine from vinyl acetate was 74%.

4.22.4 Miscellaneous methods

Berlinquet[26] synthesized DL-serine (III) in 70% yield from ethyl acetamidocyanoacetate (XLVII), which was reduced with sodium borohydride to an alcohol (XLVIII). DL-Serine (III) was obtained on hydrolysis of the product (XLVIII) with acid.

$$\underset{\text{XLVII}}{CH_3CONHCH(COOC_2H_5)CN} \xrightarrow[C_2H_5OH]{NaBH_4} \underset{\text{XLVIII}}{CH_3CONHCH(CH_2OH)CN} \xrightarrow{HCl} \underset{\text{III}}{H_2NCH(CH_2OH)COOH}$$

Sprinson and Chargaff[27] converted bromopyruvic acid (XLIX) by treatment with alkali to hydroxypyruvic acid (L), which was then treated with 2,4-diphenylhydrazine. On reduction of the resulting hydrazone (LI) with aluminium amalgam in 80% ethanol, DL-serine was obtained in 24% yield.

$$\underset{\text{XLIX}}{BrCH_2COCOOH} \xrightarrow{NaOH} \underset{\text{L}}{HOCH_2COCOOH} \xrightarrow{NO_2\text{-}C_6H_3(NO_2)\text{-}NHNH_2} \underset{\text{LI}}{HOCH_2C(=N\text{-}NC_6H_5)COOH}$$

$$\xrightarrow{Al/Hg} \underset{\text{III}}{HOCH_2CH(NH_2)COOH}$$

Sasaki[28] prepared 2-phenyl-4-hydroxymethylene-5-oxazoline (LIV) by alkaline treatment of 2-phenyl-4-ethoxymethylene-5-oxazoline (LIII), which was derived from hippuric acid (LII) and ethyl orthoformate. It was decyclized with Raney nickel, and DL-serine (III) was obtained in 28% yield.

$$\underset{\text{LII}}{C_6H_5CONHCH_2COOH} \xrightarrow[(CH_3CO)_2O]{(C_2H_5O)_3CH} \underset{\text{LIII}}{C_2H_5OCH=C\text{—}CO \;(\text{ring: } C\text{—}N=C(C_6H_5)\text{—}O\text{—}CO)}$$

$$\xrightarrow{\text{NaOH}} \underset{\text{LIV}}{\begin{array}{c} HOCH=C-CO \\ \quad\;| \quad\;\; | \\ \quad N \quad O \\ \quad\;\; \backslash\backslash \; / \\ \quad\quad C \\ \quad\quad | \\ \quad\quad C_6H_5 \end{array}} \xrightarrow[\text{Raney nickel}]{H_2} \underset{\text{III}}{\begin{array}{l} HOCH_2CHCOOH \\ \qquad\quad | \\ \qquad\; NH_2 \end{array}}$$

Knunyants and Shokina[29] treated α-phenacetylaminoacrylic acid (LV) with hydrogen bromide in acetic acid, hydrolyzed the resulting bromide derivative (LVI) with hydrogen bromide, and obtained DL-serine in 80% yield.

$$\underset{\text{LV}}{\begin{array}{l} CH_2{=}CCOOH \\ \qquad\; | \\ \quad NHCOCH_2C_6H_5 \end{array}} \xrightarrow[CH_3COOH]{HBr} \underset{\text{LVI}}{\begin{array}{l} BrCH_2CHCOOH \\ \qquad\quad | \\ \quad NHCOCH_2C_6H_5 \end{array}} \xrightarrow{HBr} \underset{\text{III}}{\begin{array}{l} HOCH_2CHCOOH \\ \qquad\quad | \\ \qquad\; NH_2 \end{array}}$$

4.22.5 Resolution of DL-serine

There have been many reports on chemical methods for resolution of DL-serine, but enzymatic and fractional crystallization methods are most practical. An enzymatic method which yields L-serine on asymmetric hydrolysis of *N*-acetyl-DL-serine by aminoacylase is favourable. Although fractional crystallization cannot be applied directly to DL-serine, it is applicable to simple derivatives such as *N*-benzoyl-DL-serine ammonium salt,[30] DL-serine paratoluene sulfonate[31] and DL-serine xylene-4-sulfonate,[32] and is considered to be useful in the commercial production of optically active serine.

References

1. E. Fischer and H. Leuchs, *Chem. Ber.*, **35**, 3787 (1902).
2. H. Leuchs and W. Geiser, *ibid.*, **39**, 2644 (1906).
3. A. Pinner, *ibid.*, **5**, 147 (1872).
4. P. Fritsch and W. Schumaker, *Ann. Chem.*, **279**, 288, 301 (1894).
5. F. D. Chataway and O. G. Backeberg, *J. Chem. Soc.*, **125**, 1097 (1924).
6. A. Lieben, *Ann. Chem.*, **146**, 180 (1868).
7. M. S. Dunn, C. E. Redemann and N. L. Smith, *J. Biol. Chem.*, **104**, 511 (1934).
8. C. E. Redemann and R. N. Icke, *J. Org. Chem.*, **8**, 159 (1943).
9. P. G. Nadeau and R. Gaudry, *Can. J. Chem.*, **27B**, 421 (1949).
10. E. Erlenmeyer and F. Stoop, *Ann. Chem.*, **337**, 236 (1904).
11. S. K. Matra, *J. Indian Chem. Soc.*, **7**, 799 (1930).
12. S. Maeda, M. Terumi and T. Suzuki, *Bull. Inst. Phys. Chem. Res.* (*Tokyo*), **17**, 267 (1938).

13. J. A. King, *J. Am. Chem. Soc.*, **69**, 2738 (1947).
14. M. Sato, K. Okawa and S. Akabori, *Bull. Chem. Soc. Japan*, **30**, 937 (1957).
15. S. Akabori, T. T. Otani, R. Marshall, M. Winitz and J. P. Greenstein, *Arch. Biochem. Biophys.*, **83**, 1 (1959).
16. L. R. Schiltz and H. E. Carter, *J. Biol. Chem.*, **116**, 793 (1936).
17. H. E. Carter and H. D. West, *Org. Syn. Coll.*, vol. 3, p. 774, 1955.
18. J. L. Wood and V. du Vigneaud, *J. Biol. Chem.*, **134**, 413 (1940).
19. H. Bretschneider, N. Karpitschka and G. Piekarski, *Monatsh. Chem.*, **84**, 1084 (1953).
20. A. M. Mattocks and W. H. Hartung, *J. Biol. Chem.*, **165**, 501 (1946).
21. H. Brockmann and H. Musso, *Chem. Ber.*, **87**, 581 (1954).
22. K. D. Gundermann and H. J. Rose, *ibid.*, **92**, 1081 (1959).
23. H. C. White and D. V. Wysong, *U.S. Patent* No. 2783274 (1957).
24. J. Wolf, J. Wojciechowski and L. Rolaczek, *Polish Patent* No. 48569 (1964).
25. H. Geipel, J. Gloede, K-P. Hilgetag and H. Gross, *Chem. Ber.*, **98**, 1677 (1965).
26. L. Berlinquet, *Can. J. Chem.*, **33**, 1119 (1955).
27. D. B. Sprinson and E. Chargaff, *J. Biol. Chem.*, **164**, 417 (1946).
28. H. Sasaki, *J. Chem. Soc. Japan*, **78**, 304 (1957).
29. I. L. Knunyants and V. V. Shokina, *J. Gen. Chem. U.S.S.R.*, **25**, 1175 (1955).
30. N. Sugimoto, I. Chibata, S. Yamada and M. Yamamoto, *Japanese Patent* No. 42–19449 (1967).
31. I. Chibata, S. Yamada, M. Yamamoto, M. Wada and T. Yoshida, *Japanese Patent* No. 43–24410 (1968).
32. I. Chibata, S. Yamada and M. Yamamoto, *Japanese Patent* No. 46–29843 (1971).

4.23 Threonine

In comparison with the discovery of other amino acids which are constituents of proteins, that of threonine was rather late, namely in 1935. However, the discovery of threonine did make it possible for the first time to grow animals by feeding an amino acid mixture instead of protein.

Rose and his co-workers and Maeda independently tried to isolate a growth-promoting factor for rats from hydrolyzates of fibrin or casein. In 1933, Maeda succeeded in obtaining a crystalline substance which was effective at 1% addition to the diet, and suggested that this amino acid might be hydroxyaminovaleric acid or hydroxyaminobutyric acid, as judged from its elementary analysis. On the other hand, Rose performed more detailed investigations and, in 1935, showed that the effective substance was in fact α-amino-β-hydroxybutyric acid. In the following year, he designated this new amino acid "*d*-(−)-threonine" since its stereochemical structure was equivalent to that of *d*(+)-threose. Following this work, West and Carter succeeded in synthesizing four isomers of threonine. Threonine has two asymmetric carbon atoms, and there are two threo type isomers, L-, D-threonine, and two erythro type isomers, L-, D-allothreonine. The threonine that constitutes protein is L-threonine; the other three isomers are nutritionally unavailable.

Ordinary synthetic methods for threonine yield a mixture of the four stereoisomers. Therefore, separation of the threo form from the allo form, and subsequent optical resolution, is necessary in order to obtain the L-threonine needed for food and drug use. Accordingly, methods which selectively afford the threo form are most advantageous as synthetic methods for threonine. Generally, however, such synthetic methods are few, and many techniques yield the allo form (which is of stable configuration) in larger amounts than the threo form. Methods which mainly yield allothreonine may be of industrial use when combined with a method which can readily convert the allo form to threo form in good yield.

For the synthesis of threonine, many techniques have been reported as follows: (1) the method via epoxide or α-halogenobutyric acid derivative, starting from crotonic acid; (2) reduction of an appropriate α-substituted acetoacetic acid ester; (3) the method via azlactone, starting from hippuric acid; (4) the method using a metal complex of glycine.

In the following sections, the method using glycine–metal complex, which is now employed industrially, and other synthetic methods having potential for industrial application are described.

4.23.1 Synthesis from glycine–metal complex

Threonine synthesis based on the condensation of a glycine–metal complex such as glycine–copper complex with acetaldehyde, was first reported by Akabori and his co-workers in 1957,[1,2] and this technique has subsequently been improved.[3-6] The process employed is very short. Furthermore, the threo form is mostly formed and can be efficiently separated from the allo form. From these reasons, this method is considered to represent one of the most advantageous methods for threonine synthesis, and is currently being employed for industrial production.

Addition of basic copper carbonate or copper sulfate and sodium hydroxide to an aqueous solution of glycine (I) gave the glycine–copper complex (II). Although the methylene group of glycine has very low reactivity, that of the glycine–copper complex is highly reactive, and by the addition of excess acetaldehyde in the presence of a basic catalyst such as potassium hydroxide in methanol, condensation occurred to form the acetaldehyde addition compound of threonine–copper complex (III). By this reaction allothreonine was also formed to some extent, in addition to threonine. However, these compounds could be separated from one another by their difference in stability and solubility.

$$\underset{\text{I}}{NH_2CH_2COOH} \xrightarrow{Cu^{2+}} \underset{\text{II}}{(CO{-}O{-}CH_2{-}NH_2)_2Cu} \xrightarrow[OH^-]{CH_3CHO}$$

$$\underset{\text{III}}{(CH_3CH(OH){-}CH(NH_2){-}CO{-}O)_2Cu} \xrightarrow{-Cu} \underset{\text{IV}}{CH_3CH(OH)CH(NH_2)COOH}$$

The copper complex (III) was dissolved in aqueous ammonia and passed through a column packed with an ammonium form cation-exchange resin (such as Dowex-50 or Amberlite IR-120) to remove the copper. Concentration of the eluate, followed by methanol addition, led to the precipitation of crude threonine, and recrystallization from aqueous methanol gave DL-threonine (IV) with over 90% threo form in a yield of about 60% based on copper glycinate (II). The effects of catalyst amount, solvent, and reaction temperature on the threonine: allothreonine ratio and on the yield have been investigated.[6] A typical flowsheet for this synthetic method is presented in Fig. 4–7.

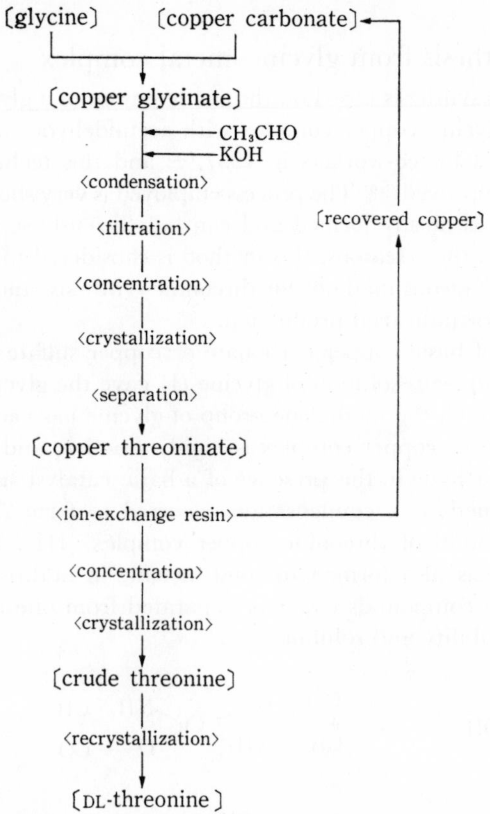

Fig. 4–7. General flowsheet for DL-threonine synthesis from glycine–metal complex.

Glycine–cobalt complex has also been used as the glycine–metal complex, in place of the above copper complex.[7)] Moreover, by employing glycine–nickel complex, the reaction can be carried out at high concentrations, and separation of the threo form from the allo form is easier due to the large difference in solubility between the two forms. According to the patent[8)] for this procedure, 31.5 g of DL-threonine having over 95% threo form was obtained from 60 g of nickel glycinate. On the other hand, glycine–copper complex has been condensed with vinylalcohol in place of acetaldehyde.[9)]

In addition to the above techniques, the condensation of *N*-salicylidene glycine–copper complex (V) with various aldehydes has been investigated. When acetaldehyde was used for the reaction, it yielded mainly the allo form, e.g. DL-threonine with 31–34% threo form in a yield of 89%.[10)]

$$\text{V (Cu complex: } CH_2\text{–C=O, N, O, HC, Cu, O, } OH_2\text{)} \xrightarrow[NaHCO_3]{CH_3CHO} \text{(Cu complex: } CH_3\overset{OH}{C}HCH\text{–C=O, N, O, HC, Cu, O, } OH_2\text{)} \xrightarrow{-Cu} \underset{HO\ \ NH_2}{CH_3CHCHCOOH}\ \text{IV}$$

4.23.2 Synthesis from acetoacetate

Ethyl acetoacetate is a readily available substance, and a number of methods for threonine synthesis starting from the ester have been reported since Adkins and Reeve[11] published their route via α-hydroxyiminoacetic acid ester.[12-16] However, in the majority of cases the main product is the allo form, and so this method is often combined with a process for conversion of the allo form to the threo form. One example is presented below.

Ethyl acetoacetate (VI) was treated with phenyldiazoniumchloride to give ethyl-α-phenylazoacetoacetate (VII). Reduction of this with zinc dust in acetic acid and acetic anhydride led to ethyl α-acetamidoacetoacetate (VIII). Hydrogenation of (VIII) with a platinum oxide catalyst under pres-

$$\underset{\text{VI}}{CH_3COCH_2COOC_2H_5} \xrightarrow[92.7\%]{C_6H_5N_2Cl} \underset{\text{VII}}{CH_3COCHCOOC_2H_5\ (N{=}N{-}C_6H_5)}$$

$$\xrightarrow[Zn]{(CH_3CO)_2O,\ CH_3COOH} \underset{\text{VIII}}{CH_3COCHCOOC_2H_5\ (NHCOCH_3)} \xrightarrow[PtO_2]{H_2} \underset{\text{IX}}{CH_3CHCHCOOC_2H_5\ (HO,\ NHCOCH_3)}$$

15~20% threo form

$$\xrightarrow{SOCl_2} \left[\underset{\text{X}}{CH_3CH\text{—}CHCOOC_2H_5\ (O\text{–}C(CH_3){=}N\ \text{ring})} \right] \longrightarrow \underset{HO\ \ NH_2}{CH_3CHCHCOOH}$$

88.7% based on VIII
83.2% threo form

$$\xrightarrow{C_2H_5ONa} \underset{\text{XI}}{\underset{HO\ \ NH_2}{CH_3CHCHCOONa}}\ (83.5\%) \xrightarrow{HCl,\ C_6H_5NH_2} \underset{\text{IV}}{\underset{HO\ \ NH_2}{CH_3CHCHCOOH}}$$

57.2% based on VI
100 % threo form

sure yielded ethyl α-acetamido-β-hydroxybutyrate (IX), which gave threonine after hydrolysis. The product contained 15–20% threonine (or 85–80% allothreonine) as determined by the microbioassay method. In order to convert the allo form to threo form, the product was treated with thionylchloride in benzene to give oxazoline ester (X). Hydrolysis of the latter, with heating, led to a product containing 83.2% threo form. To remove the remaining allo form, the product (in alcohol) was treated with sodium ethoxide to yield the sodium salt. By this reaction, the barely soluble sodium salt of DL-threonine was obtained. After being freed to amino acid, recrystallization then yielded pure DL-threonine in a yield of 57.2% based on ethyl acetoacetate.

The intermediate, ethyl α-acetamido-β-hydroxybutyrate (IX), has also been obtained in good yield by the reduction of ethyl α-hydroxyiminoacetoacetate (XII) under pressure with Raney nickel in acetic anhydride.[16)]

$$\underset{\text{VI}}{CH_3COCH_2COOC_2H_5} \xrightarrow[CH_3COOH]{NaNO_2} \underset{\text{XII}}{\underset{\|}{CH_3COCCOOC_2H_5}}\atop NOH \quad 85\sim90\%$$

$$\xrightarrow[\text{Raney nickel, }(CH_3CO)_2O]{H_2} \underset{NHCOCH_3}{CH_3COCHCOOC_2H_5} \; 85\sim93\% \xrightarrow[\text{Raney nickel, }(CH_3CO)_2O]{H_2} \underset{\text{HO NHCOCH}_3}{CH_3CHCHCOOC_2H_5} \; 97\% \; (\text{31\% threo form}) \quad \text{IX}$$

In addition, catalytic hydrogenation of the α-hydroxyimino compound (XII) with Pd-C in absolute alcohol containing hydrogen chloride gave ethyl α-aminoacetoacetate hydrochloride (XIII). Catalytic hydrogenation of this in aqueous solution with a Pt-C catalyst, and subsequent hydrolysis with hydrochloric acid, led to threonine (IV) in a good yield (98%), although the content of threo form is not known.[17)]

$$\underset{\text{XII}}{\underset{NOH}{CH_3COCCOOC_2H_5}} \xrightarrow[Pd\text{-}C,\ HCl,\ C_2H_5OH]{H_2} \underset{\text{XIII}}{\underset{NH_2\cdot HCl}{CH_3COCHCOOC_2H_5}} \quad \text{quantitative}$$

$$\xrightarrow[Pt\text{-}C]{H_2} \xrightarrow{HCl} \underset{\text{IV}}{\underset{\text{HO NH}_2}{CH_3CHCHCOOH}} \quad 98\%$$

4.23.3 Synthesis from diketene

Recently, diketene has become available in large quantities as an industrial material, and its possible use for amino acid synthesis has attracted some attention. When diketene (XIV) was reacted with isoamylnitrite and gaseous hydrogene chloride in ether, amyl α-hydroxyiminoacetoacetate (XV) was obtained.[17] Further reduction and hydrolysis of this, in the same manner as described above for ethyl α-hydroxyiminoacetate (XII), resulted in threonine.

$$\underset{\text{XIV}}{\begin{matrix} CH_2=C\text{—}O \\ | \quad\;\; | \\ CH_2\text{—}C=O \end{matrix}} \xrightarrow[\text{HCl}]{C_5H_{11}ONO} \underset{\text{XV}}{\begin{matrix} CH_3COCCOOC_5H_{11} \\ \| \\ 60\% \quad NOH \end{matrix}}$$

4.23.4 Synthesis from vinylacetate

Vinylacetate (XVI), which is a low-priced starting material, was hydroformylated by the oxo method to give α-acetoxypropionic aldehyde (XVII).[18] Subjection of this aldehyde (XVII) to the Strecker reaction and subsequent rapid heating with hydrogen cyanide and ammonia in a sealed vessel, followed by hydrolysis with hydrochloric acid, gave a product with 60% threo form in a yield of 50%.[19] Phosgene was passed through an aqueous alkaline solution of the intermediate compound, α-amino-β-hydroxybutyronitrile (XVIII), to yield 5-methyl-4-cyano oxazolidone (XIX). Subsequent alkaline hydrolysis afforded a product containing over 91% threonine in a yield of 46% based on propionic aldehyde.[20]

$$\underset{\text{XVI}}{CH_2=CHOCOCH_3} \xrightarrow[\text{tricarbonyl rhodium}]{CO,\ H_2} \underset{\text{XVII}}{\begin{matrix} CH_3CHCHO \\ | \\ 72\% \quad OCOCH_3 \end{matrix}} \xrightarrow{NH_3,\ HCN}$$

$$\left[\begin{matrix} CH_3CHCHCN \\ |\quad\; | \\ O \;\; NH_2 \\ | \\ COCH_3 \end{matrix} \longrightarrow \underset{\text{XVIII}}{\begin{matrix} CH_3CHCHCN \\ |\quad\; | \\ HO \;\; NH_2 \end{matrix}} \right] \xrightarrow{\text{HCl}} \underset{\text{IV}}{\begin{matrix} CH_3CHCHCOOH \\ |\quad\; | \\ HO \;\; NH_2 \end{matrix}}$$

$$\text{XVIII} \xrightarrow[\text{KOH}]{COCl_2,} \underset{\text{XIX}}{\left[\begin{matrix} CH_3CH\text{—}CHCN \\ |\qquad | \\ O \qquad NH \\ \diagdown \;\; \diagup \\ C \\ \| \\ O \end{matrix} \right]} \xrightarrow{\text{KOH}} \text{IV}$$

4.23.5 Conversion and separation of the allo and threo forms

As shown in several of the above examples, the main product obtained by chemical synthesis of threonine is often the allo form of the amino acid. Therefore, methods for conversion of the allo form to the threo form and separation of one form from the other have been studied. As described in the section on synthesis from ethyl acetoacetate, the treatment of *N*-acetyl- or *N*-benzoylallothreonine ester (XX) with thionyl chloride caused inversion in the *β*-position, and *trans*-oxazoline (XXII) was formed. Hydrolysis of this then led to threonine of the threo form.[21-23]

$$\underset{\substack{\text{allo form XX}}}{\begin{array}{c}CH_3\underset{|}{C}H\underset{|}{C}HCOOR\\ HO\quad NHCOR'\end{array}} \xrightarrow{SOCl_2 \text{ or } CH_3-C_6H_4-SO_2Cl} \underset{\substack{\text{XXII}\\ trans\text{ form}}}{CH_3CH-CHCOOR \text{ (O, N, C–R' oxazoline ring)}}$$

$$\text{XX} \xrightarrow{C_2H_5O(NH{=})C-C_6H_5} \underset{\substack{\text{XXI}\\ cis\text{ form}}}{\left[\begin{array}{c}CH_3CH-CHCOOR\\ O \quad N\\ C\\ R'\end{array}\right]} \longrightarrow \text{XXII}$$

$$\text{XXII} \xrightarrow{HCl} \underset{\substack{\text{threo form}\\ \text{IV}}}{\begin{array}{c}OH\\ CH_3\overset{|}{C}H\underset{|}{C}HCOOH\\ NH_2\end{array}}$$

Reaction of the *N*-benzoylallothreonine ester (XX) with *p*-toluene–sulfonylchloride gave the *O*-tosyl derivative. Treatment of this with potassium acetate yielded *trans*-oxazoline (XXII).[22] When the allothreonine ester was reacted with benzimino ethyl ester, *cis*-oxazoline (XXI) was obtained. Subsequent alkaline treatment of this then gave *trans*-oxazoline (XXII).[24] By employing the conversion of oxazoline, the four isomers of threonine can be interconverted.[23]

In the same manner as threonine synthesis by the Strecker method via *α*-acetoxypropionic aldehyde, starting from vinylacetate, treatment of allothreonine with phosgene in an alkaline medium gave *cis*-form oxazoline ester (XXIII). Treatment of this with alcoholic potassium hydroxide resulted in the formation of *trans*-oxazolidone carboxylic acid (XXIV). Subsequent hydrolysis with hydrochloric acid then led to threo type threonine.

As described under the method using glycine–metal complex, allothreonine can be separated on the basis of the difference in solubility and stability of the acetaldehyde addition compound of threonine–copper complex. In a similar fashion, allothreonine was converted to the corresponding aldehyde addition compound of the metal complex. Heating of this in the presence of sodium methoxide or potassium hydroxide in methanol, resulted in threonine free from allothreonine in a yield of 40%.[25]

Besides these methods, the separation of allo form from threo form can

$$CH_3\underset{\substack{| \\ HO}}{C}H\underset{\substack{| \\ NH_2}}{C}HCOOC_2H_5 \xrightarrow{COCl_2} CH_3CH{-}CHCOOC_2H_5$$

allo form

(cyclic: O–C(=O)–NH ring) *cis* form XXIII

$$\xrightarrow{C_2H_5OK} CH_3CH{-}CHCOOH \text{ (O–C(=O)–NH ring) } \xrightarrow{HCl} CH_3\underset{\substack{| \\ NH_2}}{\overset{\substack{OH \\ |}}{C}}HCHCOOH$$

trans form XXIV

threo form IV

be effected utilizing various salts, complexes or derivatives. One example employing *O*-acyl derivative is presented here. An aqueous solution of the *O*-acetyl derivative of a threonine-allothreonine mixture was kept at pH 6–8 by the addition of alkali. Following treatment, intramolecular conversion of *O*-acetyl group to the nitrogen of amino group occurred only in the threo form. The reaction mixture was passed through a column of cation-exchange resin. *O*-Acetylallothreonine was absorbed while *N*-acetylthreonine passed through the column. Thus, the two forms were separated from each other.[26]

4.23.6 Resolution of DL-threonine

Optical resolution of the DL-threonine obtained by the various synthetic methods described above can be effected by enzymatic resolution based on the asymmetric hydrolysis of the *N*-chloroacetyl derivative by hog kidney aminoacylase,[27,28] or of the acetyl[29-31] and benzoyl[32,33] derivatives by bacterial aminoacylase.

In the case of chemical resolution, the *p*-nitrobenzoyl[34] and toluene–sulfonyl derivatives[35] have been resolved as brucine salts. Another method, i.e. resolution of the dibenzyl derivative with L- or D-threo-*p*-nitrophenyl-2-aminopropane-1,3-diol,[36] is also available. In addition to this, as described in section 4.23.5, DL-*trans*-oxazoline[23] and the oxazoline derivative[37] have been resolved as brucine salts.

Since DL-threonine can be optically resolved in aqueous solution by direct resolution, i.e. by the preferential crystallization method, this seems advantageous for industrial application. Thus, 9 parts of DL-threonine and 1 part of L-threonine were dissolved in 30 parts of water at 80°C and cooled to 20°C. This resulted in precipitation of about 2 parts of L-threonine. After filtration of the precipitate, the same amount of DL-threonine as the precipitated product was added to the mother liquor, and the above process repeated to separate D-threonine. Recrystallization of the crystals obtained led to 1.7 parts of optically pure isomer. By repeating the above operations, the two optical isomers were obtained successively.[38,39]

For the resolution of threonine by preferential crystallization, a solution of the racemic mixture was also fed into the bottom of a column and seed crystals of the optical isomer were added to the top of the column. Preferentially crystallized product was removed from the bottom of the column, and the effluent fed into another column seeded with the optical antipode.[40]

As described above, in order to obtain L-threonine by chemical synthesis, separation of the threo form and optical resolution are essential steps. The D-isomer obtained in equal amounts to the L-isomer can be converted to the latter by the method via oxazoline. However, the D-isomer is not so easily racemized as other amino acids having one asymmetric carbon. Thus, many problems still remain before the industrial production of threonine by chemical synthesis can be established, and competition with the fermentation method is likely to be severe. On the other hand, there is one example where the synthesis and optical resolution of threonine have been successfully industrialized by combination with a fermentative method using D-threonine for the production of L-isoleucine.[41]

References

1. M. Sato, K. Okawa and S. Akabori, *Bull. Chem. Soc. Japan*, **30**, 937 (1957).
2. K. Okawa and M. Sato, *Japanese Patent* No. 34–2964 (1959).
3. I. Chibata and S. Yamada, *Japanese Patent* No. 36–1968 (1961).
4. I. Chibata and S. Yamada, *Japanese Patent* No. 36–1477 (1961).
5. K. Fujii, M. Oda, J. Arita, K. Sakai and M. Takeda, *Japanese Patent* No. 36–19562 (1962).
6. V. M. Belihov, N. L. Kuzoietsova and E. N. Safonova, *Khim. Prirod. Soedin.*, **1967**, 31.
7. Y. Ikutani, T. Okuda, M. Sato and S. Akabori, *Bull. Chem. Soc. Japan*, **32**, 203 (1959).
8. M. Tanaka, *Japanese Patent* No. 40–27732 (1965).
9. H. Kawasaki, *Japanese Patent* No. 39–17816 (1964).
10. K. Harada and J. Ohhashi, *J. Org. Chem.*, **32**, 1103 (1967).
11. H. Adkins and E. W. Reeve, *J. Am. Chem. Soc.*, **60**, 1328 (1938).
12. K. Pfister, C. A. Robinson, A. C. Shabica and M. Tishler, *J. Am. Chem. Soc.*, **71**, 1101 (1949).
13. N. F. Albertson, B. F. Tuller, J. A. King, B. B. Fishburn and S. Archer, *ibid.*, **70**, 1151 (1948).
14. Y. Izumi and S. Konishi, *Nippon Kagaku Kaishi* (Japanese), **74**, 957 (1953).
15. Y. Izumi and H. Kawasaki, *ibid.*, **78**, 660 (1957).

16. H. Kawasaki, *ibid.*, **78**, 1251, 1254 (1967).
17. T. Kato and M. Sato, *Yakugaku Zasshi* (Japanese), **87**, 1209 (1967).
18. M. Takesada, Y. Komachiya, H. Wakamatsu and E. Negoro, *Japanese Patent* No. 40–1575 (1965).
19. Y. Komachiya, M. Yakesada and S. Sakurai, *Japanese Patent* No. 40–5808 (1965).
20. M. Hara, *Japanese Patent* No. 40–11608 (1965).
21. K. Pfister, C. A. Robinson, A. C. Shabica and M. Tishler, *J. Am. Chem. Soc.*, **70**, 2297 (1948).
22. J. Attenburrow, D. F. Elliott and G. F. Penny, *J. Chem. Soc.*, **1948**, 310.
23. D. F. Elliott, *ibid.*, **1950**, 62.
24. D. F. Elliott, *ibid.*, **1949**, 589.
25. S. Kudo, K. Fujii, N. Takemoto and K. Nakano, *Japanese Patent* No. 43–24406 (1968).
26. S. Seno and T. Furuyama, *Japanese Patent* No. 40–6208 (1965).
27. V. E. Price, J. B. Gilbert and J. P. Greenstein, *J. Biol. Chem.*, **179**, 1169 (1949).
28. S. M. Birnbaum, L. Levintow, R. B. Kingsley and J. P. Greenstein, *ibid.*, **194**, 455 (1952).
29. M. Tanaka and K. Mineura, *Japanese Patent* No. 42–13447 (1967).
30. M. Tanaka and K. Mineura, *Japanese Patent* No. 42–6325 (1967).
31. K. Mineura and M. Tanaka, *Nippon Nogeikagaku Kaishi* (Japanese), **42**, 216 (1968).
32. Y. Kameda, E. Toyoura, K. Matsui, Y. Kimura, Y. Kanaya, A. Nakatani, H. Saito and K. Kawase, *Yakugaku Zasshi* (Japanese), **78**, 769 (1958).
33. Y. Kameda, E. Toyoura and K. Matsui, *Chem. Pharm. Bull.*, **7**, 702 (1959).
34. A. J. Zambito, W. L. Peretz and E. E. Howe, *J. Am. Chem. Soc.*, **71**, 2541 (1949).
35. M. Brenner, K. Rübenacht and E. Sailer, *Helv. Chim. Acta*, **34**, 2102 (1951).
36. L. Velluz, G. Amiard and R. Heymès, *Bull. Soc. Chim. France*, **22**, 201 (1955).
37. T. Inui and T. Kaneko, *Nippon Kagaku Kaishi* (Japanese), **82**, 1078 (1961).
38. L. Velluz and G. Amiard, *Bull. Soc. Chim. France*, **20**, 903 (1953).
39. G. Amiard, *ibid.*, **23**, 447 (1956).
40. K. Ito, T. Akashi and S. Tatsumi, *Japanese Patent* No. 36–17710 (1961).
41. I. Chibata, M. Kisumi, Y. Ashikaga and J. Kato, *Amino Acids* (Japanese), **5**, 76 (1962).

4.24 Tryptophan

The name "tryptophan" was given to an indole compound in putrified protein by Neumeister in 1890, although it was actually isolated later from enzymatic digests of protein by Hopkins and Cole in 1902. In 1907, Ellinger and Flamand synthesized tryptophan by condensing indole-β-aldehyde with hippuric acid, and proved that it is β-indole-α-aminopropionic acid.

Tryptophan can be prepared by extraction, fermentation or chemical synthesis. Extraction is difficult because tryptophan is easily destroyed during the acid hydrolysis of protein, and racemization occurs on alkaline hydrolysis. Fermentative methods for tryptophan production from anthranilic acid or indole have been examined. On an industrial scale, L-tryptophan is produced by DL-tryptophan synthesis followed by resolution.

The synthetic methods can be classified into 2 groups, i.e. that using indole as a starting material, and that synthesizing the indole nucleus during the synthetic process. Compared to the former method, the latter is preferable from an economic viewpoint for industrial-scale production, since indole is rather costly as a raw material. Methods belonging to both groups will be described below. Since the method using β-cyanopropionaldehyde as starting material is already used commercially, it will be presented in greater detail.

4.24.1 Synthesis from indole

A. Synthesis from indole via indole-3-aldehyde

This method is based on the principle that indole (I) is converted to indole-3-aldehyde (II), following which the aldehyde group is condensed with hippuric acid, hydantoin or thiohydantoin, and tryptophan (III) is obtained from the condensation product by reduction and hydrolysis.

$$\text{I (indole, NH)} \longrightarrow \text{II (3-CHO indole, NH)} \longrightarrow \text{III (3-}CH_2CH(NH_2)COOH\text{ indole, NH)}$$

Ellinger and Flamand,[1] in an attempt to elucidate the structure of tryptophan, synthesized tryptophan from indole-3-aldehyde (II) by the following procedure. 2-Phenyl-4-(3-indolylmethylene)-oxazolone-5 (V), a condensation product of indole-3-aldehyde (II) with hippuric acid (IV) in the presence of acetic anhydride and sodium acetate, was hydrolyzed to give

II + IV $\xrightarrow{(CH_3CO)_2O,\ CH_3COONa}$ V

NaOH

VI

i) OH^- ii) H^+

VII

NH_2OH

VIII

H_2 Raney nickel

Na-Hg

III

α-benzimido-β-indolylacrylic acid (VI) which was then reduced to tryptophan (III).

Baugness and Berg[2)] hydrolyzed the azlactone (V) with alkali to prepare β-(3-indolyl)pyruvic acid (VII), which was then condensed with hydroxylamine to give α-hydroxyimino-β-indolylpropionic acid (VIII) from which tryptophan was obtained by reduction with Raney nickel.

Majima and Kotake[3)] synthesized tryptophan (III) through condensation of indole-3-aldehyde (II) with hydantoin (IX) in the presence of acetic anhydride and sodium acetate, reduction of β-indolalhydantoin (X) with sodium amalgam, and hydrolysis of tryptophanhydantoin (XI).

II + IX $\xrightarrow{(CH_3CO)_2O,\ CH_3COONa}$ X $\xrightarrow{Na\text{-}Hg}$ XI $\xrightarrow{OH^-}$ III

Holland and Nayler[4] converted 4-(3′-indolylmethylene)-2-thio-5-thiazoline (XIII), a condensation product of indole-3-aldehyde (II) with 2-thio-5-thiazolidone (XII) in the presence of zinc chloride, to 5-(3′-indolylmethylene)-2-thiohydantoin (XIV) with ammonia, and obtained tryptophan (III) by reduction and hydrolysis.

II: indole-3-CHO + XII: CH_2—CO / NH S / CS $\xrightarrow{ZnCl_2}$ XIII: indole-3-CH=C—CO / NH S / CS

$\xrightarrow{NH_3}$ XIV: indole-3-CH=C—CO / NH NH / CS $\xrightarrow{H_2}$ XV: indole-3-CH_2—CH—CO / NH NH / CS

$\xrightarrow{OH^-}$ III: indole-3-$CH_2CH(NH_2)COOH$

The yield was improved by using piperidine[5] instead of acetic anhydride and sodium acetate as a catalyst for the condensation of indole-3-aldehyde (II) with hydantoin (IX). Another example using di- or trialkanolamine as the catalyst has been reported.[6] On hydrolysis of tryptophanhydantoin with an alkali metal hydroxide at 150°C for 40 min, tryptophan was obtained in good yield.[7]

B. Synthesis from indole via gramine

This method is based on the principle that tryptophan (III) can be synthesized via gramine (XVI), formed from indole (I), dimethylamine and formaldehyde by the Mannich reaction.

I: indole → XVI: indole-3-$CH_2N(CH_3)_2$ → III: indole-3-$CH_2CH(NH_2)COOH$

Smith and Snyder[8] treated gramine (XVI) with methyl iodide, and condensed the resulting gramine methiodide (XVII) with ethyl acetamidomalonate (XVIII) to give the product (XIX). Saponification and decarbonation yielded acetyl tryptophan (XXI), which was then converted to tryptophan (III) by alkaline hydrolysis. The yield of tryptophan based on original indole (I) was 45%.

XVII (indole-3-$CH_2\overset{+}{N}(CH_3)_3I^-$) + XVIII ($HC(COOC_2H_5)_2NHCOCH_3$) → XIX (indole-3-$CH_2C(COOC_2H_5)_2NHCOCH_3$) $\xrightarrow{OH^-}$ XX (indole-3-$CH_2C(COOH)_2NHCOCH_3$) $\xrightarrow{-CO_2}$ XXI (indole-3-$CH_2CH(NHCOCH_3)COOH$) $\xrightarrow{OH^-}$ III (indole-3-$CH_2CH(NH_2)COOH$)

Albertson *et al.*[9] also synthesized tryptophan by almost the same method. Albertson and Tuller[10] improved the yield to 71% by using acetamidocyanoacetic acid instead of acetamidomalonic acid ester.

Howe *et al.*[11] accomplished the direct condensation of gramine (XVI) with ethyl acetamidomalonate (XVIII) in the presence of sodium hydroxide. Saponifying and hydrolyzing the product (XIX) simultaneously, they obtained tryptophan in 66% yield from indole. Hellmann,[12] substituting ethyl formamidomalonate for ethyl acetamidomalonate, improved the yield to 90%.

XVI (indole-3-$CH_2N(CH_3)_2$) + XVIII ($HC(COOC_2H_5)_2NHCOCH_3$) $\xrightarrow{NaOH}$ XIX (indole-3-$CH_2C(COOC_2H_5)_2NHCOCH_3$)

Weisblat and Lyttle[13,14] treated gramine (XVI) with ethyl nitromalonate (XXII) or ethyl nitroacetate (XXIII), and obtained the condensation products (XXIV or XXV). The product (XXIV) is converted by treatment with sodium in alcohol to the product (XXV), which yields the tryptophan ester (XXVI) on reduction with Raney nickel. Tryptophan (III) is obtained by saponifying the ester (XXVI). The overall yields from ethyl nitromalonate and ethyl nitroacetate were 77% and 45%, respectively.

Indole-3-$CH_2N(CH_3)_2$ (XVI) + $HC(NO_2)(COOC_2H_5)_2$ (XXII) → Indole-3-$CH_2C(COOC_2H_5)_2NO_2$ (XXIV)

XXIV —$NaOC_2H_5$→ XXV

Indole-3-$CH_2N(CH_3)_2$ (XVI) + $CH_2(NO_2)COOC_2H_5$ (XXIII) → Indole-3-$CH_2CH(NO_2)COOC_2H_5$ (XXV)

XXV —H_2, Raney nickel→ Indole-3-$CH_2CH(NH_2)COOC_2H_5$ (XXVI) —OH^-→ Indole-3-$CH_2CH(NH_2)COOH$ (III)

Indole-3-$CH_2N(CH_3)_2$ (XVI) + $CH(COCH_3)COOC_2H_5$ (XXVII) —$NaOC_2H_5$→ Indole-3-$CH_2CH(COCH_3)COOC_2H_5$ (XXVIII)

XXVIII —HN_3, H_2SO_4→ Indole-3-$CH_2CH(NHCOCH_3)COOC_2H_5$ (XXIX)

XXVIII → Indole-3-$CH_2C(=NOH)COOH$ (VIII)

XXIX —OH^-→ Indole-3-$CH_2CH(NHCOCH_3)COOH$ (XXI) —OH^-→ Indole-3-$CH_2CH(NH_2)COOH$ (III)

VIII —H_2→ Indole-3-$CH_2CH(NH_2)COOH$ (III)

Holland and Nayler[15] achieved the condensation of gramine (XVI) with ethyl acetoacetate (XXVII) in the presence of sodium ethoxide. By Schmidt rearrangement of the condensation product (XXVIII) with hydrogen azide in the presence of sulfuric acid, they obtained the ethyl ester of *N*-acetyltryptophan (XXIX). Saponification and hydrolysis of the product (XXIX) gave tryptophan (III) with a yield from indole of 62%. By treatment of the condensation product (XXVIII) with butyl nitrite in the presence of sodium ethoxide, followed by saponification, they obtained α-hydroxyimino-β-indolylpropionate (VIII), which was then catalytically reduced to give tryptophan. The overall yield from indole was 40%.

Jones and Kornfeld[16] have also synthesized tryptophan from the tertiary ammonium salt of skatole by treatment with acylaminoxaloacetate followed by alkaline hydrolysis.

C. Synthesis by direct condensation of indole

Butenandt *et al.*[17] reported a method of synthesizing tryptophan by direct condensation of indole (I) with piperidylmethyl formamidomalonic acid ethyl ester (XXX). When the condensation was carried out in boiling xylene containing sodium hydroxide, liberation of piperidine and formation of skatoformamidomalonic acid ester (XXXI) occurred simultaneously. Tryptophan was obtained in 71% yield on hydrolysis of the ester (XXXI).

I + XXX ($C_5H_{10}N-CH_2C(COOC_2H_5)_2NHCHO$) $\xrightarrow{NaOH}$ XXXI (indolyl-$CH_2C(COOC_2H_5)_2NHCHO$) + piperidine (NH)

Snyder and MacDonald[18] prepared acetyltryptophan (XXI) in 57% yield by condensing indole (I) with α-acetamidoacrylic acid (XXXII) in the presence of acetic acid and acetic anhydride.

Behringer and Taul[19] prepared α-benzoylamino-β-[3-indolyl]acrylate (VI) by alkali and acid treatment of 2-phenyl-4-[3′-indolyl methylene]-oxazol-5-one (V), a product of condensation of indole (I) with 2-phenyl-4-chloromethyleneoxazol-5-one (XXXIII). Tryptophan (III) was obtained by hydrogenation of the product (VI) with Raney nickel followed by alkaline hydrolysis.

I + $CH_2=C(COOH)NHCOCH_3$ (XXXIII) $\xrightarrow{(CH_3CO)_2O,\ CH_3COOH}$ 3-indolyl-$CH_2CH(COOH)NHCOCH_3$ (XXI) $\xrightarrow{OH^-}$ 3-indolyl-$CH_2CH(COOH)NH_2$ (III)

I + $ClCH=C-CO$ oxazolone with C_6H_5 (XXXIII) $\longrightarrow$ 3-indolyl-$CH=C-CO$ oxazolone with C_6H_5 (V) $\xrightarrow{i)\ OH^-,\ ii)\ H^+}$ 3-indolyl-$CH=C(COOH)NHCOC_6H_5$ (VI) $\xrightarrow{i)\ H_2,\ Raney\ nickel;\ ii)\ OH^-}$ 3-indolyl-$CH_2CH(COOH)NH_2$ (III)

Okuda,[20] treating indole (I) with 1,1,5,5-tetracarbomethoxy-1,5-dinitro-3-azapentane (XXXIV) in the presence or absence of solvent at 100–200°C, prepared methyl α-methoxycarbonyl-α-nitro-β-(3-indolyl)propionate (XXXV) which gave tryptophan (III) by direct reduction or demethoxycarbonylation followed by reduction and hydrolysis.

2 I + $(CH_3OOC)_2C(NO_2)CH_2NHCH_2C(NO_2)(COOCH_3)_2$ (XXXIV) $\longrightarrow$ 2 3-indolyl-$CH_2C(NO_2)(COOCH_3)_2$ (XXXV) + NH_3

$\xrightarrow{i)\ NaOC_2H_5;\ ii)\ H_2,\ Raney\ nickel;\ iii)\ OH^-}$ 3-indolyl-$CH_2CH(COOH)NH_2$ (III)

4.24.2 Methods involving indole nucleus formation during the synthetic process

This approach was first used by Hegedüs.[21] As illustrated in the diagram, sodium ethyl acetoacetate (XXXVI) is condensed with diethylchloroethylamine first, and the product (XXXVII) is condensed with chlorobenzene diazonium salt. The condensation product (XXXVIII) is cyclized and the resulting indole compound (XXXIX) is condensed with ethyl acetamidomalonate. The final condensation product (XL) is converted by saponification and decarboxylation with sodium hydroxide to the product (XLI), from which tryptophan (III) is derived.

$COCH_3$–$CHNa$–$COOC_2H_5$ (XXXVI) $\xrightarrow{ClCH_2CH_2N(C_2H_5)_2}$ $COCH_3$–$CHCH_2CH_2N(C_2H_5)_2$–$COOC_2H_5$ (XXXVII) $\xrightarrow{C_6H_5N_2Cl}$

$CH_2CH_2N(C_2H_5)_2$, $C-COOC_2H_5$, N, N, H (XXXVIII) $\xrightarrow{CH_3COOH,\ H_2SO_4}$ $CH_2N(C_2H_5)_2$, $COOC_2H_5$, N, H (XXXIX)

$\xrightarrow{CH_3CONHCH(COOC_2H_5)_2}$ $NHCOCH_3$, $CH_2C(COOC_2H_5)_2$, $COOC_2H_5$, N, H (XL)

$\xrightarrow{NaOH}$ $NHCOCH_3$, $CH_2CHCOOH$, $COOH$, N, H (XLI) $\xrightarrow{H_2SO_4}$ $CH_2CHCOOH$, NH_2, N, H (III)

Plieninger[22] has also synthesized tryptophan from α-bromobutyrolactone.

Sakurai and Ito[23] prepared ethyl 3-(2-ethoxycarbonylindolyl)oxalacetate (XLIV) by condensing ethyl oxalate with ethyl 3-(2-ethoxycarbonylindolyl)acetate (XLIII), which was derived from α-ketoglutaric acid phenylhydrazone (XLII) by cylization. It was then converted to 2′,3′-

indo-3,4-pyr-2-one-6-carboxylic acid (XLV) by heating in the presence of *p*-toluene sulfonic acid in acetic acid. The product (XLV) was treated with sodium hydroxide solution to give β-[3-(2-carboxyindolyl)]pyruvic acid (XLVI). Sakurai and Komachiya[24-26] prepared β-[3-(2-carboxyindolyl)] α-aminopropionic acid (XLVII) either from 2′,3′-indo-3,4-pyr-2-one 6-carboxylic acid (XLV) by treatment with ammonia followed by hydrolysis and reduction, or from β-[3-(2-carboxyindolyl)]pyruvic acid (XLVI) by reduction in the presence of ammonia. Tryptophan was obtained when the product (XLVII) was decarboxylated by heating in an aqueous solution containing a strong acid.

CH₂CH₂COOH / CCOOH / N / NH (XLII) → (C_2H_5OH, H^+) → indole with $CH_2COOC_2H_5$, $COOC_2H_5$, N H (XLIII) → ($(COOC_2H_5)_2$) → indole with $COCOOC_2H_5$ / $CHCOOC_2H_5$, $COOC_2H_5$, N H (XLIV) → (p-$CH_3C_6H_4SO_3H$, CH_3COOH) → lactone with COOH, N H, C–O, O (XLV)

XLIV → (OH^-) → indole with $CH_2COCOOH$, COOH, N H (XLVI) → (NH_3, H_2, Raney nickel) → indole with NH_2 / $CH_2CHCOOH$, COOH, N H (XLVII)

XLV → (i) NH_3 ii) H_2-PdO) → XLVII → (H^+) → indole with $CH_2CHCOOH$ / NH_2, N H (III)

Warner and Moe[27] synthesized tryptophan using acrolein as a starting material, but their method is only suitable for the preparation of tryptophan on a small scale. Acrolein (XLVIII) is condensed with ethyl acetamidomalonate (XVIII) in the presence of sodium ethoxide, and the condensation product (XLIX) is treated with phenylhydrazine. The resulting hydrazone (L) is cyclized by boiling in an aqueous solution containing sulfuric acid or

$$CH_2{=}CHCHO + HC(COOC_2H_5)_2NHCOCH_3 \xrightarrow{NaOC_2H_5} [OHCCH_2CH_2C(COOC_2H_5)_2NHCOCH_3]$$

XLVIII XVIII XLIX

$$\xrightarrow{C_6H_5NHNH_2} C_6H_5NHN{=}CHCH_2CH_2C(COOC_2H_5)_2NHCOCH_3 \quad (L) \xrightarrow{H_2SO_4 \text{ or } BF_3}$$

$$\text{(indol-3-yl)}CH_2C(COOC_2H_5)_2NHCOCH_3 \quad (XIX) \longrightarrow \text{(indol-3-yl)}CH_2CH(NH_2)COOH \quad (III)$$

boron trifluoride. Tryptophan is obtained in 50% overall yield via the resulting indole compound (XIX).

Chibata and Yamada[28] used Amberlite IRA-400, an ion-exchange resin, as the catalyst for the condensation of acrolein (XLVIII) with ethyl acetamidomalonate, and cyclized the phenylhydrazone (L) by heating in an aqueous solution containing Amberlite IR-120.

Maeda and Yoshida[29] also achieved tryptophan synthesis from acrolein (XLVIII). By treatment of the acrolein with acetic anhydride, acrolein acetate (LI) was easily prepared.[30] When the acetate (LI) was hydroformylated with carbon monoxide and hydrogen in the presence of cobalt carbonyl in orthoethyl formate, 1,1-diacetoxy-4,4-diethoxybutane (LII) was obtained in high yield. The Bucherer reaction gave 5-(γ,γ-diethoxypropyl)hydantoin (LIII) or 5-ethoxy-1,3-dioxoperhydropyrid[1,2-*c*]imidazole (LIV) from 1,1-diacetoxy-4,4-diethoxybutane (LII). Tryptophan hydantoin (LV) was then obtained by Fischer's indole synthesizing reaction.

A peculiarity of this method is that it is based on the finding that 1,1-diacetoxy-4,4-diethoxybutane (LII) has two active groups of which one, the diacetylacetal group, is converted to the hydantoin while the other, the diethylacetal group, remains unchanged. This method may be suitable for the large-scale production of tryptophan, both as regards the raw material and as regards the process itself.

$$CH_2{=}CHCHO \xrightarrow{(CH_3CO)_2O} CH_2{=}CHCH(OCOCH_3)_2 \xrightarrow[\text{cobalt carbonyl}]{CO,\ H_2}$$

XLVIII LI

$$(C_2H_5O)_2CHCH_2CH_2CH(OCOCH_3)_2 \xrightarrow{NH_3,\ HCN,\ CO_2} \left[\text{LIII: } (C_2H_5O)_2CHCH_2CH_2CH{-}CO{-}NH{-}CO{-}NH \quad \text{LIV} \right]$$

LII LIII LIV

$$\xrightarrow{\text{i) } C_6H_5NHNH_2 \text{ ii) HCl}} \text{LV} \xrightarrow{OH^-} \text{III}$$

LV III

Komachiya *et al.*[31] synthesized tryptophan from β-cyanopropionaldehyde. The method is suitable for large-scale production and has already been put into practical use. β-Cyanopropionaldehyde is an intermediate in the synthesis of DL-glutamic acid, and is prepared by oxo-reaction from acrylonitrile.

$$NCCH_2CH_2CHO \xrightarrow{NH_4CN} NCCH_2CH_2CH(NH_2)CN \xrightarrow{(NH_4)_2CO_3} NCCH_2CH_2CH{-}CO{-}NH{-}CO{-}NH$$

LVI LVII LVIII

(LVI → LVIII directly: $NH_4CN,\ (NH_4)_2CO_3$)

$$\xrightarrow[\text{CH}_3\text{COOH, Pb-coated Raney nickel}]{H_2} OHCCH_2CH_2CH{-}CO{-}NH{-}CO{-}NH \xrightarrow{C_6H_5NHNH_2\cdot HCl} \text{LX}$$

LIX LX

$$\xrightarrow{HCl} \text{LV} \xrightarrow{NaOH} \text{III}$$

LV III

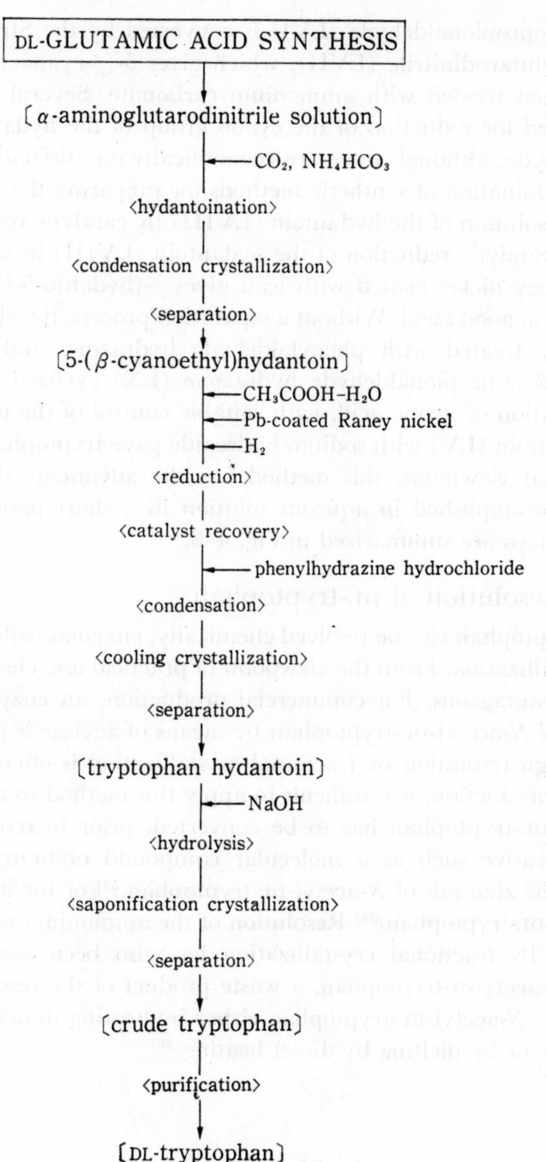

Fig. 4–8. General flowsheet for DL-tryptophan synthesis.

β-Cyanopropionaldehyde (LVI) is converted by the Strecker reaction to α-aminoglutarodinitrile (LVII), which gives 5-(β-cyanoethyl)hydantoin (LVIII) when treated with ammonium carbonate. Several methods have been reported for reduction of the cyano group of the hydantoin (LVIII) to the aldehyde, although these are economically unsatisfactory. Through a detailed examination of synthetic methods for preparing the aldehyde from an aqueous solution of the hydantoin (LVIII) by catalytic reduction, it was found that catalytic reduction of the hydantoin (LVIII) in weak acid solution by Raney nickel coated with lead gives β-(hydanto-5-yl)propionaldehyde (LIX) in good yield. Without a separation process, the aldehyde (LIX) was directly treated with phenylaldehyde hydrazine, and the resulting β-(hydanto-5-yl)propionaldehyde hydrazone (LX) cyclized by heating in a dilute solution of strong acid, with suitable control of the pH. Hydrolysis of the hydantoin (LV) with sodium hydroxide gave tryptophan (III). From a commercial viewpoint, this method has the advantage that the whole process is accomplished in aqueous solution in a short period, with good yield. The steps are summarized in Fig. 4–8.

4.24.3 Resolution of DL-tryptophan

DL-Tryptophan can be resolved chemically, enzymatically, and by fractional crystallization. From the viewpoint of practical use, chemical methods are not advantageous. For commercial production, an enzymatic method for resolving *N*-acetyl-DL-tryptophan by means of acylase is preferred.

Although resolution by fractional crystallization is often profitable for large-scale production, it is difficult to apply this method to DL-tryptophan. Therefore, DL-tryptophan has to be converted, prior to resolution, into a simple derivative such as a molecular compound of DL-tryptophan and inosine,[32] the zinc salt of *N*-acetyl-DL-tryptophan,[33] or the ammonium salt of *N*-acetyl-DL-tryptophan.[34] Resolution of the ammonium salt of *N*-acetyltryptophan by fractional crystallization has also been used in practice. Residual *N*-acetyl-D-tryptophan, a waste product of the resolution, can be converted to *N*-acetyl-DL-tryptophan either by heating in acetic acid-acetic anhydride[35] or by melting by direct heating.[36]

REFERENCES

1. A. Ellinger and C. Flamand, *Chem. Ber.*, **40**, 3029 (1907).
2. L. C. Baugness and C. P. Berg, *J. Biol. Chem.*, **104**, 675 (1934).
3. R. Majima and M. Kotake, *Chem. Ber.*, **55**, 3859 (1922).
4. D. O. Holland and J. H. C. Nayler, *J. Chem. Soc.*, **1953**, 285.

5. W. J. Boyd and W. Robson, *Biochem. J.*, **29**, 542, 546, 555, 2256 (1935).
6. J. E. Livak and M. F. Murray, *U.S. Patent* No. 2435399 (1948).
7. H. C. White, *U.S. Patent* No. 2557920 (1951).
8. C. W. Smith and H. R. Snyder, *J. Am. Chem. Soc.*, **66**, 350 (1944).
9. N. F. Albertson, S. Archer and C. M. Suter, *ibid.*, **66**, 500 (1944).
10. N. F. Albertson and B. F. Tuller, *ibid.*, **67**, 502 (1945).
11. E. E. Howe, A. J. Zambito, H. R. Snyder and M. Tishler, *ibid.*, **67**, 38 (1945).
12. H. Hellmann, *Z. Physiol. Chem.*, **284**, 163 (1949).
13. D. I. Weisblat and D. A. Lyttle, *J. Am. Chem. Soc.*, **71**, 3079 (1949).
14. D. A. Lyttle and D. I. Weisblat, *ibid.*, **69**, 2118 (1947).
15. D. O. Holland and J. H. C. Nayler, *J. Chem. Soc.*, **1953**, 280.
16. R. G. Jones and E. C. Kornfeld, *U.S. Patent* No. 2621187 (1952).
17. A. Butenandt, H. Hellmann and E. Renz, *Z. Physiol. Chem.*, **284**, 175 (1949).
18. H. R. Snyder and J. A. MacDonald, *J. Am. Chem. Soc.*, **77**, 1257 (1955).
19. H. Behringer and H. Taul, *Chem. Ber.*, **90**, 1398 (1957).
20. T. Okuda, *Bull. Chem. Soc. Japan*, **32**, 1165 (1959).
21. B. Hegedüs, *Helv. Chim. Acta.*, **29**, 1499 (1946).
22. H. Plieninger, *Chem. Ber.*, **83**, 268 (1950).
23. S. Sakurai and T. Ito, *Japanese Patent* No. 36–16026 (1961).
24. S. Sakurai and Y. Komachiya, *Japanese Patent* No. 36–19833 (1961).
25. S. Sakurai and Y. Komachiya, *Japanese Patent* No. 38–2531 (1963).
26. Y. Komachiya and S. Sakurai, *Japanese Patent* No. 38–17035 (1963).
27. O. T. Warner and O. A. Moe, *J. Am. Chem. Soc.*, **70**, 2765 (1948).
28. I. Chibata and S. Yamada, *Bull. Agr. Chem. Soc. Japan*, **21**, 58 (1957).
29. I. Maeda and R. Yoshida, *ibid.*, **41**, 2975 (1968).
30. C. W. Smith, *U.S. Patent* No. 2483852 (1949).
31. Y. Komachiya, S. Suzuki, Y. Yamada, H. Miyayashiki and S. Sakurai, *Nippon Kagaku Kaishi* (Japanese), **86**, 856 (1965).
32. Y. Suzuki, T. Akashi and T. Nakamura, *Japanese Patent* No. 44–20983 (1969).
33. S. Miyoshi, K. Harano and N. Shigematsu, *Japanese Patent* No. 40–23659 (1965).
34. I. Sasaji, K. Ohno and J. Kato, *Japanese Patent* No. 38–6183 (1963).
35. J. J. Kolb and G. Toeniss, *J. Biol. Chem.*, **144**, 193 (1942).
36. S. Kameda, K. Matsui and Y. Hotta, *Yakugaku Zasshi* (Japanese), **80**, 770 (1960).

4.25 Tyrosine

In 1864, Liebig, who had been engaged in studies on the alkaline hydrolysis of proteins, isolated an insoluble crystalline compound from an alkaline hydrolyzate of casein. In the next year, he reported that the same product could be obtained from fibrin or albumin, and named it "tyrosine" from the Greek word for cheese. Subsequently, it was demonstrated that tyrosine could in fact be isolated from various proteins by acid hydrolysis, at better yields than by alkaline hydrolysis. Determination of the chemical structure of tyrosine was accomplished somewhat later; namely, in 1883, Erlenmeyer and Lipp confirmed its structure as *p*-hydroxyphenylaminopropionic acid, based on synthesis via phenylalanine.

No efficient technique involving either chemical synthesis or fermentation has yet been developed for tyrosine production. Therefore, this amino acid is still produced chiefly by isolation from hydrolyzates of natural proteins. The isolation process can be carried out very easily since tyrosine is only barely soluble in water.

The known procedures for synthesis of tyrosine are fundamentally the same as those for phenylalanine, although the reaction sequences of the former are, due to the hydroxy group, a little longer than those of the latter. Thus, the development of procedures for tyrosine production by synthesis and/or fermentation without dependence on proteins is awaited. Since there is no established industrial procedure for tyrosine synthesis, typical methods that have been shown to yield tyrosine are described here.

4.25.1 Synthesis from acetamidomalonic acid ester

Ethyl acetamidomalonate, which is often employed for general amino acid synthesis, has been used for the synthesis of tyrosine. Initial condensation of anisyl bromide (I) with ethyl acetamidomalonate (II) in the presence of sodium ethoxide, followed by hydrolysis of the condensation product (III), gave tyrosine (IV) in an overall yield of 48%.[1)]

$$\underset{\text{I}}{CH_3O\text{-}C_6H_4\text{-}CH_2Br} + \underset{\text{II}}{HC(COOC_2H_5)_2(NHCOCH_3)} \xrightarrow{C_2H_5ONa} \underset{\text{III}}{CH_3O\text{-}C_6H_4\text{-}CH_2C(COOC_2H_5)_2(NHCOCH_3)}$$

$$\xrightarrow{HCl} \underset{\text{IV}}{HO\text{-}C_6H_4\text{-}CH_2CH(NH_2)COOH}$$

A procedure using methyl formamidomalonate instead of ethyl acetamidomalonate has also been reported.[2]

4.25.2 Synthesis via hydroxyimino acid

In a manner similar to other amino acids, tyrosine can be synthesized via hydroxyimino acid, starting from acetoacetic acid ester.[3] Treatment of ethyl acetoacetate (VI) with anisylchloride (V) resulted in the formation of ethyl α-anisylacetoacetate (VII). Transformation of this to the hydroxyimino compound (VIII) by treatment with butyl nitrite and subsequent Raney nickel-catalyzed hydrogenation, led to *O*-methyltyrosine ethyl ester (IX). Alkaline hydrolysis of the ester (IX), followed by acid hydrolysis, gave tyrosine (IV) in an overall yield of approximately 63%.

$$\underset{\text{V}}{CH_3O\text{-}C_6H_4\text{-}CH_2Cl} + \underset{\text{VI}}{CH_3COCH_2COOC_2H_5} \xrightarrow[C_2H_5ONa]{} \underset{\text{VII}}{CH_3O\text{-}C_6H_4\text{-}CH_2CH(COCH_3)COOC_2H_5}$$

$$\xrightarrow[H_2SO_4]{C_4H_9ONO} \underset{\text{VIII}}{CH_3O\text{-}C_6H_4\text{-}CH_2C(=NOH)COOC_2H_5} \xrightarrow[\text{Raney nickel}]{H_2} \underset{\text{IX}}{CH_3O\text{-}C_6H_4\text{-}CH_2CH(NH_2)COOC_2H_5}$$

$$\xrightarrow[\text{i) } OH^- \text{ ii) } H^+]{} \underset{\text{IV}}{HO\text{-}C_6H_4\text{-}CH_2CH(NH_2)COOH}$$

A method for tyrosine synthesis via hydroxyimino acid, starting from the readily available acetone and anisaldehyde (*p*-methoxybenzaldehyde), has also been developed.[4] Reduction of anisalacetone (XII), obtained by condensation of anisaldehyde (X) and acetone (XI), with Raney nickel gave α-hydroxyiminoanisyl acetone (XIV). Subsequent oxidation of (XIV) with sodium hypobromite led to the hydroxyimino acid (XV). Reduction of this imino acid with Raney nickel resulted in *O*-methyltyrosine (XVI), and subsequent alkaline hydrolysis of (XVI) yielded tyrosine (IV) in 42% yield.

$$\underset{\text{X}}{CH_3O\text{-}C_6H_4\text{-}CHO} + \underset{\text{XI}}{CH_3COCH_3} \xrightarrow[NaOH]{} \underset{\text{XII}}{CH_3O\text{-}C_6H_4\text{-}CH{=}CHCOCH_3}$$

$$\xrightarrow[\text{Raney nickel}]{H_2} \underset{\text{XIII}}{CH_3O\text{-}C_6H_4\text{-}CH_2CH_2COCH_3} \xrightarrow[HCl]{C_4H_9ONO} \underset{\text{XIV}}{CH_3O\text{-}C_6H_4\text{-}CH_2C(=NOH)COCH_3}$$

$$\xrightarrow{NaOBr} CH_3O\text{-}C_6H_4\text{-}CH_2C(=NOH)COOH \ (\mathrm{XV}) \xrightarrow[\text{Raney nickel}]{H_2} CH_3O\text{-}C_6H_4\text{-}CH_2CH(NH_2)COOH \ (\mathrm{XVI})$$

$$\xrightarrow{HCl} HO\text{-}C_6H_4\text{-}CH_2CH(NH_2)COOH \ (\mathrm{IV})$$

4.25.3 Condensation of aldehyde with active methylene

A number of reports exist for tyrosine synthesis involving the condensation of *p*-hydroxybenzaldehyde or anisaldehyde with a compound having active methylene group and nitrogen group which can be converted to amino group.

For example, tyrosine has been synthesized by Erlenmeyer's method via azlactone.[5] Initial condensation of anisaldehyde (X) with hippuric acid (XVII), followed by hydrolysis of the resulting azlactone (XVIII) with ethanolic alkali, gave *p*-methoxy-α-benzoylaminocinnamic acid (XIX). Treatment of this product with a mixture of acetic anhydride, hydriodic acid and red phosphorus, led to tyrosine (IV) in an overall yield of 60%.

$$CH_3O\text{-}C_6H_4\text{-}CHO \ (\mathrm{X}) + CH_2(NHCO\text{-}C_6H_5)COOH \ (\mathrm{XVII}) \xrightarrow[CH_3COONa]{(CH_3CO)_2O} CH_3O\text{-}C_6H_4\text{-}CH{=}C\text{—}CO\ (\text{ring: N, O, C-}C_6H_5) \ (\mathrm{XVIII})$$

$$\xrightarrow{OH^-} CH_3O\text{-}C_6H_4\text{-}CH{=}C(NHCO\text{-}C_6H_5)COOH \ (\mathrm{XIX}) \xrightarrow[(CH_3CO)_2O]{HI,\ red\ P} HO\text{-}C_6H_4\text{-}CH_2CH(NH_2)COOH \ (\mathrm{IV})$$

Synthesis of tyrosine has also been achieved by the conventional hydantoin method developed by Wheeler and Hoffman. Condensation of anisaldehyde (X) with hydantoin (XX) in the presence of acetic anhydride and sodium acetate yielded anisalhydantoin (XXI). Treatment of this hydantoin derivative with a mixture of hydriodic acid and red phosphorus led to simultaneous reduction and hydrolysis, and tyrosine (IV) was formed in an overall yield of 65%.[6]

RO–C₆H₄–CHO + H₂C–CO / HN NH / CO → ((CH₃CO)₂O, CH₃COONa) → RO–C₆H₄–CH=C–CO / HN NH / CO

X (R=CH₃) XXII (R=H) XX XXI (R=CH₃) XXIII (R=H)

H_2, Raney nickel; HI, red P

HO–C₆H₄–CH₂CH–CO / HN NH / CO (XXIV) → (OH^-) → HO–C₆H₄–CH₂CHCOOH / NH₂

When *p*-hydroxybenzaldehyde (XXII), instead of anisalaldehyde, was condensed with the hydantoin in the presence of morpholine as catalyst, *p*-hydroxybenzalhydantoin (XXIII) was obtained. Reduction of the latter with Raney nickel gave *p*-hydroxybenzylhydantoin (XXIV), and subsequent alkaline hydrolysis led to tyrosine (IV) in a overall yield of 63%.[7]

Another synthetic route for tyrosine using diketopiperazine has also been reported.[8] Condensation of anisaldehyde (X) with diketopiperazine (XXV), and simultaneous reduction and hydrolysis of the condensation product (XXVI) with hydriodic acid and red phosphorus, gave tyrosine (IV) in a yield of 48%.

CH_3O–C₆H₄–CHO (X) + CO–NH / CH₂ CH₂ / NH–CO (XXV) → ((CH₃CO)₂O, CH₃COONa)

CH_3O–C₆H₄–CH=C / CO–NH / NH–CO / C=CH–C₆H₄–OCH_3 (XXVI)

→ (HI, red P) → HO–C₆H₄–CH₂CHCOOH / NH₂ (IV)

CO–NH / CH₂ C(CH₃)(C₆H₅) / NH–CO (XXVII)

When the optically active diketopiperazine, glycyl-D-α-phenylpropionic anhydride (XXVII), was used, L-tyrosine of 35% optical purity was obtained by the same procedure as the hydantoin method.[9]

4.25.4 Resolution of DL-tyrosine

Concerning the optical resolution of synthesized DL-tyrosine, various methods such as chemical resolution and enzymatic resolution have been reported. In the case of chemical resolution, different combinations of an acyl derivative and optically active resolving agent are used to prepare diastereomeric salts, and optical resolution is achieved based on the difference in their solubility to solvents. For example, diastereomeric salts of acetyl-DL-tyrosine with brucine[10] or L-amphetamine,[11] and DL-tyrosine ethylester with D-tartaric acid,[12] have been utilized.

The main enzymatic procedure for DL-tyrosine resolution depends on enzymatic asymmetric hydrolysis of acyl-DL-tyrosine. For example, chloroacetyl-DL-tyrosine was asymmetrically hydrolyzed by crude carboxypeptidase obtained from beef pancrease, and insoluble L-tyrosine was precipitated. Recrystallization of the crude precipitate from hot water gave optically pure L-tyrosine. Concentration of the mother liquor and extraction of the concentrate with ethyl acetate yielded chloroacetyl-D-tyrosine. Hydrolysis of this with hydrochloric acid led to D-tyrosine. The overall yield, including the chloroacetylation process of DL-tyrosine, was 30% for tyrosine, and 24% for D-tyrosine.[13]

References

1. A. T. Schulgin and E. M. Gal, *J. Chem. Soc.*, **1953**, 1316.
2. J. S. Meek, S. Minkowitz and M. M. Miller, *J. Org. Chem.*, **24**, 1397 (1959).
3. K. E. Hamlin and W. H. Hartung, *J. Biol. Chem.*, **145**, 349 (1942).
4. K. Mori, *Nippon Kagaku Kaishi* (Japanese), **77**, 1011 (1956).
5. C. R. Harington and W. McCartney, *Biochem. J.*, **21**, 852 (1927).
6. H. L. Wheeler and C. Hoffmann, *Am. Chem. J.*, **45**, 368 (1911).
7. E. T. Borrows, J. C. Clayton and B. A. Hems, *J. Chem. Soc.*, Suppl. Issue, no. 1, 185 (1949).
8. T. Sakai, *Chem. Ber.*, **54**, 163 (1921).
9. G. Maeda, *Nippon Kagaku Kaishi* (Japanese), **77**, 1011 (1956).
10. R. R. Sealock, *J. Biol. Chem.*, **166**, 1 (1946).
11. J. M. Gillingham, *U.S. Patent* No. 3028395 (1962).
12. G. Lösse, *J. Prakt. Chem.*, **7**, 141 (1958).
13. J. B. Gilbert, V. E. Price and J. P. Greenstein, *J. Biol. Chem.*, **180**, 473 (1949).

4.26 Valine

Valine is an amino acid which was first found together with isoleucine in crude leucine preparations isolated from certain natural products. In 1856, von Gorup-Besanez, who was engaged in an investigation of the chemical identity of components of alcoholic extracts from various organs, found a crystalline substance different from the leucine in pancreatic extract. Thereafter, the same substance was identified in extracts from various plants, and its chemical structure was assumed to be aminovaleric acid. In 1901, Fischer isolated this amino acid from an acid hydrolyzate of casein and described it as optically active. Furthermore, in 1906, he synthesized DL-aminoisovaleric acid by bromination of isovaleric acid and subsequent amination, and also performed optical resolution of the DL-amino acid as the brucine salt of its formyl derivative. Thus, he confirmed the structure of the amino acid, and named it "valine".

Although valine occurs widely in protein hydrolyzates, no specific proteins rich in valine are known, and isolation of this amino acid in a pure state is difficult. Therefore, L-valine is at present produced by optical resolution of chemically synthesized DL-valine. In addition to the chemical method, however, production of valine by fermentation has also been developed. In fact, it seems that chemical synthesis and fermentation will compete together as methods for the mass production of valine.

A number of papers on valine synthesis have been published. The essential reactions used are very similar to those for leucine or isoleucine synthesis. Of them, those methods which are considered of interest from the viewpoint of industrial production are described here.

4.26.1 The Strecker method

There have been many reports on valine synthesis by the Strecker method, which is one of the general methods for amino acid synthesis. Isobutylaldehyde, the starting material, can be synthesized by vapor phase oxidation of isobutanol, and if isobutylaldehyde of high purity is readily available, this method appears to be one of the most advantageous techniques for industrialization. Gaudry[1)] has reported detailed studies of this method. One example from his work is described next.

Treatment of isobutylaldehyde (I) with a solution containing potassium cyanide, ammonium chloride and ammonia, gave aminoisovaleronitrile (II). Hydrolysis of this with hydrochloric acid led to valine (III) in an overall yield of 65%.

The Bucherer method is also applicable to the synthesis of valine.

Isobutylaldehyde (I) was converted to 5-isopropylhydantoin (IV) by treatment with ammonium carbonate and sodium cyanide, and subsequent alkaline hydrolysis gave valine (III) in an overall yield of 49%.[2)]

$$(CH_3)_2CHCHO\ (\text{I}) \xrightarrow[NH_4OH]{KCN,\ NH_4Cl} (CH_3)_2CHCH(NH_2)CN\ (\text{II}) \xrightarrow{HCl} (CH_3)_2CHCH(NH_2)COOH\ (\text{III})$$

$$(CH_3)_2CHCHO\ (\text{I}) \xrightarrow[(NH_4)_2CO_3]{NaCN} (CH_3)_2CH\text{-}CH\text{-}CO\text{-}NH\text{-}CO\text{-}NH\ (\text{IV}) \xrightarrow{OH^-} \text{III}$$

Although it is uncertain which synthetic method is best used in practice for the production of valine, a general flowsheet for L-valine production via the Strecker method (starting from isobutylaldehyde) and the enzymatic resolution technique described below, is shown in Fig. 4–9.

4.26.2 Synthesis via α-halogeno acid

Synthesis of valine can be accomplished by amination of α-halogeno acid; in fact, this technique constitutes one of the most advantageous methods. For example, treatment of isovaleric acid (V) with bromine in the presence of phosphorus trichloride gave α-bromoisovaleric acid (VI). Treatment of (VI) with ammonia led to valine (III) in an overall yield of 42%.[3)]

$$(CH_3)_2CHCH_2COOH\ (\text{V}) \xrightarrow[PCl_3]{Br_2} \underset{87.5\sim88.6\%}{(CH_3)_2CHCH(Br)COOH\ (\text{VI})} \xrightarrow[55\sim66\%\quad 47\sim48\%]{NH_3} (CH_3)_2CHCH(NH_2)COOH\ (\text{III})$$

$$(CH_3)_2CHBr + CH_2(COOC_2H_5)_2 \xrightarrow[70\sim75\%]{C_2H_5ONa} (CH_3)_2CHCH(COOC_2H_5)_2\ (\text{VII}) \xrightarrow{\text{i) KOH; ii) HCl; iii) } Br_2\text{; iv) heat}} \text{VI}$$

For the purpose of synthesizing the α-bromo acid, another route via isopropylmalonic acid ester (VI) is available.[4,5)]

Concerning the amination process of the α-halogeno acid, a number of

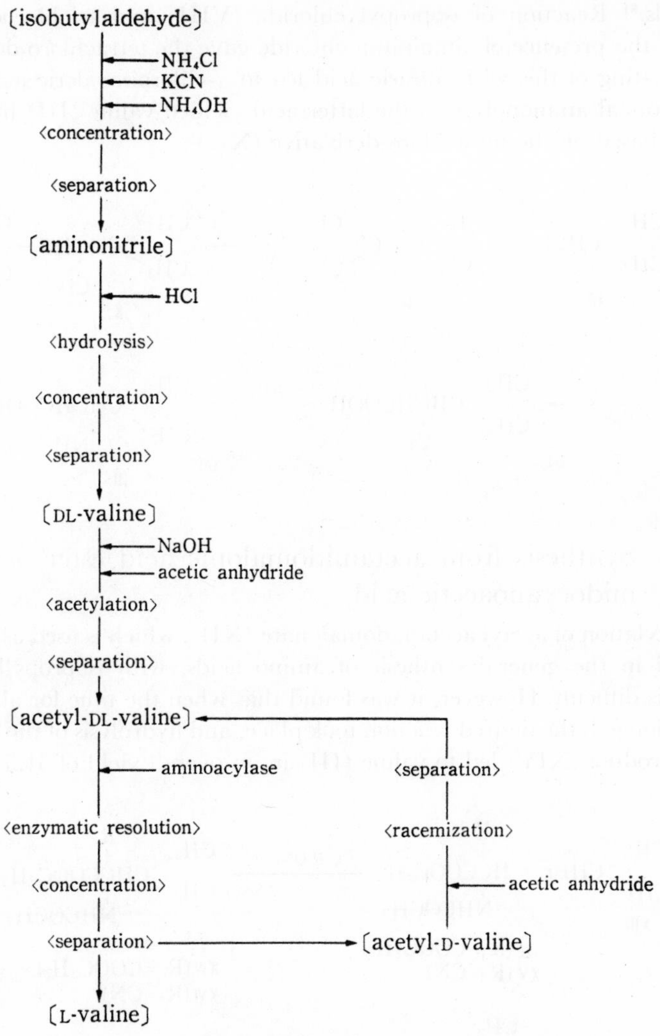

Fig. 4–9. General flowsheet for L-valine synthesis from isobutyl aldehyde.

reports have been published.[6] Treatment of the α-bromo acid (VI) with hexamethylenetetramine in dioxane–xylene, followed by decomposition of the intermediate adduct with hydrochloric acid, gave valine in a yield of 91%.[7]

Another route has been reported for the synthesis of valine via α-chloro acid prepared using trichloroethylene and isopropyl chloride as starting materials.[8] Reaction of isopropyl chloride (VIII) and trichloroethylene (IX) in the presence of aluminium chloride gave the tetrachloro derivative (X). Heating of this with sulfuric acid led to α-chloroisovaleric acid (XI). Conventional ammonolysis of the latter acid yielded valine (III) in a yield of 32% based on the tetrachloro derivative (X).

$$\underset{\text{VIII}}{(CH_3)_2CHCl} + \underset{\text{IX}}{ClHC{=}CCl_2} \xrightarrow{AlCl_3} \underset{\text{X}}{(CH_3)_2CHCH(Cl)CCl_3}$$

$$\xrightarrow[54\%]{H_2SO_4} \underset{\text{XI}}{(CH_3)_2CHCH(Cl)COOH} \xrightarrow[60\%]{NH_3} \underset{\text{III}}{(CH_3)_2CHCH(NH_2)COOH}$$

4.26.3 Synthesis from acetamidomalonic acid ester or acetamidocyanoacetic acid

Alkylation of acetyl acetamidomalonate (XII), which is used as starting material in the general synthesis of amino acids, with isopropylbromide (XIII) is difficult. However, it was found that when the time for alkylation was prolonged, the desired reaction took place, and hydrolysis of the condensation product (XIV) led to valine (III) in an overall yield of 31.5%.[9]

$$\underset{\text{XIII}}{(CH_3)_2CHBr} + \underset{\substack{\text{XII } (R=COOC_2H_5)\\ \text{XV } (R=CN)}}{HC(R)(NHCOCH_3)COOC_2H_5} \xrightarrow{C_2H_5ONa} \underset{\substack{\text{XIV } (R=COOC_2H_5)\\ \text{XVI } (R=CN)}}{(CH_3)_2CHC(R)(NHCOCH_3)COOC_2H_5}$$

$$\xrightarrow{H^+ \text{ or } OH^-} \underset{\text{III}}{(CH_3)_2CHCH(NH_2)COOH}$$

In a similar manner to that described above, ethyl acetamidocyanate (XV) was alkylated with isopropylbromide (XIII), and alkaline hydrolysis of the condensation product (XVI) resulted in the formation of valine (III) in an overall yield of 43%[10] or 25.5–28.5%.[11]

4.26.4 Other miscellaneous methods

As presented in the scheme below, valine can be synthesized via azlactone.[12] Hippuric acid (XVIII), obtained by the reaction of glycine (XVII) and benzoylchloride, was condensed with acetone in the presence of acetic anhydride and sodium acetate to yield the azlactone derivative (XIX). Simultaneous reduction and hydrolysis of this derivative by heating with red phosphorus, acetic anhydride and hydriodic acid, gave valine (III) in an overall yield of 42.5% based on glycine (XVII) (or 56% based on consumed glycine).

$$C_6H_5COCl + \underset{\text{XVII}}{\underset{NH_2}{CH_2COOH}} \longrightarrow \underset{\text{XVIII}}{\underset{NHCOC_6H_5}{CH_2COOH}} \xrightarrow[\substack{CH_3COONa\\(CH_3CO)_2O}]{CH_3COCH_3} \underset{\text{XIX}}{(CH_3)_2C{=}C{-}CO\ (\text{azlactone ring: } N{=}C(C_6H_5){-}O)}$$

57% based on XVII

$$\xrightarrow[(CH_3CO)_2O]{\text{red P, HI}} \underset{\text{III}}{(CH_3)_2CHCH(NH_2)COOH}$$

74.5%

In a similar manner to the above, condensation of acetone with 2-mercaptothiazol-5-one (XX) in the presence of morpholine, followed by reduction of the condensation product (XXI) and subsequent hydrolysis, led to valine (III) in an overall yield of 61%.[13]

$$CH_3COCH_3 + \underset{\text{XX}}{H_2C{-}C{=}O\ (N{=}C(SH){-}S\ \text{ring})} \xrightarrow{\text{morpholine}} \underset{\text{XXI}}{(CH_3)_2C{=}C{-}C{=}O\ (N{=}C(SH){-}S\ \text{ring})}$$

77%

$$\xrightarrow[CH_3COOH]{\text{red P, HI}} \underset{\text{III}}{(CH_3)_2CHCH(NH_2)COOH}$$

4.26.5 Resolution of DL-valine

For the optical resolution of synthesized DL-valine, many methods involving chemical resolution have been published. Of these, only those techniques which are relatively advantageous for industrial application are outlined below.

Optical resolution of DL-valine can be achieved by asymmetric hydrolysis of the chloroacetyl derivative of DL-valine using aminoacylase. However, for the purpose of industrial operation, it appears more advantageous to carry out asymmetric hydrolysis of acetyl-DL-valine with aminoacylase obtained from hog kidney[14] or moulds, and then separate the L-valine and acetyl-D-valine by their difference of solubility. In particular, insolubilized mould aminoacylase can readily hydrolyze acetylvaline, and continuous resolution using a column packed with the insolubilized enzyme is considered to be most suitable.[15]

DL-Valine can also be resolved by physicochemical procedures. Seeding of an optically active isomer into a supersaturated solution of DL-valine hydrochloride[16] or acetyl DL-valine ammonium salt[17] resulted in the preferential crystallization of either of the two enantiomorphs. However, resolution of DL-valine by the preferential crystallization procedure is not so easy as in the case of glutamic acid or threonine. Therefore, enzymatic resolution will probably be adopted for the industrial production of L-valine.

References

1. R. Gaudry, *Can. J. Res.*, **24B**, 301 (1946).
2. H. C. White, *U.S. Patent* No. 2642459 (1953).
3. C. S. Marvel, *Org. Syn. Coll.* vol. 3, p. 848, 1955.
4. R. Adams and R. M. Kamm, *ibid.*, vol. 1, p. 250, 1941.
5. C. S. Marvel and V. du Vigneaud, *ibid.*, vol. 2, p. 93, 1943.
6. N. D. Cheronis and K. H. Spitznueller, *J. Org. Chem.*, **6**, 349 (1941).
7. G. Hillmann and A. Hillmann, *Z. Physiol. Chem.*, **283**, 71 (1948).
8. H. Kawashima, T. Takami and Y. Yamada, *Nippon Kagaku Kaishi* (Japanese), **86**, 1190 (1965).
9. R. O. Atkinson and P. A. A. Scott, *J. Chem. Soc.*, **1949**, 1040.
10. N. F. Albertson and B. F. Tuller, *J. Am. Chem. Soc.*, **67**, 502 (1945).
11. N. F. Albertson and S. Archer, *U.S. Patent* No. 2479662 (1949).
12. P. T. Adams and B. M. Tolbert, *J. Am. Chem. Soc.*, **74**, 6272 (1952).
13. I. M. Heilbron and A. H. Cook, *British Patent* No. 666226 (1952).
14. S. M. Birnbaum, L. Levintow, R. B. Kingsley and J. P. Greenstein, *J. Biol. Chem.*, **194**, 455 (1952).
15. T. Tosa, T. Mori, N. Fuse and I. Chibata, *Biotechnol. Bioeng.*, **9**, 603 (1967).
16. S. Tatsumi, J. Sasa and H. Ohno, *Japanese Patent* No. 37–18470 (1962).
17. M. Shibasaki and T. Fukuro, *Japanese Patent* No. 39–2440 (1964).

PART 2

Amino Acid Utilization

CHAPTER **5**

Quality of Amino Acids

T. Itoh*

Due to recent advances in the technology of amino acid production and quality control, the quality of amino acids now available on the market has become very reliable. For several amino acids, national specifications for quality have been established in Japan (the *Pharmacopoeia of Japan*, *Japanese Standards of Food Additives*, etc.).

At the time when amino acids were produced mostly by extraction from protein hydrolyzates, it was difficult to avoid contamination to a greater or lesser extent with other amino acids having similar properties. However, highly pure amino acids are now readily obtainable by the synthetic and fermentative production methods. The purity (homogeneity) of such commercially produced amino acids is examined by chromatography, and the degree of contamination with impurities is also checked in detail. Concerning

* Tamio Itoh, Euro-Americas Department, Ajinomoto Co., Inc., Chuo-ku, Tokyo 104, Japan

the use of specific rotation as an index for optical purity, this should be standardized after the conditions of measurement have been strictly examined with samples of known quality determined by a variety of means.

5.1 Quality standards

Quality standards have been set for amino acids in regard to the following items: (1) homogeneity, (2) degree of contamination with impurities, and (3) safety.

5.1.1 Guarantee of homogeneity

For assessing homogeneity, measurements of purity by chemical assay, degree of contamination with other amino acids, and specific rotation can be made.

Quantitative analysis of an amino acid is carried out by (1) determination of the nitrogen content and/or (2) titration of the amino acid as a base. Nitrogen is generally determined by the Kjeldahl method. When the amino acid is decomposed by heating in sulfuric acid for several hours in the presence of potassium sulfate and copper sulfate, its nitrogen forms ammonium sulfate from which ammonia is liberated on alkalization. The nitrogen content is then calculated from the titrated value of this ammonia. Although the accuracy of the Kjeldahl method is high, the process is somewhat complicated and needs considerable skill to obtain accurate and consistent results. On the other hand, the nonaqueous titration of the amino acid as a base in glacial acetic acid is comparatively simple and accurate. This titration is carried potentiometrically with perchloric acid. Sulfur-containing amino acids such as cysteine and methionine can also be assayed by iodometry. The purities of individual amino acids set by the national standards or guaranteed by the manufacturer are generally higher than 98.5–99.0%. Moreover, the actual amino acid content may be more than this since the standards were established giving due consideration to analytical errors.

Apart from such analysis, it is of course necessary to confirm that the sample is homogeneous and does not contain appreciable amounts of other amino acids. Accurate and simple methods for detecting contaminant amino acids are paper chromatography and thin layer chromatography. The developing solvent used depends on the sample to be analyzed. For example, butanol–acetic acid–water is used for neutral amino acids, butanol–dicyclohexylamine–water for basic amino acids, and methylethyl ketone–ammonia water for threonine. Ninhydrin is commonly employed as a detecting agent. The amount of contaminants in amino acids on the market is generally less than 0.3%, although it is difficult to generalize since the kinds of possible contaminant amino acids depend rather on the method of production.

Discussion of specific rotation, an important parameter representing the optical purity of L-amino acids, is left until a later section (5.3).

5.1.2 Limits for contamination with impurities

Limits are provided for inorganic compounds which may contaminate the amino acid preparation during the course of production. For example, since amino acids are in general purified by recrystallization from water, the moisture content of the products must be determined. One common procedure is to determine the loss in weight on heat-drying under constant conditions (say, 105°C for 3 h). Since L-histidine monohydrochloride and monosodium L-glutamate, which are obtained in the form of the monohydrate, cannot be dehydrated in such a manner, their free water is determined. The established standard values for moisture content are generally less than 0.3%, although for certain amino acids they are less than 0.5–1.0%.

Chloride is determined by treatment with silver nitrate followed by comparison of the resultant turbidity with a reference prepared from silver chloride solution of known concentration. The standard values are generally less than 0.02%, although in some cases values up to 0.1% are permitted.

The ammonia present in distillates of the sample solutions to which magnesium oxide has been added is determined by Nessler's method. The standard values are less than 0.02–0.04%.

Sulfate ion is determined by comparing the turbidity formed with barium ion against a reference. The standard values are less than 0.02–0.05%.

Inorganic salts are determined as nonvolatile residues on ignition in the presence of sulfuric acid. The ignited residue is represented as sulfate, and the standard values are generally less than 0.1–0.3%.

5.1.3 Safety

From the viewpoint of safety, heavy metals and arsenic must be chemically analyzed when the amino acid is to be used as a medicine or food additive, although the amount of such contamination, if any, is generally minimal. Special biological tests should also be done with amino acid preparations to be used for injections.

Heavy metals such as lead, mercury, cadmium, copper, tin and antimony develop colors when treated with sodium sulfide under acidic conditions, and are so readily analyzed. Since they are highly toxic to man, contamination is strictly regulated, and the standard values (represented typically by the lead concentration) are generally less than 10–20 ppm.

Arsenic, another possible toxic contaminant, is determined colorimetrically by the reaction between arsine derived from the arsenic by reduction and mercuric bromide. A reference color is prepared using arsenous acid. The standard value is less than 2 ppm as arsenic trioxide.

Contamination by pyrogens is examined by biological tests in the case of amino acids to be used for injections. The pyrogens present are heat stable substances produced by contaminating microorganisms in the course of amino acid production, and consist of phospholipopolysaccharides accompanied by small amounts of nucleic acids and proteins. Pyrogens are generally assayed by measuring the temperature rise in the rectum of rabbits to which an aqueous solution of the sample has been intravenously injected. The standard set is expressed as a limiting value for the body temperature rise.

Recently, there has been a movement to regulate the number of intact cells present in pharmaceutical materials. In Sweden, the national standards provide that the intact cell concentration should be less than 100 per gram of product. Contamination by *E. coli* and other pathogenic microorganisms is prohibited.

The assay procedures for each of the above items are described in detail in the *Pharmacopoeia of Japan*[1] and *Japanese Standards of Food Additives*.[2]

5.2 NATIONAL STANDARDS FOR AMINO ACIDS

Examples where national standards for individual amino acids to be used as medicines or food additives have been established are summarized below.

5.2.1 Medicines

Japan (*Pharmacopoeia of Japan*)[1]: monosodium L-glutamate monohydrate, glycine, L-isoleucine, L-lysine monohydrochloride, DL,L-methionine, L-phenylalanine, L-threonine, L-tryptophan, L-valine.

U.S.A. (*National Formulary*)[3]: L-glutamic acid monohydrochloride, glycine, dihydroxyaluminium glycinate, DL-methionine.

U.S.S.R. (*State Pharmacopoeia of the U.S.S.R.*)[4]: L-glutamic acid.

West Germany (*Deutsches Arzneibuch*)[5]: DL-methionine.

East Germany (*Deutsches Arzneibuch*)[6]: L-cysteine hydrochloride, L-cystine, L-glutamic acid, glycine, L-leucine, DL-methionine, acetyl-L-cysteine.

France (*Pharmacopée Française*)[7]: L-cysteine hydrochloride, glycine, L-histidine monohydrochloride, DL-lysine monohydrochloride, DL-methionine, DL-tryptophan.

Switzerland (*Pharmacopoea Helvetica*)[8]: L-histidine monohydrochloride. Britain does not yet provide a standards list for amino acids.

In addition to the common amino acids shown above, national standards have been set for sodium L-thyroxine, sodium L-3,5,3′-triiodothyronine (U.S.A. and Britain), L-Dopa (Britain) and L-methyldopa (Japan and Britain).

5.2.2 Food additives

Japan (*Japanese Standards of Food Additives*)[2]: DL-alanine, L-arginine–L-glutamate, monosodium L-aspartate monohydrate, L-glutamic acid, monosodium L-glutamate monohydrate, glycine, L-histidine monohydrochloride monohydrate, L-isoleucine, L-lysine monohydrochloride, L-lysine–L-aspartate, L-lysine–L-glutamate dihydrate, DL,L-methionine, L-phenylalanine, DL,L-threonine, DL,L-tryptophan, L-valine.

U.S.A. (*Food Chemical Codex*)[9]: DL,L-alanine, L-arginine, L-arginine monohydrochloride, DL-aspartic acid, L-cystine, L-glutamic acid, ammonium L-glutamate, monosodium L-glutamate monohydrate, potassium L-glutamate, L-glutamic acid monohydrochloride, glycine, DL,L-isoleucine, DL,L-leucine, L-lysine monohydrochloride, DL,L-methionine, DL,L-phenylalanine, L-proline, DL,L-serine, L-threonine, DL,L-tryptophan, L-tyrosine, L-valine.

5.2.3 Japanese quality standards

Japanese quality standards for amino acids are shown in Table 5–1. As indicated, the values have been extracted from the *Pharmacopoeia of Japan*,[1] *Japanese Standards of Food Additives*,[2] or refer to manufacturers' standards in cases where national standards have not been set in Japan.

The *Japan Industrial Standards* list[10] and NAS–NRC standards[11] of the U.S. also provide reagent standards for 16 amino acids and 65 amino acids and related compounds, respectively.

5.3 Optical rotation

Optical rotation is important because its value is specific to each amino acid, and it can thus be considered as an index for representing the quality of amino acid preparations. Numerous data on specific rotation have been reported but all cannot be considered reliable.[12,13] Since the optical rotation value is affected by the assay conditions, it is necessary to define a suitable set of conditions by systematic examination. Standard values should be decided

TABLE 5–1
Quality standards for amino acids

Amino acid	Standard source†1	Specific rotation $[\alpha]^{20}_{D}$	Concn. (g/dl)	Solvent	Chloride (% Cl)	Ammonia (% NH_3)
DL-Alanine	F	—	—	—	<0.02	<0.03
L-Alanine	AB	+13.5∼+15.5°	10	6N–HCl	<0.02	<0.02
L-Arginine	AB	+25.0∼+27.5°	8	6N–HCl	<0.02	<0.02
L-Arginine monohydrochloride	AB	+20.5∼+23.0°	8	6N–HCl	—	<0.02
L-Asparagine	B	+33.5∼+37.8°	10	6N–HCl	<0.02	—
L-Asparagine monohydrate	A	+33.0∼+36.0°	10	3N–HCl	<0.02	<0.1
L-Aspartic acid	B	+25.5∼+26.5°	10	2N–HCl	<0.02	<0.05
Monosodium L-aspartate	F	+18∼+21°	8	6N–HCl	<0.041	<0.04
L-Citrulline	A	+24.5∼+26.5°	8	6N–HCl	<0.02	<0.02
L-Cysteine hydrochloride monohydrate	F	+4.5∼+7.0°	8	1N–HCl	—	<0.03
L-Cystine	AB	−215.0∼−225.0°	2	1N–HCl	<0.02	<0.02
L-Glutamic acid	F	+31.5∼+32.5°	10	2N–HCl	<0.1	<0.02
Monosodium L-glutamate monohydrate	F	+24.8∼+25.3°	10	2.5N–HCl	<0.1	<0.04
L-Glutamine	A	+6.3∼+7.3°	4	H_2O	<0.02	<0.1
Glycine	F	—	—	—	<0.02	<0.03
L-Histidine	A	+11.8∼+12.8°	11	6N–HCl	<0.02	<0.02
L-Histidine monohydrochloride monohydrate	F	+8.5∼+10.5°	11	6N–HCl	—	<0.02
L-Hydroxyproline	A	−74.0∼−77.0°	4	H_2O	<0.02	<0.02
L-Isoleucine	F	+39.0∼+41.5°	4	6N–HCl	<0.02	<0.02
L-Leucine	A	+14.9∼+16.5°	4	6N–HCl	<0.02	<0.02
L-Lysine monohydrochloride	F	+19.0∼+21.5°	8	6N–HCl	—	<0.02
L-Lysine acetate	B	+17.0∼+19.0°	8	6N–HCl	<0.02	<0.02
DL-Methionine	F	—	—	—	<0.02	<0.02
L-Methionine	F	+21.0∼+25.0°	2	6N–HCl	<0.02	<0.02
L-Ornithine monohydrochloride	A	+23.0∼+25.0°	4	6N–HCl	—	<0.02
L-Phenylalanine	P	−33.0∼−3.50°	2	H_2O	<0.021	<0.02
L-Proline	A	−84.0∼−86.0°	4	H_2O	<0.02	<0.02
L-Serine	AB	+13.5∼+16.0°	10	2N–HCl	<0.02	<0.02
DL-Threonine	F	—	—	—	<0.02	<0.02
L-Threonine	F	−26.0∼−29.0°	6	H_2O	<0.02	<0.02
DL-Tryptophan	F	—	—	—	<0.02	<0.03
L-Tryptophan	F	−30∼−33°	1	H_2O	<0.02	<0.02
L-Tyrosine	AB	−11.3∼−12.5°	5	1N–HCl	<0.04	<0.02
L-Valine	F	+26.5∼+28.0°	4	6N–HCl	<0.04	<0.02

NOTE: In addition to the above items, limits for contamination with arsenic acid (<2 ppm in all cases) and for the state of the amino acid aqueous solution (transparency at a given concentration) are provided.

†1 F, *Japanese Standards of Food Additives*, 3rd ed.; P, *Pharmacopoeia of Japan*, 8th rev. ed.; A, Company "A"; B, Company "B".

Sulfate (% SO_4)	Heavy metals (ppm Pb)	Other amino acids[†2] Assay method	(μg)	Loss on drying (%)	Temp (°C)	Time (h)	Residue on ignition (% SO_4)	Purity[†3] Assay method	(%)
—	<20	P	10	<0.3	105	3	<0.2	T	>98.5
<0.03	<10	T	10	<0.2	105	3	<0.1	T	>98.5
<0.03	<10	T	10	<0.6	105	3	<0.1	T	>98.5
<0.03	<10	T	10	<0.2	105	3	<0.1	T	>98.5
<0.03	<20	<1%		<1.0	130	3	<0.1	T	>98.5
<0.03	<10	<1%		11.5~12.5	130	3	<0.1	T	>98.5
<0.03	<10	T	10	<0.2	105	3	<0.1	T	>98.5
—	<20	P	10	<0.3	†4	5	—	T	>98.0
<0.03	<10	T	10	<0.4	105	3	<0.1	T	>98.5
—	<20	P	10	8.5~12.0	†4	20	<0.2	I	98.0~100.0
<0.03	<10	T	10	<0.4	105	3	<0.1	T	>98.5
—	<10	P	10	<0.3	105	3	<0.2	T	>99.0
—	<10	P	10	<0.3	100	5	—	T	>99.0
<0.03	<10	<1%		<0.2	80	3	<0.1	T	>98.5
—	<20	P	10	<0.3	105	3	<0.1	T	>98.5
<0.03	<10	T	10	<0.2	105	3	<0.2	T	>98.5
—	—	P	10	<0.3	98	3	<0.1	K	98.0
<0.03	<10	T	10	<0.2	105	3	<0.1	T	>98.5
—	<20	P	10	<0.3	105	3	<0.1	T	>98.5
<0.03	<10	<2%		<0.2	105	3	<0.1	T	>98.5
—	<10	P	30	<1.0	105	3	<0.3	T	>98.5
<0.03	<20	P	100	<0.3	105	5	<0.2	T	>98.5
<0.048	<20	—	—	<0.5	105	4	<0.1	I	>98.5
<0.048	<20	—	—	<0.5	105	4	<0.1	I	>98.5
<0.03	<10	T	10	<0.2	105	3	<0.1	T	>98.5
<0.028	<20	P	10	<0.3	105	3	<0.1	T	>98.5
<0.03	<10	T	10	<0.4	105	3	<0.1	T	>98.5
<0.03	<10	T	10	<0.2	105	3	<0.1	T	>98.5
—	<20	P	10	<0.3	105	3	<0.1	T	>98.5
—	<20	P	10	<0.3	105	3	<0.1	T	>98.5
—	<20	P	10	<0.3	105	3	<0.1	T	>98.0
—	<20	P	10	<0.3	105	3	<0.1	T	>98.5
<0.03	<30	T	10	<0.2	105	3	<0.1	T	>98.5
—	<20	P	10	<0.3	105	3	<0.1	T	>98.5

†2 T, Thin layer chromatography; P, paper chromatography. "10 μg" indicates that no other amino acid is detectable when a 10 μg sample is analyzed.

†3 T, Nonaqueous titration; K, Kjeldahl method; I, iodometry. The upper maximum is 101.0%.

†4 Reduced pressure.

by analyzing the many available data statistically, giving due consideration to technical error.

The wavelength of the light, temperature, concentration of sample, and solvent are generally specified. The D-line of a sodium lamp is most commonly used. Optical rotation is measured at 20°C according to the Japanese national standards but at 25°C according to the U.S. standards. The temperature coefficient indicating the increase in specific rotation on 1°C rise in temperature is specific for different amino acids under given conditions.

Either water or hydrochloric acid is used as solvent. The specific rotation varies with the ionic form of the amino acids. At the isoelectric point, L-amino

TABLE 5–2

Specific rotation and temperature coefficients of amino acids

Amino acid	Specific rotation, $[\alpha]_D^{20}$	Concn. (g/dl)	Solvent	Temp. coefficient (°/°C)
L-Alanine	+14.7°	10	6N–HCl	−0.07
L-Arginine	+27.3°	8	6N–HCl	−0.04
L-Arginine monohydrochloride	+22.5°	8	6N–HCl	−0.03
L-Aspartic acid	+26.4°	10	2N–HCl	−0.08
L-Cystine	−222.5°	2	1N–HCl	+2.20
L-Glutamic acid	+32.0°	10	2N–HCl	−0.07
L-Glutamic acid hydrochloride	+25.6°	10	conc.HCl,16.6ml $+H_2O \rightarrow$100ml	−0.06
L-Histidine	+12.5°	11	6N–HCl	+0.19
L-Histidine monohydrochloride monohydrate	+ 9.2°	11	6N–HCl	+0.19
L-Hydroxyproline	−75.9°	4	H_2O	+0.07
L-Isoleucine	+41.1°	4	6N–HCl	−0.09
L-Leucine	+15.2°	4	6N–HCl	+0.06
L-Lysine monohydrochloride	+21.1°	8	6N–HCl	−0.031
L-Methionine	+23.8°	2	6N–HCl	+0.02
L-Phenylalanine	−34.4°	2	H_2O	+0.10
L-Proline	−85.2°	4	H_2O	+0.005
L-Serine	+15.2°	10	2N–HCl	−0.08
L-Threonine	−28.5°	6	H_2O	+0.05
L-Tryptophan	−31.5°	1	H_2O	+0.06
L-Tyrosine	−11.7°	5	1N–HCl	+0.25
L-Valine	+28.1°	8	6N–HCl	−0.03

acids exhibit a minimum value which shifts to the plus side by adding acid or alkali, but the value returns to a constant level when acid or alkali is present in excess over a certain specific level (Lutz's law). Water is a suitable solvent for L-phenylalanine, L-tryptophan, L-threonine, L-proline and L-hydroxyproline, since these amino acids exhibit a levorotation in aqueous solutions; the absolute values in acidic solutions are smaller. The rotation of L-cystine and L-tyrosine, which exhibit a levorotation in water, is measured with highly concentrated solutions in hydrochloric acid since the solubilities in water of these amino acids are so low that analytical errors easily result. The rotation of an aqueous solution of L-glutamine is levo but its value is rather small. The rotation of a solution in hydrochloric acid is dextro and the value greater, but it varies showing a gradual decomposition. Water is therefore used as solvent for measuring the optical rotation of glutamine. Amino acids other than those described above exhibit dextrorotation of suitable degree in hydrochloric acid, which is commonly employed as the solvent. It is necessary to decide the assay conditions, such as concentration of hydrochloric acid and concentration of sample, in order to obtain suitable and consistent values.[14)] Standard values for the specific rotation and temperature coefficients of various amino acids of high purity are listed in Table 5–2.[15)]

References

1. *The Pharmacopoeia of Japan*, 8th ed., part I, II, Soc. Japan. Pharmacopoeia, Yakuji Nippo Ltd., 1973.
2. *The Japanese Standards of Food Additives*, 2nd ed., Japan Food Hygiene Assoc., 1970.
3. *National Formulary*, 13th ed., American Pharm. Assoc., 1970.
4. *State Pharmacopoeia of the U.S.S.R.*, 9th ed. (Engl. ed.), Ministry of Health, U.S.S.R., 1961.
5. *Deutsches Arzneibuch*, 7 Ausgabe, 5 Bde, Academie Verlag, 1964–69.
6. *Deutsches Arzneibuch*, 7 Ausgabe, Deutscher Apotheker-Verlag, 1968.
7. *Pharmacopée Française*, 8th ed., Ministère de la Santé Publique et de la Population, 1965; supplément, 1968.
8. *Pharmacopoea Helvetica*, Editio Quinta, Drucksachen und Materialzentrale, 1953–62.
9. *Food Chemical Codex*, 2nd ed., National Academy of Sciences–National Research Council, 1972.
10. *Japan Industrial Standards*, Agency of Industrial Science and Technology, Japan, 1961.
11. *Specifications and Criteria for Biochemical Research*, 2nd ed., National Academy of Sciences–National Research Council, 1967.

12. J. P. Greenstein and M. Winitz, *Chemistry of the Amino Acids*, p. 1724, Wiley, 1961.
13. *Ed.* S. Akabori and S. Mizushima, *Tampakushitsu Kagaku* (Japanese), vol. 1, p. 511, Kyoritsu Shuppan, 1954.
14. K. Hayashi, Y. Fujii, R. Saito, H. Kanao and T. Hino, *Agr. Biol. Chem.*, **30**, 1221 (1966).
15. Ajinomoto Co., Inc., *unpublished.*

CHAPTER

6

Utilization of Amino Acids

T. Itoh*[1] et al.

*1 Tamio Itoh, Euro-Americas Department, Ajinomoto Co., Inc., Chuo-ku, Tokyo 104, Japan

*2 Katsuyuki Toki, Sumitomo Chemical Co., Ltd., Takarazuka-shi, Hyogo-ken 665, Japan

*3 Ichiro Chibata, Research Laboratory of Applied Biochemistry, Tanabe Seiyaku Co., Ltd., Higashi-Yodogawa-ku, Osaka 532, Japan

*4 Ryonosuke Yoshida, Product Development Laboratories, Ajinomoto Co., Inc., Kawasaki-shi, Kanagawa-ken 210, Japan

Industrialization of amino acid production was initiated in 1909, following the extensive investigations on amino acids carried out since the birth of modern chemistry in the 1800's. The first step was the production and use of monosodium L-glutamate as a tasty substance for Japanese traditional soup stock. However, it is only in the past 10–15 yr that the technology of amino acid production has been diversified, mainly in Japan. The present annual production of amino acids in the world has an estimated value of about 100 billion yen.

Monosodium glutamate, a well-known flavor enhancer, is still the predominant product, with a worldwide annual production of about 200,000 tons. L-Glutamic acid has been produced by extraction from protein hydrolyzates for about 50 yr, but this method has the drawback of yielding a large volume of by-products. In the technological revolution following World War II, new methods of production were sought and in 1956 the fermentation method was successfully developed. Within a few years almost all glutamic acid was being obtained by the new method. The success of glutamic acid fermentation opened the door to the fermentative production of comparatively complex compounds by artificially controlling the metabolism of microorganisms, and fermentation methods for the production of most amino acids are now available.

Methods of chemical synthesis for glutamic acid and methionine were industrialized in 1960. Even though the chemical synthesis of glutamic acid requires an optical resolution step to obtain the L-isomer, this method remains competitive with the fermentation processes. In the case of methionine, optical resolution is unnecessary, since D-methionine has the same nutritional value as the L-isomer, and thus the synthetic method is particularly advantageous in this case. The annual production of methionine as a feed additive is over 50,000 tons.

Lysine is produced on the next largest scale, of nearly 20,000 tons/yr, and is used mainly as a feed additive. Although methods of chemical synthesis have been studied intensively, the fermentation method remains the cheapest. Demand is expected to increase more rapidly for this amino acid than for any other.

Although the above three amino acids are produced on a relatively large scale, demand for other amino acids is small, amounting to between several tons and several hundred tons per year. Most amino acids are now produced by chemical synthesis or fermentation, and the extraction method is scarcely used, except for the production of cystine. In general, the prices of such amino acids are high, though many of them have applications in the medical field, and as new applications are developed prices may fall.

Amino acids are constituents of proteins, and are essential nutrients for living organisms. Demand for amino acids is therefore based in such areas as medicines, food additives and additives for animal feed. The potential demand for amino acids as food and feed additives is enormous, but is centered on only a few essential amino acids such as lysine and methionine. However, in the medical field, mixtures of many kinds of amino acids for administration either orally or by injection are in use throughout the world. Some amino acids act as important physiologically active substances, such as hormones, during the course of normal metabolism. Thus, a disorder involving the metabolic pathway of a particular amino acid may require administration of that amino acid or its metabolite. For instance, Dopa has recently been used as a treatment for Parkinson's disease.

Amino acids are also useful in relation to food flavors. Glutamic acid is a well-known flavor enhancer, and glycine is employed as a sweetener. Various mixtures of amino acids producing a complex taste are also used.

Thus, the utilization of amino acids is centered on their nutritional value and taste, and on their physiological activity, involving the food and animal feed industries and the pharmaceutical industry. The growth of demand will therefore depend on progress in production technology to reduce costs, and also on the possible application of amino acids as chemical raw materials. Since amino acids are naturally occurring substances, there is little danger of industrial pollution or danger to human health in the development of the amino acid industry.

6.1 Utilization for food

Amino acids occur mainly as protein constituents, and also in the free state to some extent in foods. They are important factors in evaluating the nutritive value and taste of foods. The nutritive value of food depends not only on its calory content but also on the quantity and type of amino acids which it contains. Several amino acids are of particular importance from a nutritional point of view. In addition, the flavor of food is due to a combination of palate and the sense of smell, and it is known that the role of amino acids is important. In fact, amino acids may be essential flavor-forming substances. The development of the amino acid industry stemmed from the commercialization of glutamic acid as a flavoring agent.

The amino acids used as nutritional additives are essential amino acids such as lysine and methionine which are often present in insufficient amounts in foods. These amino acids are not only employed directly in food but are also added to animal feed. Amino acids used as flavoring agents include alanine and glycine as sweeteners, in addition to glutamic acid which improves the natural taste of various foods by acting as a flavor enhancer.

Several other amino acids are also utilized in food processing for certain specific properties.

The production of amino acids for use in foods (including animal feed) is much greater than that for medicinal purposes. The worldwide annual production of monosodium L-glutamate monohydrate is about 200,000 tons, produced mainly by fermentation. DL-Methionine, which is mainly used as an additive for animal feed, is wholly produced by chemical synthesis and its worldwide annual production is more than 50,000 tons. About 20,000 tons of L-lysine monohydrochloride, which is used in food, feed and medicines, is obtained annually by fermentation, and demand is expected to increase rapidly. The worldwide annual production of glycine, which is obtained by chemical synthesis, is 2000 tons.

6.1.1 Utilization of amino acids as essential nutrients

A. World protein problem and amino acids

The world population, which was about 1.6 billion in the early 1900's, increased to 2.5 billion in 1950 and 3.6 billion in 1970, and is estimated to reach over 6 billion at the end of the century. Developing countries in

TABLE 6–1

Supply of protein in the world (estimation for 1975)

	Developing countries	Developed countries	Total	Increase from 1960
Population (billions)	2.9	1.1	4.0	1.0
Animal protein (million tons)	14.6	19.0	33.6	12.2
Meat	6.7	9.4	16.1	5.6
Fish	2.7	1.3	4.0	2.0
Eggs	1.1	1.1	2.2	0.8
Milk products	4.1	7.2	11.3	3.8
Vegetable protein (million tons)	51.6	26.4	78.0	30.7
Cereals	33.0	10.6	43.6	11.0
Beans	13.1	11.8	24.9	17.1
Fruits and vegetables	3.1	2.1	5.2	1.9
Root vegetables	2.4	1.9	4.3	0.7
Total (million tons)	66.2	45.4	111.6	43.0
Proportion of animal protein (%)	22	42	30	

Southeast Asia, Africa, and Central and South America, possessing the majority of the world population, are suffering from permanent malnutrition, and the gap between the population growth and agricultural production in these countries is a difficult problem. A continuous lack of protein represses not only the children's physical growth but also their mental development, and it makes the adults lethargic. It causes a vicious cycle of economic and social underdevelopment and malnutrition.

The United Nations, having paid special attention to the nutritional problems throughout the world, proposed the "Freedom from Hunger" campaign in 1960. In 1968, the Protein Advisory Group, consisting of experts from various fields in the United Nations, presented their recommendations to the Economic and Social Council on the increased production and utilization of protein foods.[1] This study has quantified the problem by expressing the relationship between the population increase and food shortage, especially that of protein foods, in terms of a "protein gap", and proposed 14 plans to supply a large quantity of protein of high nutritive value in a form readily acceptable to consumers. New protein sources which were listed in the recommendations included oilseed protein, fish protein concentrate, proteins from genetically improved vegetables, and single-cell protein. The use of amino acids for the fortification of protein was also discussed.[2]

B. Fortification of vegetable protein with amino acids

According to estimates by the FAO of the United Nations, the total protein consumption in the world in 1975 will be 112 million tons (34 million tons of animal protein and 78 million tons of vegetable protein)[3] (Table 6–1). Since meat and eggs are produced by using feed containing vegetable protein, more than 100 million tons of vegetable protein is necessary to produce 8 million tons of meat and eggs proteins. The total requirement for vegetable protein, including that for direct consumption as food, thus reaches 180 million tons. It may be said that the key to the problem of food protein in the world is the rational use of vegetable protein sources such as rice, wheat, corn and soybean. This depends on how efficiently they are produced and utilized.

The purpose of amino acid fortification is to achieve increased production of animal protein by efficient utilization of vegetable protein. Amino acid fortification is based on the finding that the nutritional efficiency of protein depends on the balance of its constituent amino acids, especially the essential amino acids, and on the experimental finding that the nutritional value of protein can be improved through modification of the amino acid balance by the addition of the amino acids which are the limiting factors in the nutritional value of the protein.

FAO proposed an ideal pattern of essential amino acids in 1957, based on experimental data on the requirements for essential amino acids obtained by Rose during the 1940's and 1950's.[4] Since then, the FAO model has been widely applied, and in 1965 it was modified to the present pattern, which is identical to the composition of essential amino acids in whole egg or human milk.[5]

The nutritional value of proteins can be roughly evaluated by comparing the amino acid composition with the standard values for essential amino acids in whole egg or human milk, but it can be more critically evaluated by biological tests with rats or mice (growth tests or nitrogen-balance tests). Biological tests also reveal which amino acid is the limiting factor, and how much the nutritional value is improved by the addition of any particular amino acid.

TABLE 6–2

Fortification of cereals with amino acids

Cereal	Amino acids added	PER
Rice	none	1.50
	L-lysine monohydrochloride, 0.2%; DL-threonine, 0.2%	2.61
Wheat flour	none	0.65
	L-lysine monohydrochloride, 0.2%	1.56
	L-lysine monohydrochloride, 0.4%; DL-threonine, 0.3%	2.67
Corn	none	0.85
	L-lysine monohydrochloride, 0.4%	1.08
	L-lysine monohydrochloride, 0.4%; DL-tryptophan, 0.07%	2.55
Sorghum	none	0.69
	L-lysine monohydrochloride, 0.2%	1.77
Barley	none	1.66
	L-lysine monohydrochloride, 0.2%; DL-threonine, 0.2%	2.28

The experimental data of Howe *et al.*[6,7] on the fortification of protein with amino acids are illustrated in Table 6–2. The protein efficiency ratio (PER) is the ratio of weight gain to the total feed protein administered during the experimental period. The PER of casein, which has a comparatively high nutritional value, is used as a reference standard for experiments on protein nutrition. Its value is 2.5, while the value for cereals is as low as 1. The limiting amino acid in cereals is lysine, without exception, and the PER is greatly improved by supplementing lysine. Supplementing threonine (rice and wheat) or tryptophan (corn), the next most limiting amino acids to lysine, further improves the nutritional values of cereals. The amino acids valuable in practice for the fortification of cereal proteins are lysine, threonine and tryptophan.

Protein from oilseeds such as soybean is important as a source for animal feed. The limiting amino acid in soybean meal is methionine. The addition of lysine and threonine to cottonseed meal, lysine to sesame meal, and lysine, threonine and methionine to peanut meal is effective.[8)]

C. Present status and prospects for amino acid fortification

The effectiveness of the fortification of protein with amino acids has been confirmed and a basis for the practice of fortification has been provided by the development of the amino acid industry. The current status of fortification will be reviewed from the standpoint of food and animal feed manufacture.

(i) Fortification of feed

Amino acid fortification has been practiced in the manufacture of formula feeds for domestic animals and poultry, effectively converting vegetable protein into animal protein. Following the increase in the consumption of methionine as an ingredient of formula feed in recent years, the use of lysine, for which the potential demand may be greater, has been increasing rapidly.

The importance of vegetable protein as a protein source in feed has long been recognized. Vegetable protein is economically favorable compared with animal protein such as fish meal, the resources of which are limited. Protein sources for formula feed include grains and oilseed meals such as soybean and cottonseed meal. Protein from soybean meal is useful but it is deficient in methionine. All other grains and oilseed meals are primarily deficient in lysine.

Material obtainable locally at a low price is usually chosen as a base for feed, and formulation is based on the concept that the feed should be economic and satisfy nutritional requirements. Methionine and lysine are generally essential additives.

There is no difference as regards nutritional considerations between amino acid fortification of feed and direct utilization of amino acids for food. However, in the case of food there are many complex factors to be considered, whereas with animal feed only economic factors need be considered.

This is why the fortification of animal feed was developed first. Since feed is already essential for the production of animal protein such as meat and eggs, it may be said that the economical supply of food protein increasingly involves the effective utilization of amino acids. The use of amino acids as animal feed additives will be dealt with in detail in a separate section.

(ii) Fortification of food with amino acids

The fortification of food cereals with amino acids is obviously more effective from the viewpoint of direct utilization of food resources than the fortification of formula feed for the production of animal protein. However, amino acid fortification of human foods requires careful study in relation to eating habits, cost and social acceptability.

Amino acids which might be used for fortification of food include lysine (cereals), threonine (rice) and tryptophan (corn). This is the easiest and most efficient way to enhance the nutritional value of protein in cereals, because at least a minimum amount of cereals is ingested as an energy source, even in developing coutries where the protein intake is quantitatively and qualitatively insufficient. However, fortification of these cereals with amino acids should not involve any changes in the taste of the food, and requires a consideration of cooking methods and other food-related habits.

The fortification of wheat flour products is easy because powdered L-lysine monohydrochloride can be mixed with the flour during its distribution. In Japan, the first lysine-fortified bread appeared on the market in 1958, and in 1962, the Ministry of Welfare permitted the use of the term "special nutritional food" for wheat flour products fortified with a certain amount of L-lysine. In 1968, the Japan Essential Amino Acids Association experimentally confirmed the effectiveness of the lysine-fortified bread in improving children's growth, and proposed to the Ministry of Education that the bread eaten with school lunches should be fortified with lysine. Since then, the suggestion has been taken up and at present 7 million school children are eating lysine-fortified bread at school.

As for the fortification of rice, amino acids should be processed in granules like rice and mixed with polished rice to avoid any problems with the traditional manner of consumption. The addition of 0.2% L-lysine monohydrochloride and 0.1% L-threonine is desirable from the nutritional point of view for the fortification of rice. Such amino acid-containing rice has been developed in Japan and the nutritional testing has been carried out in Thailand with the cooperation of Japan and the U.S.

Various programs for fortification with lysine have been examined for commercial and political reasons in developing countries. Lysine-fortified bread has come onto the market or its nutritional value has been examined in India, Peru and Tunisia. In such developing countries, infants under school age, but after weaning, are often subject to protein malnutrition. Graham *et al.*[9,10] investigated weight and height gains, nitrogen balance and serum protein fluctuations after feeding badly undernourished Peruvian infants with a diet in which lysine-enriched wheat flour was the only protein source, and reported that their nutritive status was greatly improved by wheat flour fortified with 0.2–0.4% L-lysine monohydrochloride. This finding that protein malnutrition can be prevented by feeding cereal protein fortified with amino acids without intake of animal protein suggests that the requirement for protein is almost satisfied by fortified cereals if the cereal intake is sufficient for energy requirements. Productivity has been considerably improved for cereals in developing countries. It can undoubtedly be said that the fortification of cereals with amino acids is a valuable step towards solving the world protein crisis, both from the nutritional and economical points of view. The only remaining problems are consumer education, product distribution, and political decisions, based on the national economy.

D. Utilization of amino acids for chemically defined diets

The idea of a chemically defined diet consisting of amino acids and other nutrients appeared in the 1950's, and a series of investigations was carried out by Greenstein *et al.*[11] In the initial stage of development it was intended to utilize this diet as a food for astronauts,[12] and later medical applications were studied. All nutrients were to be quantitatively provided in a chemically defined diet consisting not only of essential amino acids but also of non-essential amino acids as a nitrogen source, carbohydrates such as glucose and dextrin as an energy source, vitamins, essential fats and inorganic salts, but excluding any high molecular compounds. The daily requirement for adult nutrition can be satisfied with about 500 g of such a chemically defined diet although the amount required depends of course on body weight. During physiological and psychological investigations on adult males to whom only this chemically defined diet had been administered for 6 months, it was found that no abnormality occurred except for a decrease in excretion due to the absence of indigestible matter in the diet[13] and a decrease in gut microflora.[14] The results of this experiment suggest the possibility of nutritional substitution of an amino acid mixture for protein. Although such a diet may not satisfy the appetite of a healthy man for a long period, it may be useful for patients with functional disorders of the digestive tract requiring strict dietary treatment, or before and after surgical operations.

6.1.2 Use of amino acids in food processing

A. Taste of amino acids

In 1908, Ikeda found that L-glutamic acid was the taste substance in "*konbu*" (*Laminaria japonica*), which is used traditionally for preparing soup stock in Japan, and he suggested the importance of amino acids as taste factors. It is well known that the utilization of monosodium L-glutamate as a flavoring in the food industry has been the basis of the amino acid industry, and has accelerated technological development relating to the production and utilization of amino acids.

Although it has long been known that the taste of soy sauce and soy paste, the traditional flavorings in Japan, arises from amino acids which are formed from soybean protein by the hydrolytic action of microbial enzymes, studies on the contribution on individual amino acids to taste and on the tastes due to combinations of several amino acids are new. The composition of free amino acids in various foods has been investigated by advanced techniques such as chromatography which were extensively developed after World War II, and it has been shown that certain amino acids are important elements in the taste of food. However, the taste of food is a complex phenomenon and it is very difficult to analyze the taste factors chemically.

Taste substances other than amino acids include mononucleotides such as inosinic and guanylic acids, originating from nucleic acids, organic acids, sugars and inorganic salts. Peptides and amino acid metabolites such as betaine and creatine cannot be overlooked, because they contribute to the complexity of taste sensations. It is known that the essential taste substances are present in extracts excluding polymer compounds. Although polymers such as protein and polysaccharide are important from a tactile point of view, they probably have no direct relation with taste. In general, the main taste substances in meat and meat products are glutamic acid and inosinic acid. Other constituents such as organic acids, peptides and other amino acids may modify the taste. The taste of vegetable foods is mainly due to glutamic acid and also to other amino acids, organic acids and sugars. The participation of organic acids and sugars is greater in fruit taste.

The tastes of individual amino acids are shown in Table 6–3.[15)] These taste sensations can be classified into sweetness, bitterness, sourness and tastiness, but the classification is not simple and varies with the concentration. A typical example of a sweet amino acid is glycine, which exhibits a refreshing sweetness. Amino acids of low molecular weight, such as L-alanine, L-serine and L-proline, are generally sweet. On the other hand, L-leucine, L-phenylalanine, L-tyrosine and L-tryptophan are bitter. However,

TABLE 6–3

Taste of amino acids

Aminoacid	Threshold (mg/dl)	Sweetness	Bitterness	Tastiness	Sourness	Saltiness
Sweet amino acids						
Glycine	130	+++				
Alanine	60	+++				
Serine	150	+++			+	
Threonine	260	+++	+		+	
Proline	300	+++	++			
Hydroxyproline	50	++	+			
Glutamine	250	++		+	+	
Bitter amino acids						
Valine	40	+	+++			
Leucine	190		+++			
Isoleucine	90		+++			
Methionine	30		+++	+		
Phenylalanine	90		+++			
Tryptophan	90		+++			
Arginine	50		+++			
Arginine hydrochloride	30	+	+++			
Histidine	20	+	+++			
Lysine	50	++	++	+		
Acidic amino acids						
Histidine	5		+		+++	
Asparagine	100		+	+	++	
Aspartic acid	3		+		+++	
Glutamic acid	5		+	++	+++	
Tasty amino acids						
Sodium aspartate	100			++		+
Sodium glutamate	50			+++		

the D-isomers of bitter amino acids are sweet. The bitterness of L-tryptophan is about half that of caffeine, and the sweetness of D-tryptophan is 35 times as strong as that of sucrose.[16)] Although L-aspartic acid and L-glutamic acid are acidic, in neutralized solutions they are tasty. The D-isomers of these amino acids are tasteless. From this relation between the stereostructure and taste of amino acids, it has been presumed that the taste of an amino acid is due to stereospecific "three-point adsorption" of the amino acid on the taste receptor on the surface of the tongue, as seen in enzyme-substrate complex formation.[17)] It is known that there is a synergistic action between amino acids and mononucleotides as regards taste.[18)] That is, the tastiness of sodium inosinate, which was identified as a taste substance of "*katsuobushi*" (dried bonito), is not very strong in itself, but is greatly enhanced by the addition of monosodium L-glutamate. The synergistic action of mononucleotides is also seen with monosodium L-aspartate.

The mechanism of action of peptides in food taste has not been clarified, but they are considered essential for increasing the "body" (complexity) of the taste. The taste of peptides seems to be independent of the types of constituent amino acids but dependent on the overall structure of the peptide. Dipeptides consisting of sweet amino acids are generally tasteless, whereas those containing bitter amino acids are bitter without exception.[19)] Carnosine (β-alanyl-L-histidine) and anserine (β-alanyl-L-methylhistidine), the major peptides in meat and fish, and glycyl-L-leucine, occurring in rice wine, are all bitter.

L-Aspartyl-L-phenylalanine methyl ester was found to have a very strong sweetness (more than 150 times that of sucrose) in the course of research on the synthesis of gastrin, a hormone stimulating gastric juices. This peptide is expected to find applications as an artificial sweetener.

B. Free amino acids in food

There have been many studies on the taste substances in natural foods.[20)] Free amino acids occurring in food which are thought to be responsible for the natural taste will be described briefly in this section. Fish with dark flesh contains high levels of histidine, while fish with white flesh contains less histidine and much more glycine and alanine. Although it is difficult to connect amino acid content to fish taste, it is well known that the amino acid content of fish becomes higher in the "season". Clams and snails contain large quantities of sweet amino acids such as glycine, alanine and proline. Proline content is high in squid, and proline, glycine and alanine occur at high levels in shrimp. In particular, the glycine content in shrimp reaches 1% and it is said to be a factor in the sweetness of shrimps. After glycine, arginine also occurs in large quantities but it seems to have no effect on the taste. Sea urchins, a delicacy in Japan, contain various amino acids. Glutamic acid exhibits tastiness, valine produces bitterness, and methionine is considered to be essential to give the specific taste of sea urchins.

On the other hand, the amount of free amino acids is less in animal meat, the total content of free amino acids being as low as 0.1–0.3%, while that in dark-colored fish flesh is 0.3–0.8%. Among the amino acids, glutamic acid, alanine and histidine are present at comparatively high levels. Beef extract which is essential for the manufacture of various foods such as soup stock is prepared by extracting beef at high temperature, and the amino acid content is reduced by the Maillard reaction with sugars during processing.

Cheese contains amino acids resulting from the hydrolysis of milk protein, and the peculiar taste of cheese arises from glutamic acid as well as bitter amino acids such as valine, leucine and methionine.

In vegetables and fruits, asparagine, glutamic acid, glutamine and serine are found at high levels. The content of citrulline is high in melons. In green tea, levels of glutamic acid, aspartic acid and arginine are high. The peculiar taste of green tea is due to teanine (L-glutamyl-γ-ethylamide) as well as glutamic acid. As for marine algae, glutamic acid is a major component (2–4% of dry weight) in the taste of *Laminaria*, and alanine (1%) as well as glutamic acid (0.6%) in *Porphyra*.

Soy sauce contains almost all amino acids, originating from soybean protein, and includes 20% glutamic acid, followed in amount by aspartic acid, lysine, leucine and proline. Soy paste contains glutamic acid as well as leucine and lysine. There are various amino acids in rice wine which give "body" to the taste. Proline occurs at a high level in beer.

As noted above, the relation between free amino acids and taste depends on the type of food. In certain foods, amino acids may determine the taste completely. Glycine in shrimp and methionine in sea urchins are examples where one or more amino acids form a special taste in combination with other taste substances. On the other hand, the complex taste of beef extract is due not to a particular amino acid but to a combination of various amino acids. The resulting taste is not simple but is universally acceptable so that it can be used as a basic flavoring for many types of food. Soy sauce may also belong to this category.

C. Utilization of amino acids as flavoring

Although the effect of amino acids in food taste is not simple, as noted above, it is clear that amino acids are important substances affecting taste. As the shelf life of foods has been improved recently and dependence on cooked foods has increased, the utilization of amino acids as food flavoring has also been extended. However, amino acids which are used separately are limited to monosodium L-glutamate and glycine. To obtain more complex tastes, products containing various amino acids have been developed for practical use to substitute for natural flavoring such as beef extract.

(*i*) *Monosodium L-glutamate*

Monosodium L-glutamate, which was proven to be a taste substance in *Laminaria*, has been considered to represent an essential taste factor not belonging to any of the four basic tastes; sweetness, sourness, bitterness and saltiness. It has recently been found that glutamic ions stimulate the taste receptor electrophysiologically.[21] This suggests that the concept of tastiness is physically rational, and monosodium L-glutamate is often defined as a flavor enhancer or flavor potentiator in the United States. That is, glutamate is considered to enhance or potentiate the natural flavor of foods. Actually, free glutamic acid occurs universally in large quantities in both animal and vegetable foods, so that it forms not a taste as such but a basis of food taste.

The taste threshold of monosodium L-glutamate is as low as 0.03%, while the values for sugar and salt are 0.5 and 0.2%, respectively. This clearly shows that glutamate exerts its effect even in very low concentrations. Monosodium L-glutamate also affects the taste of other constituents. It softens saltiness, for instance. In general, a suitable amount of sodium L-glutamate to be added to a food is 10–20% of the salt contained in it. Sourness and bitterness are softened and repressed by glutamate, while it enhances the "body" of the food by cooperation with sweetness and saltiness.

This action of monosodium L-glutamate is enhanced by the inosinic or guanylic acid occurring in food, as already mentioned. The effect of a compound flavoring consisting of 88% monosodium L-glutamate and 12% sodium inosinate is six times stronger than that of monosodium L-glutamate alone.

Monosodium L-glutamate was produced initially by hydrolysis of glutamate-rich and inexpensive proteins such as wheat gluten and soy-bean protein. The by-product, an amino acid mixture, may be called a chemically prepared soy sauce and has been effectively utilized as supplementary material for fermented soy sauce. However, the extraction method, relying on natural protein resources, is not suitable for the production of L-glutamic acid at low cost on a large scale.

After World War II, the fermentation technology being developed for antibiotics production offered the possibility of producing various naturally occurring physiological substances by fermentation. Since the possibility of L-glutamic acid production from glucose by fermentation was proposed in 1955, the fermentation method has all but completely replaced the extraction method. By using glucose, molasses, or acetic acid as carbon sources, and ammonia as a nitrogen source, L-glutamic acid can be accumulated efficiently in the medium.[22] The bulk of the annual worldwide production of monosodium L-glutamate, which is estimated at 200,000 tons, is produced by the fermentation method. Countries producing L-glutamate include Japan, the United States, Italy, France, Peru, Thailand, the Phillippines, Malaysia, Indonesia, Korea and China. Production in Japan is about 100,000 tons, of which 20% is exported. The production in other countries is mostly based on technology developed in Japan.

It is noteworthy that L-glutamic acid is also commercially produced from acrylonitrile by chemical synthesis based on the oxo reaction, the Strecker reaction, and preferential crystallization. In practice, chemical synthesis may be convenient if the plant is located in a petrochemical complex where raw materials are readily available. However, chemical synthesis is only economic on a large scale. For the production of L-glutamic acid, it may well be competitive with the fermentation method.

About 40% of the consumption of monosodium L-glutamate in Japan is for home use and 60% for business use. Sodium inosinate or guanylate is mixed with the product for home use to improve the flavor effect. As for business use, monosodium L-glutamate is extensively used in the manufacture of various foods such as fish paste products, instant noodles, fish sausages, meat products, soup stocks, mayonnaise, special fish delicacies, seasoned fish products and curry paste. In Southeast Asia, a considerable amount is consumed in domestic cooking, whereas it is mostly used for manufacturing foods such as dehydrated soup in Western countries.

(ii) Glycine

As glycine is refreshingly sweet, it has been used as a sweetening agent in food manufacture. When glycine was discovered in 1820, it was found to be so sweet that it was thought to be a sugar. The sweetness of 1% glycine solution is almost equal to that of 1% sucrose solution. Glycine exhibits a light bitterness and tastiness as well as sweetness. As noted above, glycine occurs at a high level in white fish, shrimps and clams, and it is recognized to be an important component of food taste.

The first example of the use of glycine in food was in a compound rice wine. When glycine was added to a compound rice wine containing a smaller amount of extract, an improved taste, full of "body", was obtained. At present, glycine is widely used in the manufacture of various foods such as soft drinks, soup stocks, pickles, fish paste products, seasoned fish products and fish delicacies. The amount added depends on the kind of food but is generally 0.1–1%. The purpose of adding glycine is not only to sweeten but also to give "body", or to improve the taste by softening saltiness and bitterness. When 1–2% glycine is added as a flavoring to fish paste products, deterioration by slime-forming bacteria is suppressed to some extent. It is thought that this effect is due to the repression of bacterial growth resulting from inhibition of biosynthesis of the cell wall in bacteria such as *Bacillus subtilis*, which cause slime formation.[23)]

Glycine is also used as an antioxidant for emulsifiers such as mono- or diglycerides. The antioxidant effect of some amino acids has been observed in model experiments using methyl linoleate, and a strong effect was observed with histidine and tryptophan.[24)] A similar effect on dry milk has also been proven in experiments using histidine.[25)]

Glycine is synthesized from formaldehyde by the Strecker reaction or from monochloroacetic acid by amination. As the synthetic process is simple and the production volume is large, it is obtainable at a low price. Annual production is about 3000 tons in Japan, the United States, France and the Netherlands. Demand for glycine comes from the medical and industrial fields as well as food manufacture.

(*iii*) *Amino acid mixtures*

The various preparations of amino acids which are produced by the hydrolysis of protein are widely used as flavorings for food manufacture. Hydrolyzed vegetable protein (HVP) and yeast extracts are examples. Products in which amino acid composition is artificially adjusted or to which various compounds other than amino acids are added have been developed recently.

Soy sauce, a traditional flavoring in Japan, and meat extracts, the base for processed foods in Western countries, contain various amino acids. Amino acid mixtures which are prepared by the hydrolysis of protein may be expected to substitute for such flavorings. It was known that yeast autolyzate could be employed as a flavoring from the end of the 19th century, and a flavoring prepared from the hydrolyzate of soybean protein was first used 50 yr ago. However, such products came into practical use all over the world after World War II.

The flavoring which is prepared from the hydrolyzate of vegetable protein is called HVP. Wheat gluten as well as soybean protein is used as raw material. Because hydrochloric acid and sodium hydroxide are used for hydrolysis and neutralization, respectively, the product contains a considerable amount of salt. There are various types of product, such as liquid, paste, powder and granules.

Yeast extract is prepared by the autolysis of brewery yeast or *Torula* yeast followed by separation and concentration of the extract. As it contains a lot of peptides and vitamins, it produces a stronger "body" than HVP, while its tastiness is weaker. HVP is produced and utilized mainly in the United States and Japan, while the demand for yeast extracts is greater in European countries. The annual production of these substances is estimated at 30,000 tons, calculated in terms of dry material. They are widely used to enhance taste and to give "body" to almost all processed foods, together with monosodium L-glutamate.

The effects of HVP and yeast extracts are increased by combination with amino acids and other organic acids. Amino acid additives are selected for their flavoring effect from the amino acids occurring in meat extracts and soy sauce. The product of reaction of cysteine and sugars is often added to substitute meat extracts as a special flavor constituent.

D. Special applications of amino acids in food processing

(*i*) *Amino acid and odor*

It has long been known that the odors resulting from the heat processing of food are due to degradation products arising from the reaction of amino acids and sugars, and the reaction is called the aminocarbonyl reaction, the Maillard reaction, or the browning reaction from the names of reacting groups, the investigator's name, or the effect (brown discoloration), respectively. This reaction is important in relation to the application of amino acids in food processing.

Odor substances formed during the heat treatment of amino acids include amines arising from the decarboxylation of amino acids and aldehydes produced by Strecker degradation, and such odors are unpleasant compared to those produced on heating amino acids in the presence of sugars. Amino acids react easily with sugars even at a comparatively low temperature, yielding not only various volatile compounds which are the major odor constituents of cooked foods but also the brown substance responsible for the coloration of baked foods.

It was recognized that aldehydes are formed by Strecker degradation when various amino acids are heated under a nitrogen stream in the presence of glucose.[26)] Of course, there are many other degradation products and the odor actually perceived is very complex and delicate. Odors such as caramel, bread and potato are produced from glycine, proline and methionine, respectively.[27)] In addition, the particular odor of each food is produced as a result of the formation of various volatile compounds by complex reactions of amino acids and sugars and/or fats in the food.

The odor of bread is formed by fermentation and baking, and in particular, the crust odor is due to the Maillard reaction during the baking process. The amino acid producing the bread-like odor is proline. It is known that a mixture of proline and glucose[28)] or proline and glycerine produces a substance having a strong bread-like or cracker-like odor. Odor substances such as 1-azabicyclo[3,3,0]oct-4-one and *N*-methyl-2-acetopyrrolidine have been isolated.[29)] It may be possible to use these compounds to enhance the odor of bread; and valine and phenylalanine are already used to improve the odor of rice crackers.

The Maillard reaction of cysteine is applied for the preparation of artificial cooked meat odor, as described in the following section.

(*ii*) *Cysteine*

As cysteine possesses an SH group in the molecule, it shows specific physicochemical properties not seen with other amino acids and it has a special effect on food odor, so that it is useful in food processing. In Japan, cysteine is permitted to be added to natural fruit juices as an antioxidant or to bread as a dough improver. Since free cysteine is easily oxidized, it is used in the form of the hydrochloride, which is comparatively stable.

Cysteine is effective as an antibrowning agent for food. It represses the formation of a colored substance, melanoidin, by reducing an intermediate in the browning reaction (cysteine itself), or by inactivating the intermediate (free radical formed from cysteine).[30] When natural fruit juice is stored, ascorbic acid is oxidized and eventually browning and reduction of the nutritive value occurs. In fact, ascorbic acid has antibrowning activity due to its reducing power, but browning proceeds when the ascorbic acid is completely oxidized by enzymes in fruit juice. This can be prevented by the addition of cysteine hydrochloride.

The reductive power of cysteine can also be used to improve the baking process and bread quality. As a result of examination of various agents in attempts to make bread of good quality in a shorter time, it was found that the use of a combination of cysteine hydrochloride as a reducing agent and potassium bromide as an oxidizing agent is advantageous, especially in a continuous baking process. This is now done in the United States, United Kingdom and Japan.[31] It is considered that cysteine produces increased dough flexibility by reacting rapidly with gluten to cause the opening of S–S linkages, while potassium bromide produces rigidity by slowly forming new S–S linkages. The amount of cysteine hydrochloride added is usually about 50 ppm to wheat flour.

So-called "meaty flavor" is due to various sulfur compounds. Studies aimed at the preparation of artificial meaty flavor by the heat treatment of amino acids and sugars have been carried out, and it was recognized that cysteine can produce a strong meaty flavor. As for sugars, pentoses, such as ribose and xylose are more reactive than hexoses. For example, a beef-like flavor is produced by heating a mixture of L-cysteine hydrochloride, glucose and ribose, followed by further treatment after soybean protein hydrolyzate and inosinic acid are added.

L-Cysteine hydrochloride is prepared by the electrolytic reduction of L-cystine. As L-cystine is difficult to prepare by either fermentation or chemical synthesis, it is produced by extraction from the insoluble fraction of human hair hydrolyzate in Japan, China and Germany. Annual production is estimated at about 100 tons.

References

1. The Advisory Committee on the Application of Science and Technology to Development, *Feeding the Expanding World Population: International Action to Avert the Impending Protein Crisis*, United Nations, 1968.
2. *Ed.* N. S. Scrimshaw and A. M. Attschul, *Amino Acid Fortification of Protein Foods*, MIT Press, 1971.
3. F. Wokes, *Plant Foods Human Nutr.*, **1**, 23 (1968).
4. *FAO Nutritional Studies*, no. 16, FAO, 1957.
5. *WHO Technical Report Series*, no. 301, WHO, 1965.
6. E. E. Howe, G. R. Jansen and G. W. Gelfillan, *Am. J. Clin. Nutr.*, **16**, 315 (1965).
7. E. E. Howe, G. R. Jansen and M. L. Anson, *ibid.*, **20**, 1134 (1967).
8. E. E. Howe, E. W. Gilfillan and M. Milner, *ibid.*, **16**, 321 (1965).
9. G. G. Graham, R. P. Placko, G. Acevedo, E. Morales and A. Cordano, *ibid.*, **22**, 1459 (1969).
10. G. G. Graham, E. Morales, A. Cordano and R. P. Placko, *ibid.*, **24**, 200 (1971).
11. J. P. Greenstein and M. Winitz, *Chemistry of the Amino Acids*, p. 245, Wiley 1961.
12. M. Winitz, J. Graff, N. Gallagher, A. Narkin and D. A. Seedman, *Nature*, **205**, 741 (1965).
13. M. Winitz, D. A. Seedman and J. Graff, *Am. J. Clin. Nutr.*, **23**, 25 (1970).
14. M. Winitz, R. F. Adams, D. A. Seedman, P. N. Davis, L. G. Tayks and T. A. Hamilton, *ibid.*, **23**, 546 (1970).
15. M. Yoshida, T. Ninomiya, S. Ikeda, S. Yamaguchi, T. Yoshikawa and M. Ohara, *Nippon Nogeikagaku Kaishi* (Japanese), **40**, 295 (1966); S. Ikeda, S. Yamaguchi and T. Yoshikawa, *Hinshitsukanri* (Japanese), **17**, (12), 69 (1966).
16. L. Solms, I. Vuataz and R. H. Egli, *Experientia*, **12**, 692 (1965).
17. T. Kaneko, *Kagaku to Kogyo* (Japanese), **24**, 864 (1971).
18. A. Kuninaka, *Nippon Nogeikagaku Kaishi* (Japanese), **34**, 489 (1960).
19. T. Ninomiya, N. Katsuya, J. Kirimura, A. Shimizu and A. Kimizuka, *J. Agr. Food Chem.*, **17**, 689 (1969).
20. M. Ohara, *Shokuhin no Aji* (Japanese), Korin Shoin, 1966.
21. A. Adachi, *Nippon Seirigaku Zasshi* (Japanese), **24**, 607 (1962).
22. S. Motozaki, *Kagaku Chomiryo* (Japanese), Korin Shoin, 1969.
23. K. Komagata, H. Ogawa, K. Fukushima and T. Ito, *Shokuhin Eiseigaku Zasshi* (Japanese), **9**, 289 (1968).
24. R. Marcuse, *J. Am. Oil Chemists Soc.*, **39**, 97 (1962).
25. H. Mitsuda, K. Yasumoto and K. Iwami, *Eiyo to Shokuryo* (Japanese), **18**, 217 (1965).
26. N. Kobayashi and M. Fujimaki, *Agr. Biol. Chem.*, **29**, 1059 (1965).
27. W. J. Herz and R. S. Shallenberger, *Food Res.*, **25**, 491 (1960).
28. T. Morimoto and J. A. Johnson, *Cereal Chem.*, **43**, 627 (1966).
29. I. R. Hunter, M. K. Walden, W. H. McFadden and J. W. Pence, *Cereal Science Today*, **11**, 493 (1966).
30. P. S. Song and C. O. Chichester, *J. Food. Sci.*, **32**, 107 (1967).
31. R. G. Henika, *Cereal Science Today*, **10**, 420 (1965); H. Fujiwara, *New Food Ind.* (Japanese), **11**, 50 (1969).

6.2 Utilization for animal feed

In section 6.1, the growing gap between the increase of world population and that of protein supply was mentioned, and the possibility of increasing the efficiency of protein utilization by fortification of cereals with amino acids was put forward. Amino acid fortification has already been put into practice in the manufacture of animal feed for raising livestock in the production of animal protein.

Food proteins can be divided broadly into two groups, animal and vegetable proteins, and the resources of the former, except for fish and shellfish which feed on plankton (although the production of fish by aquaculture has been increasing recently), are produced by biological conversion of the latter under artificial control in advanced nations. As the scale of livestock raising has been expanded and its efficiency improved, formula feed of homogenous quality and of high nutritive efficiency, consisting mainly of grains, fish meal and soybean meal as the calory and protein sources and supplemented by vitamins and minerals, is required in large quantities at low price.

Although animal feed manufacture has expanded rapidly in the advanced nations since the end of World War II and is now at the stage of maturity, such facilities will also be required in the future in developing nations. The annual production of formula feed in Japan was 16.5 million tons, worth 500 billion yen in 1972, and world production is probably 10 times greater than the production in Japan in terms of completely formulated feed.

The utilization of amino acids in feed manufacture has grown from year to year for formula feed. The kinds of amino acids utilized are limited but there is no doubt that feed manufacture will represent the biggest market for amino acids.

6.2.1 Demand for animal protein and formula feed

Eating habits in Japan have become fairly Westernized, and the ratio of animal protein in the diet has been increasing. According to a survey by the Ministry of Agriculture and Forestry, the supply of animal protein from livestock is 15 g per capita per day, corresponding to about 20% of the total protein consumption in 1970. This is not high, but the growth rate is bigger than that of fish and shellfish. Animal protein supplied by raising livestock in 1970 was three times greater than in 1960, whereas the increase for fish and shellfish was only 10%. Since there is a limit to fish and shellfish resources and the preference of Japanese consumers has been changing toward animal meat, the demand prospects for animal protein are good.

In Western countries, 60–70% of the total consumption of protein per capita per day is animal protein and the relative importance of fish and shellfish is very low. In other words, about 50 g of protein per capita per day is supplied as animal protein food.

The annual demand for animal protein in the world is estimated at 10 million tons. However, the turnover rate of feed protein to edible animal protein is as low as 20% (28% for eggs, 25% for milk, 19% for broilers, 18% for pork, and 12% for beef). Accordingly, an enormous amount of formula feed must be supplied to produce the necessary amount of animal protein, even if milk, beef and mutton are mainly produced from natural grass.

Although the production of formula feed in Japan in 1970 was 15.08 million tons, an increase of 13% over the previous 10 years (8.15 million tons in 1960), the growth rate has declined since. In 1970, 47% of the total production consisted of feed for laying hens (however, this proportion is tending to decrease), 25% for pig feed, and 10, 6 and 11% for broiler, beef and dairy cow feeds, respectively.

The production of formula feed in the United States and other Western countries is estimated at 60 million and 40 million tons, respectively. However, in these countries, farmers dealing with both agriculture and livestock raising often prepare formula feed for themselves, so more feed might actually be produced.

6.2.2 Feed formulation and amino acids

The development of formula feed is the most remarkable among a series of technical improvements involving breeding and feeding management. For the improvement of feed, a knowledge of the qualitative and quantitative requirements for nutrients of domestic animals and poultry in relation to the nutrient content of the raw materials for feed is essential. Since such studies are well advanced, it is possible to produce formula feed in a rational manner.

The formulation of feed which has optimum nutritional value and low cost is readily done by electronic computer. The formulation which completely satisfies the nutritional requirements and which has minimum cost is computed by the linear programming method.

(Condition)

(Condition)	total formulation	$X_1+X_2+\cdots+X_n=100$
	crude protein	$CP_1X_1+CP_2X_2+\cdots+CP_nX_n\geqq CP$
	lysine	$L_1X_1+L_2X_2+\cdots L_nX_n\geqq L$
	$\vdots$	$\vdots$
		$X_1\geqq 0,\ X_2\geqq 0,\cdots,\ X_n\geqq 0$

(Function to be solved) price $Y=P_1X_1+P_2X_2+\cdots+P_nX_n \rightarrow \min.$

X_i, formulation of each ingredient; CP, requirement for crude protein; CP_i, content of crude protein in each ingredient; L, requirement for lysine; L_i, content of lysine in each ingredient; P_i, price of each ingredient.

As for nutritional requirements, the standard values of the NAS–NRC (National Academy of Sciences–National Research Council) of the United States are commonly used.[1)] However there is still room for further improvements, and many studies have been carried out in Japan and other countries. As for the composition of amino acids in the feed materials, the tables of Block,[2)] the NAS–NRC[3)] and Comb[4)] are commonly used.

Protein is available from the following materials in formula feed:

Grains: These are not only a main energy source but also an important protein source since they are mixed into the feed in large quantities. Corn and sorghum, which are mainly produced in the United States, are important all over the world. Japan imports most of her requirements from the United States. Barley and wheat are also used in European countries and in Canada.

Oilseed meals: These are relatively important as protein sources for feed. Soybean meal, a by-product of soybean oil manufacture, is a universally important feed material. The main soy-producing countries are the United States and Brazil. Cottonseed meal (U.S. and India), peanut meal (India), sunflower meal (U.S.S.R. and Argentine), rapeseed meal (Canada), and sesame meal (India and Venezuela) also have local importance.

Fish meal: This is a major source of animal protein and has a great influence upon the economics of feed manufacture. Peruvian anchovy meal is the major factor in the world market, but other fish meals are also produced in the North Atlantic and North Pacific.

Since fish meal was once supplied in large quantities at a low price and the protein requirement was standardized at a rather higher level, a large quantity of fish meal was added to formula feed. As the protein content in fish meal is high and its amino acid composition is similar to the pattern required by domestic animals and poultry, amino acid deficiency has not been a problem when feed containing a large quantity of fish meal has been used. However, it is known that the amino acid balance is more important than the amount of protein, and excess protein is not necessary if the amino acid balance is satisfactory. Besides, shortages of fish meal often occur with the increase in formula feed production, and much feed has been switched to a vegetable-rich type from a fishmeal-rich type for economic reasons. Consequently, the amino acid composition in grains and oilseed meals has become a problem from the nutritional viewpoint. Grains are lacking in lysine without exception and soybean meal, in which the lysine content is comparatively high, contains less methionine. Lysine is also a limiting amino acid in other oilseed meals. Lysine and methionine are therefore added to such vegetable-based feeds.

TABLE 6–4

Essential amino acids in feed materials

Material	Crude protein (%)	Lysine (%)	Methionine (%)	Tryptophan (%)	Threonine (%)
Barley	9.0	0.29	0.12	0.10	0.30
Corn†	9.2	0.28	0.21	0.09	0.39
Sorghum	10.0	0.21	0.13	0.12	0.36
Soybean meal†	49.7	3.16	0.73	0.74	2.00
Cottonseed meal	43.0	1.53	0.55	0.48	1.20
Sesame meal	47.5	1.37	1.48	0.82	1.71
Peanut meal	47.0	1.45	0.37	0.41	1.10
Fish meal (anchovy meal)†	64.6	4.37	2.22	0.75	2.45

† The lysine and methionine were bioassayed to show "availability".[5)] Other data were obtained by chemical analysis.

6.2.3 Methionine

Fish meal and soybean meal are the most important protein sources for feed. The annual supply of fish meal worldwide has been over 4 million tons. Peru, which had an annual production of 2 million tons in the past, has made drastic restrictions on fishing to preserve fish resources. Reduced or static production is also expected in Norway, Chile and Denmark. Japan is currently importing 100,000 tons annually because local production of fish meal in the coastal waters and North Pacific is insufficient.

With the increased production of feed, the supply of fish meal has become inadequate, and the need to reduce the fish meal content in feed has resulted in the practice of supplementing feed with methionine. Since as long as 10 yr ago, there has been an increasing trend towards the use of vegetable protein in place of fish meal. The main source of this vegetable protein is soybean meal, which is highly nutritious and is available in large quantities at a low price. However, soybean meal is deficient in methionine (Table 6–4). If methionine is available in large quantities at a low price, its use as a feed supplement will be feasible. Techniques of methionine synthesis from acrolein, methylmercaptan, and hydrogen cyanide have been already established, and the development of the petrochemical industry has provided an industrial basis for methionine synthesis. Around 1965, full-scale industrialization of DL-methionine production was achieved successively in Japan, France and Germany. Subsequently, production has gradually increased, although there were some fluctuations. The worldwide annual production is over 50,000 tons at present. In addition to methionine, several thousand tons of the calcium salt of DL-methionine hydroxyl analogue, a derivative of methionine in which the amino group is replaced by a hydroxyl group, are produced in the United States.

It can be said that methionine is now essential in the formulation of feed as a regular ingredient, regardless of the price fluctuations of fish meal and soybean meal. Conversely, the value of soybean meal as a protein source for feed has been firmly established by its use in combination with methionine.

It is a matter of course that the equal nutritional values of D-methionine and L-methionine favored the practical use of DL-methionine. Methionine is the only amino acid which can be completely utilized in the D-form by animals, including human beings.

6.2.4 Lysine

Lysine is the next most important amino acid for feed manufacture after methionine. As the relative importance of vegetable feed materials has become greater and the amino acid composition has become an important factor in the formulation of feed, attention has been paid to lysine, for which the requirement is high but the content in the feed material is low.

Since the content of grains which are considered to be the energy source is as high as 60–80% of the feed, grains are also important as a source of protein. However, the lysine content in grains and also in oilseed meals, except for soybean meal, is very low (Table 6–4). Therefore, if fish meal is partly replaced by such oilseed meals, lysine becomes a limiting factor. Miscellaneous meals are comparatively cheap and form a useful vegetable protein source which may rank with soybean meal according to regional circumstances. A typical example of the use of lysine is seen in feed containing high levels of miscellaneous oilseed meals.

Short-term fattening of animals, especially pigs, has been achieved by the use of energy-rich, low-protein feed, and feed of high protein efficiency has been developed with due regard to amino acid composition. Since then, the addition of lysine to feed has also been put into practice.

An important point in considering the lysine content in feed is the nutritional availability of lysine. Among the amino acids in protein, only lysine possesses an amino group at the side chain. If compounds having carbonyl groups are present in the feed materials, the ε-amino group of lysine may react with the carbonyl group to form a Schiff base, and the lysine is no longer nutritionally available. Therefore, formulation must be carried out with due consideration to the amount of lysine actually available in the feed.

A major factor causing loss of lysine availability is heat treatment, which is carried out to improve the protein utilization efficiency. During this process, vegetable protein reacts with reducing sugars, and fish meal protein reacts with oxidation products of fats. It is known that the availability of methionine is also reduced by heat treatment. Although there have been many studies on assay methods for available lysine and methionine,[5)] the value fluctuates depending on the sample and it is very difficult to formulate feed on the basis of exact availability. However, to produce feed of higher efficiency, more attention must be paid to the availability of lysine, and consequently the requirement for lysine may be further increased.

Lysine is produced by fermentation technology developed in Japan. Unlike methionine, only L-lysine is physiologically active, D-lysine being inactive. Thus, synthetic methods for lysine production are at a disadvantage. Several years ago, a plan for the industrialization of lysine synthesis which had been examined for a long period was abandoned in the Netherlands. This was an apparently reasonable process using caprolactam as a starting material, but it was reported that there were technical problems in the optical resolution process.

Among amino acid fermentations, lysine fermentation has been industrialized on the next largest scale to glutamic acid fermentation. The annual production of lysine is expected to exceed 20,000 tons in the near future.

As described above, artificially produced methionine and lysine are utilized in formula feeds to supplement the methionine and lysine in the natural feed materials. The production of these amino acids for use as feed additives is certain to increase in the future, since the cost, availability and quality of natural feed materials fluctuate annually, whereas there is no such fluctuation with lysine and methionine produced industrially.

In this section, only amino acid balance has been taken into consideration, but there have also been many studies on amino acid utilization in animal feed from the physiological point of view. After methionine and lysine, it is expected that tryptophan and threonine may be utilized for feed in the future. Tryptophan is a limiting amino acid in corn, a representative feed grain. Its content is high in soybean meal but low in cottonseed and peanut meals. Threonine content is low in all grains and oilseed meals except for soybean meal. Therefore, providing economic and other factors permit, tryptophan and threonine may be used as additives for animal feed in the future. If the demand for these amino acids increases, techniques for their production by synthesis or fermentation at low cost will be developed.

References

1. NAS–NRC Publication 1192, *Nutrient Requirements of Swine*, 1964; NAS–NAC Publication 1345, *Nutrient Requirements of Poultry*, 1966.
2. R. J. Block and K. W. Weiss, *Amino Acid Handbook*, Thomas, 1966.
3. NAS–NRC Publication 1232, *Joint United States–Canadian Table of Feed Composition*, 1964.
4. G. F. Comb and H. Nott, *Feedstuffs*, p. 36 (1967–10–21).
5. G. F. Comb, E. H. Bossard and G. R. Childs, *ibid.*, p. 36 (1967–10–21).

6.3 Utilization as medicines and cosmetics

Recently, the production of pharmaceuticals has continued to grow year by year, and one outstanding feature is the increase in amino acid preparations. This wide use of amino acids as medicines is probably due in part to the fact that physiological, biochemical and nutritional research on proteins and amino acids has rapidly advanced, so revealing the important role of amino acids in maintaining life. A second reason for the increase is undoubtedly the technological progress made in recent years in the commercial production of amino acids. In addition to medicines, amino acids have also been utilized for their characteristic physicochemical properties in cosmetics.

6.3.1 Nutritional preparations

In the 1930's, Rose and others established the concept of "essential (indispensable) amino acids" and demonstrated that a diet including a mixture of essential amino acids as a nitrogen source could be used in place of proteins to grow rats[1] and to maintain nitrogen equilibrium in man.[2] After the work of Rose and his co-workers, research on protein and amino acid nutrition advanced rapidly, and our knowledge of not only essential amino acids but also of nonessential (dispensable) amino acids increased, providing wide scope for the beneficial use of amino acids in the nutritional and pharmaceutical fields.

A. Amino acid infusion

Orally ingested dietary proteins are broken down in the gastrointestinal tract to amino acids. These amino acids are absorbed into the portal blood stream, and finally utilized in various tissues and organs. The living body is in a dynamic state where synthesis and degradation of body proteins occur continuously, and thus an external supply of protein or amino acids is essential for maintenance of the nitrogen balance. Therefore, when oral ingestion of an adequate diet is impossible because of gastrointestinal disorders or other causes, amino acids should be supplied by parenteral administration.

Madden *et al.*[3,4] showed that an intravenously injected casein hydrolyzate was utilized in the living body, since plasma proteins were regenerated in protein-depleted dogs after infusion of the hydrolyzate. Following this, Madden *et al.*[5,6] and Silver *et al.*[7] demonstrated that a mixture containing the 10 essential amino acids for dogs determined by Rose, plus glycine, was effectively utilized in the same manner as casein hydrolyzate when administered intravenously to dogs. In the case of protein hydrolyzates, it is difficult to obtain a product of a certain desired amino acid composition. Moreover, protein hydrolyzates often contain impurities such as peptides and ammonia which are considered to be one cause of the nausea or vomiting that may accompany parenteral infusions. When chemically pure crystalline amino acids are used instead of protein hydrolyzates for infusion, a mixture having the desired amino acid composition can easily be prepared. Thus, the use of mixtures of pure amino acids in place of protein hydrolyzates has lately become widespread. Apart from this, Madden *et al.*[8] suggested that one of the causes of the vomiting accompanying infusion of protein hydrolyzates might be the presence in them of glutamic acid and aspartic acid. It has in fact been experimentally confirmed that an excess of these acidic amino acids can induce vomiting.[9-11]

All amino acids which occur in proteins are in the L-form, and the final form of amino acids utilized in the living body is also of the L-configuration. Although it is well known that certain D-amino acids are nutritionally available, they are utilized only after being transformed to the L-form. Therefore, a mixture of L-amino acids is considered to have greater availability than a mixture of DL-amino acids. In fact, Howe *et al.*[12] demonstrated that a decrease in the use of racemic amino acids and an increase in the use of the corresponding L-antipodes improved both the tolerance to and effectiveness of amino acid infusions. Based on these facts, and also as a result of the recent progress in amino acid production, mixtures containing only L-form amino acids have come into wider use for parenteral infusion.

Formerly, essential amino acids were the main components of the mixture for infusion, and only glycine was added as a nonessential amino acid to increase the nitrogen content. However, recent progress in the study of amino acid metabolism in the living body has disclosed the important physiological role of nonessential amino acids.[13-16] The nutritional significance of these amino acids has also stimulated the development of amino acid infusions containing various dispensable amino acids in addition to the essential amino acids,[17] and many such preparations are now commercially available. However, the optimum pattern of essential amino acids (A/E pattern) and the ratio of essential/nonessential amino acids (E/N ratio) for parenteral nutrition have not yet been defined. At present, the amino acid pattern of human milk protein, FAO reference pattern and Rose's minimum requirement pattern, which show high nutritive value when ingested orally, are used for parenteral infusions. However, some question still remains as to whether the optimum amino acid pattern for parenteral infusion should be taken as identical with that for oral administration. Recent studies indicate a broad coincidence between them, and further extensive research is anticipated to provide formulas for specific infusions having the most suitable amino acid pattern according to age, sex, and disease of the patient.

Calory intake influences amino acid metabolism. For example, dietary carbohydrates and fats have a suppressing action on the expenditure of protein as an energy source. This action is nutritionally well known and is termed "protein-sparing action".[18,19] In fact, it has been demonstrated that the addition of glucose to an amino acid infusion may improve the utilization of amino acids. Glucose is, however, highly reactive to amino acids (the Maillard reaction) and it is difficult to prepare a stable amino acid infusion that contains glucose. Recently, therefore, sugar-alcohols such as xylitol and sorbitol, which are stable in amino acid solutions, have often been used[20-22] as an energy source in the infusion preparation. Since these sugar-alcohols are incorporated into cells without insulin action,[23] are utilized efficiently as an energy source and have an antiketogenic action,[24-26] they can be employed favorably as an energy source for diabetic patients. The use of ethanol as the energy source of amino acid infusions has also been investigated,[20,27] and has already been successfully employed in practice.

Fat is a most effective energy source, having a higher caloric value than carbohydrates. Technical problems such as difficulties in the preparation of fat emulsions and of side-effects have been partially overcome. Fat emulsions are now on the market, and the combined use of amino acid and fat infusions is very effective for supplying adequate calories and nitrogen.

To increase the efficiency of the infused amino acids, preparations with added vitamins and minerals have also been placed on the market. In particular, when parenteral infusion is the only means for meeting nutritional requirements over a long period, nutritionally complete infusions containing amino acids, a calory source, vitamins, minerals and essential fatty acids are considered to be more efficient. With the solution of technical problems in their preparation, such kinds of complete parenteral infusions are expected to be used widely in practice.

B. Other nutritional preparations

Many kinds of amino acid preparations for oral administration are marketed as tablets, solutions and in powder form, and often contain vitamins and minerals.

Greenstein *et al.*[28-34] showed that nutritionally complete solution diets containing amino acids, a calory source, vitamins, minerals and essential fatty acids could ensure the normal growth, life-span and reproduction of experimental rats, and that these diets could maintain human beings in a positive nitrogen balance. Winitz *et al.*[35] obtained good results in feeding experiments of adult men over a 19-week period using this kind of chemically defined diet. Characteristics of such defined diets are (1) high nutritive efficacy, (2) complete water solubility, (3) variable composition of nutrients, and (4) complete digestibility without the assistance of digestive enzymes, adsorption at the upper section of the intestine and a resulting reduction in faecal elimination.[29] These characteristics have prompted the use of chemically defined diets as space foods, synthetic diets and special foods for patients with gastrointestinal diseases.

From the viewpoint of preventing contamination by microorganisms and pyrogens and other side-effects, oral infusions are much more readily prepared and administered than parenteral ones. Thus, since the oral administration of a chemically defined diet is safer and more natural, it seems that for some patients parenteral infusion can be replaced by oral infusion, and that increasing amounts of oral preparations will be used in general. Also, for clinical examinations, the pre-use of a chemically defined diet having a constant composition of nutrients is considered very advantageous, since its serves to standardize the nutritional and physiological condition of the patient.

In addition to the complete amino acid preparations described above, various other nutritional preparations containing special amino acids such as lysine, threonine and tryptophan, which may show a tendency to be deficient, are commercially available.

6.3.2 Therapeutic agents

From the pure nutritional viewpoint, amino acids are of major importance as protein constituents. However, clarification of the specific physiological role of individual amino acids has also brought about their therapeutic application in several cases, as follows.

A. Arginine

Although arginine is classified as a dispensable amino acid, it is one of the most important amino acids for human nutrition and was formerly recognized as an indispensable amino acid. It plays a physiologically significant role as a member of the urea cycle.

Administration of arginine increases arginase activity in the liver and results in a reduction of blood ammonia levels by enhancing the production of urea.[36] Thus, arginine is effective in the alleviation of hyperammonemia and liver disorders. In order to increase these effects, arginine is often used as a salt with aspartic acid or glutamic acid. In the same manner, ornithine, another member of the urea cycle, is also employed for ammonia detoxification.

Apart from this, arginine is known to be the main component of spermatic proteins and arginine deficiency causes hypospermia.[37] For this reason arginine administration is effective for the sterilitas virilis that results from arginine deficiency.

B. Aspartic acid

Aspartic acid was one of the first amino acids to be discovered, and it has received considerable attention as a medicine since the report on its physiological and therapeutic importance by Laborit and his co-workers[38] in 1958. A number of pathways exist for the metabolism of aspartic acid. It reacts with α-keto acids by transamination to form the corresponding amino acids used for protein synthesis. By this reaction, aspartic acid is converted to oxaloacetic acid, a member of the TCA cycle, and accelerates the cycle by linking with metabolic intermediates of carbohydrates and fats.[39] Aspartic acid is also concerned in the urea cycle, and exhibits a detoxifying action towards blood ammonia, where it is especially effective in combination with ornithine.[40] Aspartic acid is also used for the biosynthesis of precursors of nucleic acids such as orotic acid.

In recent years, the significance of the K^+ and Mg^{2+} in myocardial cells has attracted much interest, and aspartic acid was found to be an ideal "carrier substance" for them, active transport of the ions being shown to occur. Thus, potassium-magnesium aspartate is widely employed to assist recovery from fatigue, heart failure, liver disease or diabetes. Other salts of aspartic acid such as potassium aspartate are used in the treatment of hypopotassemia, calcium aspartate is used for alleviating the leg cramp syndrome, and ferrous aspartate for treating anemia.

C. Cysteine, Cystine

Although cysteine and cystine are dispensable amino acids, they possess sulfur atoms and can display a sparing action in place of methionine.[41] The physiological action of cysteine may be ascribed chiefly to its sulfhydryl group (–SH), and it protects SH-enzymes in the liver from attack by enzyme inhibitors. This amino acid also exhibits a detoxifying activity by the formation of mercapturic acid derivatives,[42] etheral sulfate combination with phenolic hydroxy group and conversion of cyanides to thiocyanates. It is considered that the radioprotective action of cysteine is due to the trapping of radicals by the sulfhydryl group[43] or the formation of disulfide bonds (–S–S–) with the free sulfhydryl group in proteins.[44] As cysteine preparations, the amino acid is often combined with sodium thiosulfate, sodium phosphate, sodium carbonate and sodium sulfite, and these preparations are used for the treatment or prevention of radiation-damage, leukocytopenia and medical toxicosis.

Derivatives of cysteine, such as *N*-acetyl-L-cysteine[45] and L-cysteine methyl ester, are employed as mucolytic agents in the treatment of bronchitis or nasal catarrh based on the ability of the sulfhydryl group of cysteine to reduce the viscosity of mucus. *N*-Acetyl-DL-homocysteine-thiolactone is employed for the treatment of hepatitis, radiation damage, allergic dermatitis, etc.

D. Glutamic acid

Glutamic acid is an important amino acid in the metabolism of the brain. Although both glutamic acid and γ-aminobutyric acid (GABA) can only scarcely pass through the blood-brain barrier, the former exists within the brain and is transformed to the latter on decarboxylation.[46] Further oxidation of the GABA leads to γ-amino-β-oxybutyric acid (GABOB).[47] Both GABA and GABOB play a physiologically significant role in the functioning of the brain.

Certain analogues of glutamic acid are thus used as medicaments for various neuropathic conditions. For example, dimethylaminoethanol acetylglutamate is employed in the treatment of the motor or memory disturbance resulting from cerebral vascular damage, encephalitis or the Down syndrome. GABA is employed in the treatment of memory and speech disturbances, hemiplegia and hypertension. GABOB is used for treating hemiplegia, memory and speech disturbances, essential neurogenous hypertension, epilepsy and arrested mental development. Glutamic acid is also linked with the TCA cycle, in addition to apartic acid, and can thus be effective in the treatment of hepatic coma.[48]

The amide of glutamic acid, glutamine, is employed as a medicament for gastric ulcers since this amino acid is concerned with the biosynthesis of hexosamine, a constituent of mucopolysaccharides, and is effective in the reparation of ulcerous tissue.

E. Glycine

Although glycine is itself rarely used as a drug, it is often applied as an improving agent for certain types of medicines. Thus, ferrous glycinate is used as a ferruginous preparation. Also, since glycine is an amphoteric electrolyte, it may be applied as an antacid in the form of its basic alminium salt.

Glycine was for a time used in the treatment of muscular dystrophy. It has also been reported that a combination of glycine, alanine and glutamic acid is effective for improving the symptoms of benign prostatic hypertrophy such as delayed micturition, residual urine or pollakiuria.[49]

F. Histidine

Histidine is a nonessential amino acid for adults, but an essential amino acid for infants.

It has been shown in animals that the administration of an amino acid mixture lacking in histidine may cause ulcerative enterocolitis, and that this disease may be rectified by the injection of histidine. Thus, this amino acid is used as a remedy for gastric and duodenal ulcers.[50)] The mechanism of the therapeutic action of histidine on ulcers is considered to be a combined effect of stimulation of the proliferation of mucous epithelia via the autonomic nervous system, reduction of motility in the gastrointestinal musculature, antacid action, regeneration of mucosa and hematosis.

Histidine is indispensable in the regeneration of hemoglobin, and it is well known that anemia may arise as a result of histidine deficiency.[51)] Recently, Gerber *et al.* have also reported that large oral doses of histidine may relieve the symptoms of rheumatoid arthritis.[52)]

G. Methionine

Methionine is a sulfur-containing amino acid and is known to play an important role in the metabolism of various sulfur-containing compounds in the body. The labile methyl group of methionine is closely concerned with choline formation by the mechanism of transmethylation.[53)] Choline itself is converted to phospholipid by binding with liver fat and shows lipotropic action. Methionine also detoxicates various toxic substances by methylation, and is therefore used in the treatment of hepatic failure, including cases of chronic and acute hepatitis, hepatocirrhosis and intoxication by arsenic, chloroform, carbon tetrachloride and benzene derivatives. Especially when it is employed as a lipotropic drug, methionine is often combined with inositol or choline.

Methionine sulfonium chloride (vitamin U) has a methyl group of high bond energy and is so a very effective methyl donor. While histamine is known to induce gastric ulcer formation, vitamin U is considered to inactivate histamine by methylation, and thus shows an antiulcer effect.[54)] It is in fact widely used in the treatment of peptic ulcers.

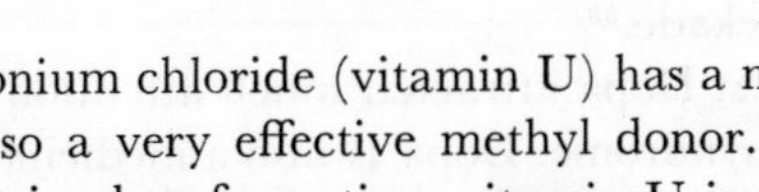

$$\left[(CH_3)_2 {}^{+}SCH_2CH_2\underset{\displaystyle NH_2}{\underset{|}{C}}HCOOH \right] Cl^-$$

vitamin U

H. Other amino acids and derivatives

(1) Methyldopa and Dopa

The recent development of research on the pharmacology and biochemistry of catecholamines led to the birth of a number of important therapeutic agents derived from aromatic amino acids. One such agent is L-α-methyl-3,4-dihydroxyphenylalanine, or "methyldopa", which is now widely employed as an antihypertensive medicament having relatively minor side-effects.

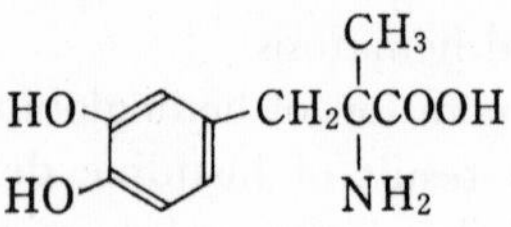

CH_3 / HO, HO / CH_2CCOOH / NH_2 — methyldopa

HO, HO / $CH_2CHCOOH$ / NH_2 — Dopa

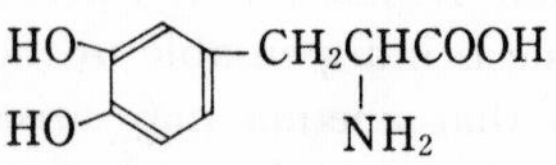

Methyldopa was first studied as an inhibitor of the enzymatic decarboxylation of dihydroxyphenylalanine (Dopa).[55] Then, based on the relationship between catecholamines and hypertension, the possibility that this inhibitor might display an antihypertensive effect was anticipated, and methyldopa was developed for clinical application.[56]

A number of hypotheses have been put forward concerning the mode of action of methyldopa. Its hypotensive action may not necessarily be related to any inhibitory effect towards aromatic amino acid decarboxylase.[57] On the other hand, certain metabolites of methyldopa have been implicated in its antihypertensive action.[58] An additional function of α-methyldopa is known whereby it reduces vascular resistance by a mechanism other than adrenergic nerve blockade.[59]

In the late 1960's, Dopa attracted much attention due to its efficacy in treating Parkinson's syndrome. Dopa is also a medicine related to catecholamine metabolism, i.e. in addition to α-methyldopa. The concentration of dopamine is significantly low in parts of the basal ganglia, particularly the striatum of patients with Parkinsonism. However, dopamine administered to raise the level in the brain does not reach the central nervous system since it cannot pass through the blood-brain barrier. Dopa, the immediate precursor of dopamine, was therefore investigated, and it has been widely used clinically. According to the report of Cotzias *et al.*,[60] administration of L-Dopa to patients with Parkinsonism improved symptoms such as akinesia and rigidity, and some patients responded dramatically. However, side-effects such as Huntington's chorea, psychotic disturbances, anorexia, nausea and orthostatic hypotension have been observed following the administration of L-Dopa. For L-Dopa treatment, large doses are required due to the fact that considerable amounts of the administered L-Dopa are degraded by tissue decarboxylase. The resulting degradation products also seem to cause side-effects. To solve this problem, a peripheral decarboxylase inhibitor, DL-α-methyldopahydrazine, is administered together with L-Dopa. By such combined treatment, the necessary dosage of L-Dopa is effectively reduced and the unwanted side-effects correspondingly minimized.

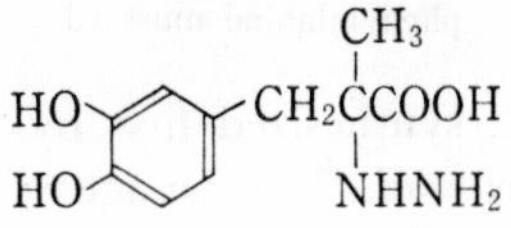

$$(HO)_2C_6H_3-CH_2C(CH_3)(NHNH_2)COOH$$

α-methyldopahydrazine

(2) Cycloserine

Cycloserine is an antibiotic which may be isolated from fermentation broths of *Streptomyces orchidaceus*[61] and *S. garyphalus*.[62] Its chemical structure has been determined as D-4-amino-3-isoxazolidone.[63,64] Cycloserine is used in the treatment of tuberculosis, but in large doses can cause mental disturbance. On the other hand, it has been demonstrated that the side-effects may be depressed by combined administration with DL-methionine or L-glutamic acid.

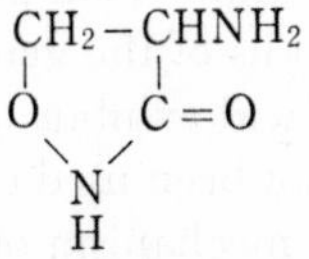

$$\begin{array}{ccc} CH_2 & - & CHNH_2 \\ | & & | \\ O & & C=O \\ & \diagdown N \diagup & \\ & H & \end{array}$$

cycloserine

The serine derivative, *O*-carbamyl-D-serine, has also been shown to have antibiotic activity. It exhibits a marked synergistic effect in combination with cycloserine.[65]

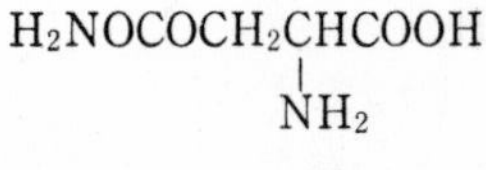

$$H_2NOCOCH_2\underset{\displaystyle NH_2}{\underset{|}{C}}HCOOH$$

O-carbamyl-D-serine

(3) Antitumor agents

The effectiveness of nitrogen mustard as an antitumor agent has been recognized for a long time, but its toxicity is very high. In order to reduce this toxicity, or to vary the selectivity of the drug against different kinds of tumors, binding of the nitrogen mustard to an amino acid has been studied. For example, phenylalanine mustard has been successfully used in practice for the treatment of myelomas.

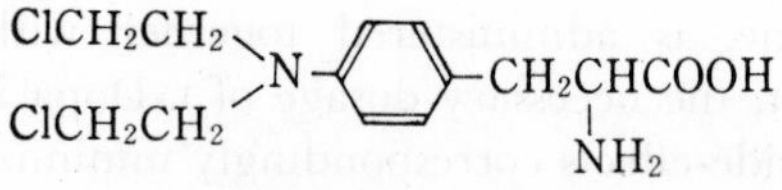

$$(ClCH_2CH_2)_2N-C_6H_4-CH_2CH(NH_2)COOH$$

phenylalanine mustard

Tryptophan mustard, synthesized first by DeGraw and Goodman,[66] has been shown by Robert *et al.* [67] to be active against plasmacytic myelomas, although it produces side-effects such as nausea and vomiting. They recommended the use of tryptophan mustard in the case of patients who failed to respond to treatment with L-phenylalanine mustard and cyclophosphamide.

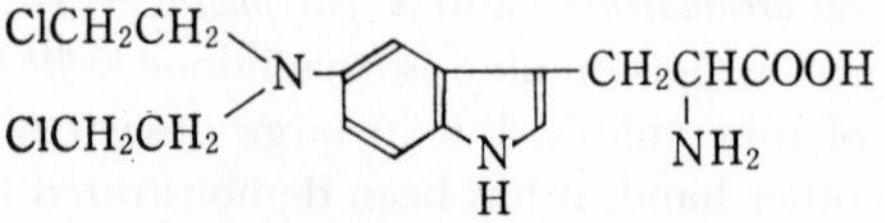

$$(ClCH_2CH_2)_2N-(\text{indolyl})-CH_2CH(NH_2)COOH$$

tryptophan mustard

Azaserine is an antitumor amino acid possessing a diazo group and was discovered in fermentation broths of the genus *Streptomyces* by Stock *et al.*[68] in 1954. Although this amino acid exhibits a very strong antitumor activity in animal experiments, it has not been used clinically due to its high toxicity. It has been suggested that the mechanism of antitumor activity is based on an inhibition of purine biosynthesis. As a structurally analogous substance, 6-diazo-5-oxonorleucine (DON) has been found to possess antitumor activity.

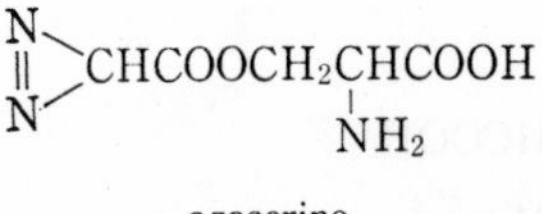

$$N_2{=}CHCOOCH_2CH(NH_2)COOH \qquad N_2{=}CHCOCH_2CH_2CH(NH_2)COOH$$

azaserine 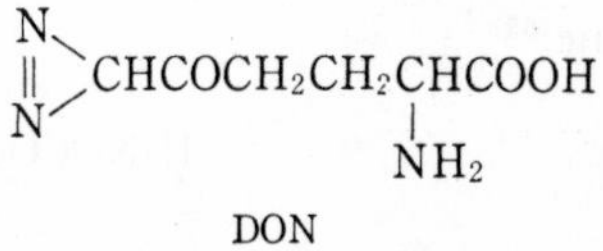 DON

In addition, based on the observation that L-asparaginase inhibits the growth of certain tumor cells that require asparagine, investigations on the antitumor activity of structurally related compounds of L-asparagine were undertaken, and *S*-carbamyl-L-cysteine was found to exhibit antitumor activity in animal experiments.[69]

$$\underset{\displaystyle NH_2}{H_2NOCCH_2\overset{|}{C}HCOOH}\qquad \underset{\displaystyle NH_2}{H_2NOCSCH_2\overset{|}{C}HCOOH}$$

L-asparagine S-carbamyl-L-cysteine

Thus, elucidation of the requirement of particular tumor cells for specific amino acids, and subsequent testing of the corresponding amino acid analogues as antitumor agents, can be expected to provide interesting and valuable information for the development of new and effective antitumor agents.

(4) ε-Aminocaproic acid

Plasmin is a kind of proteolytic enzyme formed in plasma from the inert precursor, plasminogen. The presence of plasmin may bring about destructive effects on proteins in the blood, tissues and cells. Plasmin also causes anaphylactic shock by the release of histamine or bradykinin.

Okamoto *et al.*[70] undertook the screening of a large number of compounds involving amino acids in a search for plasmin inhibitors, and found that basic amino acids such as lysine showed the greatest inhibitory effect. Further, they also tested the effect of lysine analogues, and from the results it was discovered that ε-aminocaproic acid (ε-ACA) was a very strong plasmin inhibitor. ε-ACA acts principally to inhibit the activation of plasminogen, but is less effective in inhibiting plasmin itself. ε-ACA is now widely employed as an antiplasminic agent.

$$H_2NCH_2CH_2CH_2CH_2CH_2COOH$$

ε-aminocaproic acid

Apart from this, it was also found that the inhibitory effect of *trans*-aminomethylcyclohexane carboxylic acid (*t*-AMCHA), synthesized as an analogue of ε-ACA, was some 8 times greater than that of ε-ACA.[71]

$$NH_2CH_2-\langle H \rangle-COOH$$

AMCHA

6.3.3 Peptide preparations

A large number of biologically active peptides has been isolated from various tissues and glands. Structural elucidation of many of them has been achieved by total synthesis. From studies of structure-activity relationships in these peptides, some products having stronger activity or reduced side-effects, when compared with the natural peptides, have been found among synthetic peptides with a simpler structure than the natural compounds or with certain of their component amino acids replaced by other amino acids. As a result of the recent improvements in techniques for synthesizing peptides, several new products have come into use as drugs in place of natural compounds. Accordingly, amino acids are expected to be increasingly employed for the synthesis of such peptides.

A. Cellular peptides

Among the various cellular peptides synthesized and stored in cells and extractable from tissues, only those of interest as medicines are discussed below.

(1) Oxytocin

Oxytocin is produced in the hypothalamus and stored in the pituitary portal system in addition to vasopressin. The main actions of oxytocin consist of a uterus-contracting effect and a milk-releasing effect. In particular, the former effect is employed clinically in the case of patients requiring early vaginal delivery of uterine inertia or for the purpose of inducing or shortening the delivery process.

Cys – Tyr – Ile – Glu(NH_2) – Asp(NH_2) – Cys – Pro – Leu – Gly(NH_2)

oxytocin

Oxytocin was first synthesized by duVigneaud *et al.*[72] in 1953. Recently, its synthesis has become possible by a solid phase procedure, and the synthetic oxytocin obtained has already been used as a remedy.

Of the great number of oxytocin analogues that have been prepared, 4-threonine-oxytocin,[73] in which a threonine residue replaces the glutamine residue at the 4-position, and 1-deamino-oxytocin (1-[β-mercaptopropionic acid]-oxytocin)[74,75] have been shown to be more potent oxytocic agents than oxytocin itself. The former is particularly advantageous for clinical use since its antiduretic action, which is one of the undesirable side-effects of oxytocin, is weak. Another oxytocin analogue, 4-deamino-oxytocin (4-glutamic acid oxytocin), acts mostly as a milk-ejecting agent, and its uterus-contracting effect is very weak. This analogue may therefore have application as a lactagogue with minimal side-effects.[76]

(2) Vasopressin

Vasopressin is a nonapeptide whose amino acid sequence is essentially similar to that of oxytocin, except that the two amino acids at positions 3 and 8 are different. In general, mammals have 8-arginine-vasopressin, although pigs and hippopotamuses excrete lysine-vasopressin, where the arginine is replaced by lysine.

Cys – Tyr – Phe – Glu(NH_2) – Asp(NH_2) – Cys – Pro – Arg – Gly(NH_2)

arginine-vasopressin

Vasopressin shows a hypertensive effect by its action on arteriolae and and capillary vessels. It also exhibits an antidiuretic action by increasing water reabsorption from renal tubules, promotes ACTH excretion and has an oxytocic effect.

Both lysine-vasopressin[77)] and arginine-vasopressin[78)] have been synthesized by duVigneaud and his co-workers. Since the former can be prepared more easily than the latter, synthetic forms of lysine-vasopressin or its analogues are generally used for treatment. In particular, 1-deamino-8-lysine-vasopressin shows a greater antidiuretic action and lesser pressor activity. It is therefore suitable for treating diabetes insipidus.[79,80)]

Other analogues having considerably higher antidiuretic action, such as 1-deamino-2-phenylalanine-8-arginine-vasopressin,[80,81)] 1-deamino-8-D-γ-aminobutyric acid-vasopressin,[82)] 1-deamino-8-D-arginine-vasopressin and 4-α-aminobutyric acid-8-lysine-vasopressin,[83)] are known. On the other hand, vasopressin is rarely applied as a vasocontrictor due to the accompanying coronary constriction. 2-Phenylalanine-8-lysine-vasopressin[84,85)] and 2-phenylalanine-8-ornithine-vasopressin[86)] both show a selective pressor effect and may be employed for the treatment of shock.

B. Circulating peptides

Circulating peptides are found in the circulating fluids of the body, mainly the blood, and may be divided into two groups, hypertensive peptides and hypotensive peptides.

(1) Angiotensin

Angiotensin I is produced from the angiotensionogen that occurs in serum globulin by the action of renin released by the kidney. Angiotensin I is itself inactive, but the action of a peptidase known as "converting enzyme" releases the two *N*-terminal amino acids (His–Leu) to form the octapeptide, angiotensin II.

Asp – Arg – Val – Tyr – Ile – His – Pro – Phe

angiotensin II

Angiotensin II is a strong vasopressor substance which causes greatest constriction in the arteriolae. It is also known that almost all smooth muscles, such as of the intestine and uterus, are constricted by angiotensin II, in addition to the blood vessels. Angiotensin II is thus clinically effective for maintaining blood pressure during anesthesia and surgery due to its hypertensive effect, and synthetic products such as 1-asparagine-angiotensin II may also be used.[87] Since the unnatural compound (in which L-aspartic acid is replaced by the corresponding D-isomer) resists the action of peptidase, this product is about 50% more active and its pressure effect 2–3 times longer than natural angiotensin II.[88]

(2) Plasmakinins

Kinins released from plasma proteins by the action of proteinases (kallikrein, plasmin, snake venom, trypsin, etc.) are generally known collectively as plasmakinins. The most representative kinins are bradykinin, kallidin and methionyl-lysylbradykinin, which resemble each other as regards their amino acid sequences. That is to say, kallidin and methionyl-lysylbradykinin are nonapeptides where lysine or methionyllysine, respectively, is bound in the *N*-terminal position of bradykinin.

Arg – Pro – Pro – Gly – Phe – Ser – Pro – Phe – Arg
bradykinin

These plasmakinins all show similar physiological action, causing the contraction of extravascular smooth muscles such as of the ileum and uterus. They also have a powerful vasodilator effect and act hypotensively by increasing vascular permeability.

In addition to the above three peptides, eledoisin, fisalamin and caerulein are known to be active peptides which show similar activity. In particular, eledoisin obtained from the posterior salivary glands of the Mediterranean eight-armed cephalopod, *Eledon*, shows a strong hypotensive effect and stimulant action on smooth muscle. Numerous analogues of eledoisin have been synthesized due to the ease of preparation of such peptides in a pure state. Among them, the nonapeptide formed by removal of the *N*-terminal dipeptide from eledoisin shows a more potent effect or longer activity than the natural compound.[89]

There are no recorded examples of direct clinical application of plasmakinin action, although it is known that kinin levels increase during anaphylactic shock. Thus, in order to prevent shock, antikinic agents are being developed from various standpoints, based on antagonism to kinin, activation of kininase, or inactivation of kinin. It may also be possible in the future to utilize various synthetic peptides that resemble kinin in structure as medicines.

C. Peptide antibiotics

Since the discovery of penicillin, research on antibiotics has rapidly advanced and a great number of microbially produced antibiotics are now used for the treatment of various bacterial infections. Many of these antibiotics are peptide antibiotics which generally possess D-amino acids or special amino acids not occurring in proteins. They also often possess a substituting group at the *C*- or *N*-terminal position or have a cyclic structure. Penicillin can be considered as a kind of peptide, and several varieties of semi-synthetic penicillin having a different side chain structure are commercially available.

More typical peptide antibiotics such as gramicidin S, colistin A and polymyxin B are produced by microorganisms of the genus *Bacillus*. They are widely used in combatting bacterial infections according to the respective antibacterial spectrum. These antibiotics are also effective for preventing postoperative infections. The peptide sequences comprise about ten amino acids and the structure is relatively complicated. They are therefore produced by fermentation at present. However, if synthetic procedures become more economical to operate than fermentative ones, due to technical advances in peptide synthesis, it can be expected that the synthetic compounds will be used medicinally. By the synthesis of analogues of peptide antibiotics, structure-activity relationships will become clearer, and new synthetic peptide antibiotics having different characteristics from the natural ones produced by fermentation will undoubtedly be found. They may include effective peptides which are readily synthesized and of simple structure, or peptides showing a wider antibacterial spectrum, stronger activity or reduced side-effects.

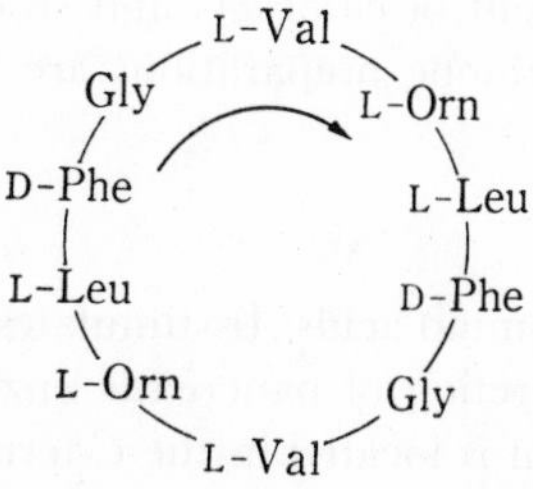

```
           L-Val
    Gly            L-Orn
 D-Phe                L-Leu
 L-Leu                D-Phe
    L-Orn          Gly
           L-Val
```

5,10-glycine-gramicidin S

```
           D-Val
    Gly            D-Orn
 L-Phe                D-Leu
 D-Leu                L-Phe
    D-Orn          Gly
           D-Val
```

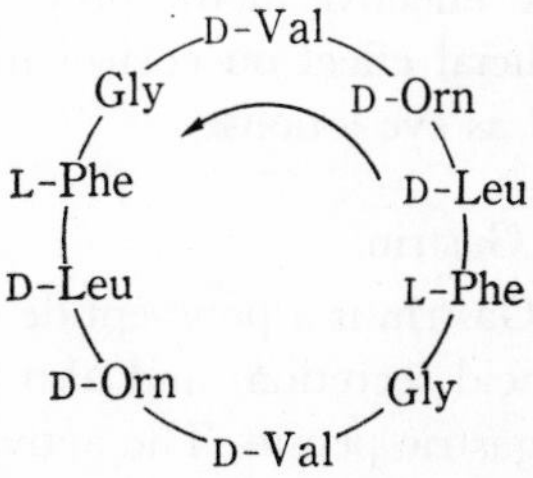

retroenantio-5,10-glycine-gramicidin S

Studies along these lines have already been extensively carried out, and it has been found for example that the peptide with the two proline residues of gramicidin S replaced by glycine is more active against *Bacillus subtilis*[90] than is natural gramicidin S. Further, it has been demonstrated that both retro and retroenantio forms of 5,10-glycine-gramicidin S show a high antimicrobial activity.[91] These facts may well provide new insights for elucidating the mode of peptide antibiotic action.

D. Other peptides

(1) Glutathione

The discovery of glutathione (γ-L-glutamyl-L-cysteinylglycine) dates back to 1935, when it was successfully synthesized by Harington and Mead.[92] Glutathione occurs widely in all living tissues and blood, and its physiological function covers a wide spectrum.

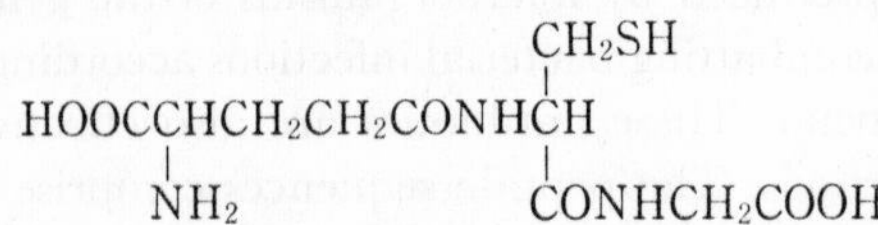

glutathione

The main action of glutathione depends on its sulfhydryl group, and it shows cysteine-like action in the activation and stabilization of enzymes, protection of enzymes from inhibitors, etc. On the other hand, it displays an antipoisoning effect based on enzymatic conjugation with certain toxic substances by the sulfhydryl group and their excretion as mercapturic acid derivatives. Glutathione is also related to the oxidation-reduction systems of the living body.

Since glutathione possesses the various physiological actions described above, it is produced by chemical synthesis or extraction from yeasts and used as a drug in the treatment of liver diseases, medical toxicosis, gestosis and allergic diseases, and for protection against radiation damage. Since it is also effective in the prevention and treatment of cataracts and shows a beneficial effect on cornea metabolism, glutathione preparations are marketed as eye lotions.

(2) Gastrin

Gastrin is a polypeptide consisting of 17 amino acids. It stimulates gastric acid secretion, and also promotes the secretion of pancreatic enzymes and gastric pepsin. The active center of gastrin is located in the *C*-terminal tetrapeptide (Trp–Met–Asp•Phe–NH_2).[93]

Pentapeptides obtained by combining this tetrapeptide with *tert*-butyloxycarbonyl-β-alanyl residue also exert gastrin-like activity.[94] Such synthetic pentagastrin is not used at present as a therapeutic agent, although the peptic acid secretion rate induced by it is greater than that induced by gastrin, and the response in the former case is highly reproducible. Moreover, its adverse side-effects are considered to be extremely few. Therefore, pentagastrin is currently employed as a diagnostic reagent in assessing the patient's ability to excrete gastric juices.

Gastrin has been identified in tumor cells of patients with the Zollinger-Elison syndrome. Injection of gastrin may cause duodenal ulcers due to its effect on gastric acid excretion. Thus, for the purpose of developing therapeutic agents for treating duodenal ulcers, substances which can depress the excretion of gastric acid have been screened. As a result, gastron, which occurs widely in the gastric mucosa, was discovered. Its possible application as a medicine is now being investigated.

Besides the above substances, ACTH, insulin and glucagon are known to be physiologically active peptides. Their chemical structures have been confirmed and total synthesis of each of them has already been achieved. The latter is difficult to carry out economically, however, since these substances are polypeptides with many constituent amino acids. On the other hand, as in the case of other physiologically active peptides, it seems possible that synthetic products having rather simpler structures may perhaps have applicability if they exhibit the same activity as the natural products or some new, characteristic activity. For example, although ACTH is a peptide composed of 39 amino acids, related peptides having an identical amino acid sequence to the first 19 or more residues (counting from the *N*-terminal of ACTH) exert similar activity to natural ACTH.[95] In fact, the synthetic peptide composed of 24 amino acids is already used for treating asthma, collagen diseases, skin diseases and burns.

6.3.4 Cosmetics

Amino acids are an important factor in the development of effective beauty treatment. For example, amino acids exhibit a buffering action that helps maintain normal skin function by regulating pH, and a protective action against bacteria. The amino acids used in hair cream or hair tonic act as hair nutrients and so increase hair lustre and softness.

Recently, the so-called natural moisturizing factor (NMF) has also attracted much attention in connection with cosmetics. One former theory was that skin became dry due to the loss of fat materials from the corneum layer, and therefore superfatting has been undertaken in cosmetic products for rectifying dry skin. However, recent progress in cosmetics chemistry has shown that dry skin is due to the loss of water-soluble material (named NMF) from the corneum layer. This NMF plays an important role in maintaining water in the corneum layer, and its main components are free amino acids such as glycine, threonine, alanine, aspartic acid, glutamic acid, serine and pyrrolidonecarbonic acid.

As mentioned above, amino acids are closely concerned in the regulation of normal skin function, and so various amino acids, their derivatives, and protein hydrolyzates have been utilized in cosmetics. Many methods have been patented, such as (1) the application of enzymatic hydrolyzates of lactalbumin[96)] or plankton[97)] to cosmetics, (2) the use of condensation products obtained by the reaction of amino acids with sugars as moisturizing agents for cosmetics,[98)] and (3) the addition of cystine dimethyl ester, ethyl glycinate and diethyl glutamate as stabilizers for cosmetic products containing amino acids or protein hydrolyzates as active ingredients.[99)] In addition to this application of amino acids as moisturizing agents for cosmetic products such as skin cream and tooth-paste, they are also used in remanufactured cellulose films, paper products and tobacco.

One of the weak points in the application of amino acids for cosmetics is that absorption of the amino acids from the skin is difficult due to their hydrophilic properties. In an attempt to solve this problem, a method employing amino acids and alkaloids together was proposed as an effective way to facilitate absorption of the amino acids via the skin.[100)]

Cysteine exhibits a significant effect on hair, and so many studies on the application of this amino acid to cosmetics have been carried out and a number of useful derivatives developed. Thioglycolic acid has been employed as a reducing agent for the cold-waving of hair, and cysteine or its derivatives may be used for a similar purpose.[101,102)] Other preparations reported include an antidandruff shampoo containing cysteine or cysteamine derivatives as the main ingredient,[103)] an oral preparation for seborrhoea containing cysteine derivatives[104)] and a preparation that combines the keratinizing action of acetylcysteine with the antiseborrhoeic action of pyridoxine.[105)] In addition to this, glutathione (as well as cysteine) inhibits the formation of melanin pigment. The mechanism of inhibition is considered to be due to complex formation between the sulfhydryl group and Dopa, a precursor of melanin. This effect may thus mark the first step in the use of glutathione for cosmetics.

Asparagine and aspartic acid derivatives are also used as ingredients of cosmetics. For example, the employment of aspartic acid derivatives as surface-active agents,[106,107)] and of the aspartic acid salt of a photosensitive pigment for the treatment of keratodermatitis,[108)] has been reported. Addition of aspartic acid or its derivatives, together with vitamin B_6, to cosmetic bases has been suggested to be effective in preventing ageing of the skin or in revitalizing old skin.[109] The relationship of aspartic acid to the TCA cycle has been described above.

6.3.5 Use as synthetic materials for medicines

A. Pyridoxine (vitamin B_6)

Pyridoxine (I) is one of the vitamins related to amino acid metabolism, and it is now produced by chemical synthesis. In 1962, Harris *et al.* discovered a new synthetic method for pyridoxine starting from alanine,[110] although the method was at the time considered unsuitable for industrial application due to difficulty in the synthesis of the oxyazole derivative (II). However, various improvements of this method have been published recently and its industrialization has been realized.[111]

$CH_3CH(NH_2)COOH$ (alanine) ⟶ $CH_3CH(NHCHO)COOC_2H_5$ $\xrightarrow{P_2O_5}$ $CH_3-C=C-OC_2H_5$ (N, O, CH ring) II

$\xrightarrow{CHCOOC_2H_5 \,\|\, CHCOOC_2H_5}$ HO–, CH_3–, N pyridine with $COOC_2H_5$, $COOC_2H_5$ $\xrightarrow{\text{reduction}}$ HO–, CH_3–, N pyridine with CH_2OH, CH_2OH I

The most advantageous method suitable for industrialization[112] utilizes oxazole derivative obtained by the cyclization of ethoxyallylalanine ester (III) with phosgene and triethylamine as a means of avoiding the difficult oxazoline synthesis of the previous method. Synthesis of an oxazole deriva-

$CH_3CH(NH_2)COOH$ $\xrightarrow{C_2H_5OH,\ (COOH)_2}$ $CH_3CH(NHCOCOOC_2H_5)COOC_2H_5$ (III) $\xrightarrow{COCl_2,\ (C_2H_5)_3N}$ $CH_3-C=C-OC_2H_5$ (N, O, C ring with $COOC_2H_5$)

$\xrightarrow{\text{i) hydrolysis; ii) } CH_2-O-CH=CH-CH_2-O \text{ ring with } CHCH(CH_3)_2}$ HO–, CH_3–, N pyridine with CH_2OH, CH_2OH I

tive from aspartic acid instead of alanine, followed by Diels-Alder reaction to form the pyridine ring and subsequent decarboxylation, also yields pyridoxine.

B. Pteroylglutamic acid (folic acid)

Pteroylglutamic acid is a vitamin that is effective against pernicious anemia. The following method for its synthesis has been reported.[113]

$$HOOCCH(CH_2CH_2COOH)NHCO\text{-}C_6H_4\text{-}NH_2 + BrCH_2CHBrCHO + \text{2,4,5-triamino-6-hydroxypyrimidine}$$

$$\longrightarrow HOOCCH(CH_2CH_2COOH)NHCO\text{-}C_6H_4\text{-}NHCH_2\text{-(2-amino-4-hydroxypteridin-6-yl)}$$

pteroylglutamic acid

C. Pantothenic acid

Pantothenic acid is a derivative of β-alanine and occurs widely in both animal and plant tissues. It is employed as an additive for feed and is synthesized in large quantities as follows.[114]

$$\text{(pantolactone: } CH_2C(CH_3)_2CH(OH)CO\text{, ring-closed through O)} + NH_2CH_2CH_2COOC_2H_5$$

$$\longrightarrow HOCH_2C(CH_3)_2CH(OH)CONHCH_2CH_2COOH$$

D. Ergotamine

Ergotamine is used as a drug since it exerts a uterus-constricting action, accelerates delivery and has a hemostatic effect on intrapartum hemorrhage. Ergotamine possesses L-prolyl-L-phenylalanine lactam as its intramolecular structure, and this lactam is used as a starting material in the synthesis of ergotamine.[115]

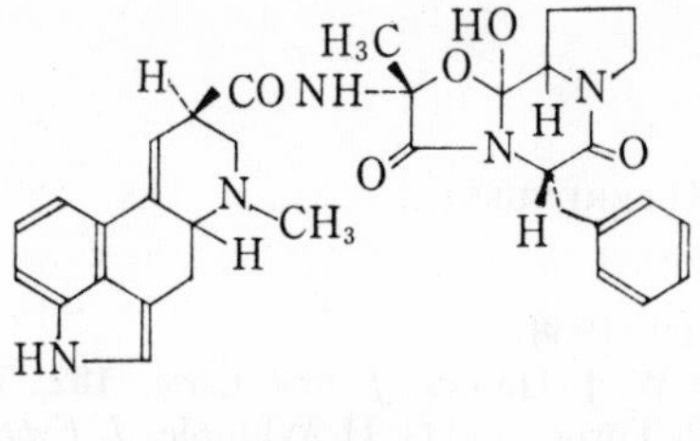

ergotamine 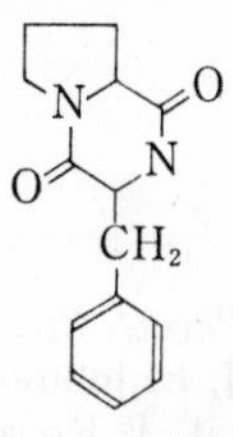 L-prolyl-L-phenylalanine lactam

The above account has summarized the principal areas of utilization of amino acids and their derivatives for medicines and cosmetics, covering both present and possible future applications. The use of amino acids as medicines is expected to broaden further as the specific physiological functions of individual amino acids become clearer based on new research into protein and amino acid metabolism in various diseases. For example, clarification of amino acid metabolism in bacteria, tumor cells and their hosts will facilitate the development of medicines composed of amino acids or their derivatives for use as metabolic antagonists. It is also expected that specific infusions containing the most suitable amino acid pattern for various particular diseases will be available in the future.

On the other hand, based on technological advances in peptide synthesis, peptide hormones and physiologically active peptides of relatively low molecular weight have already been placed on the market. Particularly in the case of peptide hormones, it is anticipated that many more active substances will be synthesized and that products having specific, higher or long-lasting activity will be found.

As one method of utilizing amino acids for medical purposes, combination with other known drugs is being studied, as seen in the case of certain anticancer drugs. Such combinations may not only alleviate the effects of drug toxicity but also provide the drugs with enhanced selectivity or specificity.

Application of the characteristics of amino acids by combining them into cosmetic preparations has been widely attempted in recent years. However, the overall number of amino acid derivatives examined for their possible use in cosmetics is still limited, and their application in this field seems very promising.

References

1. W. C. Rose, *Physiol. Rev.*, **18**, 109 (1938).
2. W. C. Rose, J. E. Johnson and W. J. Haines, *J. Biol. Chem.*, **182**, 541 (1950),
3. S. C. Madden, L. P. Rowe, A. P. Turner and G. H. Whipple, *J. Exptl. Med.*, **73**, 727 (1941).
4. S. C. Madden, A. A. Kattus Jr, J. R. Carter, L. L. Miller and G. H. Whipple, *ibid.*, **82**, 181 (1945).
5. S. C. Madden, J. R. Carter, A. A. Kattus Jr., L. L. Miller and G. H. Whipple, *ibid.*, **77**, 277 (1943).
6. S. C. Madden, R. R. Woods, F. W. Shull and G. H. Whipple, *ibid.*, **79**, 607 (1944).
7. R. H. Silver, A. O. Seeler and E. E. Howe, *J. Biol. Chem.*, **164**, 639 (1946).
8. S. C. Madden, R. R. Woods. F. W. Shull, J. H. Remington and G. H. Whipple, *J. Exptl. Med.*, **81**, 439 (1945).
9. L. W. Roth, R. K. Richards and F. R. Steggerda, *Proc. Soc. Exptl. Biol. Med.*, **62**, 284 (1946).
10. S. Levey, J. E. Harroun and C. J. Smyth, *J. Lab. Clin. Med.*, **34**, 1238 (1949).
11. C. O. Rice and J. H. Strickler, *Am. J. Clin. Nutr.*, **5**, 11 (1957).
12. E. E. Howe, K. Unna, G. Richards and A. O. Seeler, *J. Biol. Chem.*, **162**, 395 (1946).
13. W. C. Rose, M. J. Oesterling and M. Womack, *ibid.*, **176**, 153 (1948).
14. D. V. Frost, *Protein and Amino Acid Requirements of Mammals* (ed. A. A. Albanese), p. 50, Academic Press, 1950.
15. L. H. Breuer Jr., W. G. Pond, R. G. Warner and J. K. Loosli, *J. Nutr.*, **82**, 499 (1964).
16. V. R. Young and J. Zamora, *ibid.*, **96**, 21 (1968).
17. P. Juergens and D. Dolif, *Klin. Wochschr.*, **46**, 131 (1968).
18. D. H. Calloway and H. Spector, *J. Nutr.*, **56**, 533 (1955).
19. W. S. Thomson and H. N. Munro, *ibid.*, **56**, 139 (1955).
20. W. Griem and K. Lang, *Klin. Wochschr.*, **38**, 336 (1960).
21. W. Griem and K. Lang, *ibid.*, **38**, 951 (1960).
22. K. H. Baessler, W. Toussaint and G. Stein, *ibid.*, **44**, 212 (1966).
23. K. H. Baessler and W. Prellwitz, *ibid.*, **42**, 94 (1964).
24. N. L. Edson, *Biochem. J.*, **30**, 1862 (1936).
25. R. K. Haydon, *Biochim. Biophys. Acta*, **46**, 598 (1961).
26. K. H. Baessler and G. Dreiss, *Klin. Wochschr.*, **41**, 12 (1963).
27. C. O. Rice, B. Orr and I. Enquist, *Ann. Surg.*, **131**, 289 (1950).
28. J. P. Greenstein and M. Winitz, *Chemistry of the Amino Acids*, vol. 3, p. 245, Wiley, 1961.
29. J. P. Greenstein, S. M. Birnbaum, M. Winitz and M. C. Otey, *Arch. Biochem. Biophys.*, **72**, 396 (1957).
30. S. M. Birnbaum, J. P. Greenstein and M. Winitz, *ibid.*, **72**, 417 (1957).
31. S. M. Birnbaum, M. Winitz and J. P. Greenstein, *ibid.*, **72**, 428 (1957).
32. M. Winitz, S. M. Birnbaum and J. P. Greenstein, *ibid.*, **72**, 437 (1957).

33. J. P. Greenstein, M. C. Otey, S. M. Birnbaum and M. Winitz, *J. Natl. Cancer Inst.*, **24**, 211 (1960).
34. R. B. Couch, D. M. Watkin, R. R. Smith, L. E. Rosenberg, M. Winitz, S. M. Birnbaum, M. C. Otey and J. P. Greenstein, *Fed. Proc.*, **19**, 13 (1960).
35. M. Winitz, J. Graff, N. Gallagher, A. Narkin and D. A. Seedman, *Nature*, **205**, 741 (1965).
36. J. S. Najarian and H. A. Harper, *Am. J. Med.*, **21**, 832 (1956).
37. J. Tanimura, *Bull. Osaka Med. School*, **13**, 84 (1967).
38. H. Laborit, R. Moynier, R. Coirault, J. Thiebault, G. Guist, P. Niaussat, J. M. Jouany and C. Baron, *Presse Méd.*, **66**, 1307 (1958).
39. A. Meister, H. A. Sober, S. V. Tice and P. E. Fraser, *J. Biol. Chem.*, **197**, 319 (1952).
40. F. Salvatore and V. Bocchini, *Nature*, **191**, 4789 (1961).
41. W. C. Rose and R. L. Wixon, *J. Biol. Chem.*, **216**, 763 (1955).
42. J. A. Stekol, *ibid.*, **138**, 225 (1941).
43. D. G. Doherty, W. T. Burnett Jr. and R. Shapira, *Radiation Res.*, **7**, 13 (1957).
44. L. Eldjarn and A. Pihl, *ibid.*, **9**, 110 (1958).
45. A. L. Sheffner, *Ann. N. Y. Acad. Sci.*, **106**, 298 (1963).
46. E. Roberts and S. Frankel., *J. Biol. Chem.*, **190**, 505 (1951).
47. S. Seo, *Med. J. Osaka Univ.*, **7**, 833 (1957).
48. J. M. Walshe, *Lancet*, **264**, 1075 (1953).
49. F. Damrau, *J. Am. Geriat. Soc.*, **10**, 426 (1962).
50. A. G. Weiss and E. Aron, *Compt. Rend. Soc. Biol.*, **112**, 1530 (1933).
51. E. S. Nasset and V. H. Grate-Wood, *J. Nutr.*, **53**, 163 (1954).
52. D. A. Gerber and N. Y. Brooklyn, *Clin. Res.*, **17**, 352 (1969).
53. J. A. Stekel, E. I. Anderson and S. Weiss, *J. Biol. Chem.*, **233**, 425 (1958).
54. R. Suzue, *Bull. Inst. Chem. Res. Kyoto Univ.*, 201 (1967).
55. T. L. Sourkes, *Arch. Biochem. Biophys.*, **51**, 444 (1954).
56. J. A. Oates, J. Gillespie, S. Udenfriend and A. Sjoerdsma, *Science*, **131**, 1890 (1960).
57. C. C. Porter, J. A. Totaro and C. M. Leiby, *J. Pharmacol. Exptl. Therap.*, **134**, 139 (1961).
58. M. D. Day and M. J. Rand, *Brit. J. Pharmacol.*, **22**, 72 (1964).
59. S. Mohammed, T. E. Gaffney, A. C. Yard and H. Gomez, *J. Phamacol. Exptl. Therap.*, **160**, 300 (1968).
60. G. C. Cotzias, P. S. Papavisiliov and R. Gellene, *New Engl. J. Med.*, **280**, 337 (1969).
61. R. L. Harned, P. H. Hidy and E. K. LaBaw, *Antibiot. Chemother.*, **5**, 204 (1955).
62. D. A. Harris, M. Ruger, M. A. Reagan, F. J. Wolf, R. L. Peck, H. Wallich and H. B. Woodruff, *ibid.*, **5**, 183 (1955).
63. F. A. Kuehl Jr., F. J. Wolf, N. R. Trenner, R. L. Peck, R. H. Buhs, I. Putter, R. Ormond, J. E. Lyons, L. Chaiet, E. Howe, B. D. Hunnewell, G. Downing, E. Newstead and K. Folkers, *J. Am. Chem. Soc.*, **77**, 2344 (1955).
64. P. H. Hidy, E. B. Hodge, V. V. Young, R. L. Harned, G. A. Brewer, W. F. Phillips, W. F. Runge, H. E. Stavely, A. Pohland, H. Boaz and H. R. Sullivan, *ibid.*, **77**, 2345 (1955).
65. N. Tanaka and H. Umezawa, *J. Antibiotics (Tokyo) Ser. A*, **17**, 8 (1964).
66. J. DeGraw and L. Goodman, *J. Org. Chem.*, **27**, 1395 (1962).
67. A. K. Robert, P. P. Carbon, J. J. Lynch, A. H. Owens, G. Costa, R. T. Silver,

J. Cuttner, J. B. Harley, L. A. Leone, B. I. Shnider and J. F. Holland., *Cancer Res.*, **27**, 510 (1964).
68. C. C. Stock, D. A. Clarke, H. C. Reilly, C. P. Rhoads and S. M. Buckley, *Nature*, **173**, 71 (1954).
69. R. H. Adamson, *ibid.*, **217**, 751 (1968).
70. S. Okamoto, S. Oshiba, H. Mihara and U. Okamoto, *Ann. N. Y. Acad. Sci.*, **146**, 414 (1968).
71. S. Okamoto, S. Saito, Y. Takada and U. Okamoto, *Keio J. Med.*, **13**, 177 (1964).
72. V. duVigneaud, C. Ressler, J. M. Swan, C. W. Roberts, P. G. Korsoyannis and S. Gordon, *J. Am. Chem. Soc.*, **75**, 4879 (1953).
73. M. Manning and W. H. Sawyer, *Nature*, **227**, 715 (1970).
74. D. Jarvis and V. duVigneaud, *Science*, **143**, 545 (1964).
75. B. M. Ferrier, D. Jarvis and V. duVigneaud, *J. Biol. Chem.*, **240**, 4264 (1965).
76. I. Photaki and V. duVigneaud, *J. Am. Chem. Soc.*, **87**, 908 (1965).
77. M. F. Bartlett, A. Jöhl, R. Roeske, R. J. Stedman, F. H. C. Stewart, D. N. Ward and V. duVigneaud, *ibid.*, **78**, 2905 (1956).
78. V. duVigneaud, D. T. Gish, P. G. Katsoyannis and G. P. Hess, *ibid.*, **80**, 3355 (1958).
79. R. D. Kimbrough, W. D. Cash, L. A. Branda, W. Y. Chan and V. duVigneaud *J. Biol. Chem.*, **238**, 1411 (1963).
80. R. L. Huguenin and R. A. Boissonnas, *Helv. Chim. Acta*, **49**, 695 (1966).
81. E. Stürmer, R. L. Huguenin, R. A. Boissonnas and B. Berde, *Experientia*, **21**, 583 (1965).
82. M. Zaoral, J. Kolc and F. Šorm, *Collection Czech. Chem. Commun.*, **32**, 1250 (1967).
83. D. Gillessen and V. duVigneaud, *J. Biol. Chem.*, **242**, 4806 (1967).
84. R. A. Boissonnas and S. Guttmann, *Helv. Chim. Acta*, **43**, 190 (1960).
85. A. Tsakiris and A. Buehlmann, *Deut. Med. Wochschr.*, **88**, 46 (1963).
86. R. L. Huguenin, *Helv. Chim. Acta*, **47**, 1934 (1964).
87. W. Rittel, B. Iselin, H. Kappeler, B. Riniker and R. Schwyzer, *ibid.*, **40**, 614 (1957).
88. B. Riniker and R. Schwyzer, *ibid.*, **47**, 2357 (1964).
89. L. Bernardi, G. Bosisio, F. Chillemi, G. de Caro, R. de Castiglione, V. Erspamer, A. Glaesser and O. Goffredo, *Experientia*, **20**, 306 (1964).
90. H. Aoyagi, T. Kato, M. Ohno, M. Kondo and N. Izumiya, *J. Am. Chem. Soc.* **86**, 5700 (1964).
91. M. M. Shemyakin, Y. A. Ovchinnikov, V. T. Ivanov and I. D. Ryabova, *Experientia*, **23**, 326 (1967).
92. C. H. Harington and T. H. Mead, *Biochem. J.*, **29**, 1602 (1935).
93. H. J. Tracy, *Nature*, **207**, 1356 (1965).
94. S. Konturek and M. I. Grossman, *Gastroenterology*, **50**, 650 (1966).
95. C. H. Li, J. Meienhofer, E. Schnabel, D. Chung, T. B. Lo and J. Ramachandran, *J. Am. Chem. Soc.*, **82**, 5760 (1960); **83**, 4449 (1961).
96. Nestlé Products, *British Patent* No. 1050756 (1966).
97. J. S. L. Marissal, *French Patent* No. 1536017 (1967).
98. Kolmar Cosmetics A. G., *British Patent* No. 1004774 (1965).
99. Shiseido Co., Ltd., *German Patent* No. 1492121 (1969).
100. A. Yamada and Y. Hagihara, *Japanese Patent* No. 40–14317 (1965).
101. Gillette Co., *German Patent* No. 958501 (1957).
102. Mead Johnson and Co., *Belgian Patent* No. 649282 (1964).

103. L'OREAL, *Belgian Patent* No. 727534 (1969).
104. L'OREAL, *Belgian Patent* No. 711049 (1968).
105. Société d'Études et de Recherches Pharmaceutiques, *French Patent* No. 1477452 (1965).
106. G. Kalopissis, G. Vanlerberghe and A. Visut, *Parfum. Cosmet. Savons*, **10**, 293 (1967).
107. G. Kalopissis, G. Vanlerberghe and A. Visut, *Arch. Biochim. Cosmetol.*, **9**, 87 (1966).
108. K. Suzue, T. Yamashiki, T. Banno and S. Yasui, *Japanese Patent* No. 43–3884 (1968).
109. T. Taniguchi, *Japanese Patent* No. 40–24640 (1965).
110. E. E. Harris and R. A. Firestone, *J. Org. Chem.*, **27**, 2705 (1962).
111. F. Hoffmann-La Roche and Co., *Netherlands Patent Appl.* No. 6508673 (1966).
112. I. Maeda, M. Takehara, T. Togo, S. Asai and R. Yoshida, *Bull. Chem. Soc. Japan*, **42**, 1435 (1969).
113. C. W. Waller, B. L. Hutchings, J. H. Mowat, E. L. R. Stockstad, J. H. Boothe, R. B. Angier, *et al.*, *J. Am. Chem. Soc.*, **70**, 19 (1948).
114. R. J. Williams, H. K. Mitchell, H. H. Weinstock and E. E. Snell, *ibid.*, **62**, 1784 (1940).
115. P. A. Stadler, A. J. Frey, H. Ott and A. Hoffmann, *Helv. Chim. Acta.*, **47**, 1911 (1964).

6.4 Utilization in the chemical industry

Utilization of amino acids in the chemical industry as raw materials for various compounds has been studied from many viewpoints. However, little use has been made of amino acids as yet, and real development remains for the future.

In the chemical industry, the bifunctional nature of amino acids is an important feature. For example, a compound having anionic surface-active properties can be prepared by introducing a lipophilic group on the amino group. Another example of the utilization of bifunctionality is the synthesis of polyamino acids. Amino acids have chelating action and also act as antioxidants. Typical chelating agents such as EDTA and NTA are structural analogues of amino acids.

6.4.1 Surface-active agents derived from amino acids

A surface-active agent is a compound having a lipophilic group and hydrophilic group in the same molecule. As an amino acid has two hydrophilic groups, i.e. the amino and carboxyl groups, two kinds of surface-active compounds can be prepared by introducing lipophilic groups on one of the functional groups. *N*-Acylamino acids which are prepared by the reaction of an amino acid with a chloride of a higher fatty acid are anionic surface-active agents, while amino acid esters prepared by the reaction of amino acids with a higher alcohol are cationic surface-active agents. One example of such a surface-active agent which is now widely used is sodium laurylsarcosinate. It is added to tooth-paste and shampoo since it has a strong bacteriocidal action as well as a foaming action.[1]

A. Anionic surface-active agents

Various types of *N*-acylamino acids (anionic surface-active agents) can be prepared, as properties such as pH, surface tension, foaming power and emulsifying power can be modified by changing the number of carbon atoms in the acyl group and the amount of base used to neutralize the carboxylic group. Among such agents, acylglutamate (I) has excellent properties.[2] The lauryl group is often used as an acyl group and sodium hydroxide or triethanolamine can be used for neutralization.

$$\mathrm{HOOCCH_2CH_2\underset{\displaystyle NHCOR}{\underset{|}{C}}HCOOH} \qquad \mathrm{RCO = C_8 \sim C_{18}\ acyl\ group}$$

I

Sodium acylglutamate has excellent washing, foaming and emulsifying properties, is less irritative and has a softer "feel" because the pH of its aqueous solution is 5.5–6.0, which is almost the same as that of human skin, whereas the pH of soap solution is weakly alkaline. It is also more effective with hard water than soap. Since its oral toxicity is low and it is easily decomposed by biological action, it is very safe and does not cause environmental pollution. Sodium acylglutamate is used as a pressed "syndet" bar, and as a base or ingredient of shampoos, detergents and cosmetics. The addition of acylglutamate to detergents synthesized from mineral oil or alcohol is useful to protect human skin from chapping by kitchen detergents, since the irritating action of the detergents is softened by acylglutamate.[3)]

B. Cationic surface-active agents

It is known that some cationic surface-active agents derived from basic amino acids have strong bacteriocidal or bacteriostatic power. For example, palmitoyl-L-lysyl-L-lysine methyl ester hydrochloride (II), a dipeptide of lysine, is known to exhibit a broad antibacterial action.[4)] It was first studied as a model of polymyxin, a peptide antibiotic.

$$C_{15}H_{31}CONH\underset{\substack{|\\(CH_2)_4\\|\\NH_2}}{C}HCONH\underset{\substack{|\\(CH_2)_4\\|\\NH_2}}{C}HCOOCH_3\cdot 2HCl$$

II

Arginine derivatives (III) prepared by acylation of the α-amino group followed by esterification of the carboxylic group with a lower alcohol, e.g. lauroyl-L-arginine ethyl ester hydrochloride, are excellent antiseptics for foods and cosmetics.[5)]

$$\begin{matrix} HN \\ H_2N \end{matrix}\!\!>\!C\!-\!NHCH_2CH_2CH_2\underset{\substack{|\\NHCOR_2}}{C}HCOR_1\cdot HCl$$

III

$R_1 = -NH_2,\ -OCH_3,\ -OC_2H_5$

$COR_2 = C_8 \sim C_{18}$ acyl group

Higher alcohol γ-esters of glutamic acid (IV) are also said to be useful antiseptics having antibacterial action.[6)] This is an example of an intramolecularly neutralized detergent.

$$ROOCCH_2CH_2\underset{\substack{|\\NH_3^+}}{C}HCOO^-$$

IV

$R = C_8 \sim C_{18}$ alkyl group

6.4.2 Utilization of amino acids for polymer synthesis

Polyamino acids (VI) are obtained when *N*-carboxylated anhydrides (V) of amino acids, prepared by the reaction of phosgene and the amino acid, are polymerized by a suitable procedure.

$$H_2NCHRCOOH \xrightarrow{COCl_2} \underset{\text{V}}{\begin{matrix} CHR-CO \\ | \qquad\quad \\ NH-CO \end{matrix}\!\!>\!O} \xrightarrow{\text{polymerization}} \underset{\text{VI}}{H-[HNCHRCO]_n-OH}$$

Although polyamino acids have a skeletal structure similar to that of protein, their properties are different in many respects because they contain only one kind of amino acid, unlike proteins. (Copolymers can also be synthesized, but are not practical.) However, the structures of polyamino acids are similar to those of natural protein fibers such as silk and leather. They have therefore been considered as materials for synthetic fibers or leather and have been successfully developed for practical use.

Diaminomonocarboxylic acids such as lysine can be used as diamines for polymer synthesis. For example, lysine is converted to the diisocyanate and employed as a material for polyurethane synthesis.

A. Polyamino acids

The utilization of polyamino acids as materials for synthetic fibers or film and for coating vinyl leather was proposed by Woodward 20 yr ago. Subsequent studies on poly-L-alanine by Du Pont[7] and on poly-L-glutamate ester by Courtaulds[8] were of great interest. They hoped to produce a substitute for silk but unfortunately these materials did not appear to be economically viable, and studies were discontinued. Since then, a reexamination of polyglutamate has resulted in its use as a coating agent for synthetic leather in Japan.

Polymethyl L- or D-glutamate (PMG) (VII) is synthesized by the polymerization of *N*-carboxyl-γ-methyl-L- or D-glutamate anhydride (VIII), which is prepared by the treatment of γ-methyl-L- or D-glutamate with phosgene[9] in the presence of a suitable polymerization initiator in a siutable solvent.[10]

$$\underset{\text{VIII}}{\begin{matrix} H_3COOCCH_2CH_2-CH-CO \\ | \qquad\quad \\ NH-CO \end{matrix}\!\!>\!O} \xrightarrow{\text{polymerization}} \underset{\text{VII}}{H-[NH\overset{CH_2CH_2COOCH_3}{\overset{|}{C}H}CO]_n-OH}$$

PMG synthesized in this way has a highly regular linear structure and dissolves in halogenated hydrocarbons to form a right-handed α-helical structure in the case of L-PMG and a left-handed one with D-PMG. On the other hand, DL-PMG does not form a structure of high regularity and does not give a polymer of high molecular weight. As PMG becomes insoluble after separation from the solvent, it is necessary to choose a suitable solvent and to use the polymer solution directly for molding.

A wet spinning method is preferred for producing PMG fibers. The "feel" and appearance of PMG fibers are very similar to those of silk rather than those of other synthetic fibers. The equilibrium moisture content is low but the moisture-absorbing capacity and moisture-releasing rate are high. The physical properties remain unchanged even on heating at 150°C and the light resistance is excellent. However, such fibers have not yet been produced on an industrial scale.[11)]

PMG film has been put to practical use as a coating agent for synthetic leather. When a 10% solution of PMG in 1,2-dichloroethane–perchloroethylene is spread over the surface of vinyl leather, a synthetic leather with a "feel" as good as that of natural leather is obtained. Since its resistance to heat, water, chemicals and dirt, as well as its moisture penetration and dyeing properties, are excellent, synthetic leather treated with PMG has been widely used for bags, furniture, clothes and footwear.

The practical use of polyamino acids has only just started, and thus there is considerable scope for the future. Polyamino acids are expected to become one of the major growth points in the industrial utilization of amino acids.

B. Polyurethane

Lysine and ornithine may be useful as raw materials for polyamides or for diisocyanate from which polyurethane is prepared. Polyurethane is usually obtained by treatment of toluenediisocyanate with various diols and is widely used in polyurethane foam, adhesive agents and paints. Various diisocyanates have been examined in attempts to improve the properties of polyurethane. The diisocyanates prepared from esters of lysine and ornithine are examples. Methyl-2,6-diisocyanate caproate (X), for example, is a liquid of high boiling point obtained by the treatment of L-lysine methyl ester dihydrochloride (IX) with phosgene in *o*-dichlorobenzene at 150°C, followed by distillation.[12)]

$$\underset{\textbf{IX}}{H_2NCH_2CH_2CH_2CH_2\underset{\displaystyle NH_2\cdot 2HCl}{\underset{|}{C}}HCOOCH_3} \xrightarrow{COCl_2} \underset{\textbf{X}}{OCNCH_2CH_2CH_2CH_2\underset{\displaystyle NCO}{\underset{|}{C}}HCOOCH_3}$$

The resulting isocyanate has a smaller toxicity, due to its lower vapor pressure, than toluenediisocyanate. Physical properties such as solubility and dyeability are controlled by selective saponification of the ester groups in polyurethane, and the resulting product is a colorless material, and is not degraded even after exposure to ultraviolet light for a long period.

It has also been reported that moisture retention and dyeability are improved by the uptake of amino acid residues into nylon-6 polymer, if amino acid is added during the caprolactam polymerization. As amino acids are available at a fairly low price, studies on the utilization of amino acids in polymer synthesis are expected to continue.

C. Utilization as additives to synthetic resins

Some amino acid derivatives have been used as additives to synthetic resins in an attempt to stabilize the resins or to prevent the build-up of static charge. Synthetic fibers and resins easily acquire static charges, a characteristic which causes problems in processing due to dust adsorption and discomfort during use of the products. The utilization of amino acid derivatives having surface-active action has been examined in an attempt to prevent charge build-up on synthetic fibers and resins. Higher alkylphosphates of dialkyl esters of aspartic or glutamic acid provide examples of such derivatives. When fibers such as nylon or polyester are air-dried after treatment with a 0.4% methanolic solution of dibutylaspartate monolaurylphosphate (XI), the electric resistance and frictional charge build-up are greatly reduced.[13]

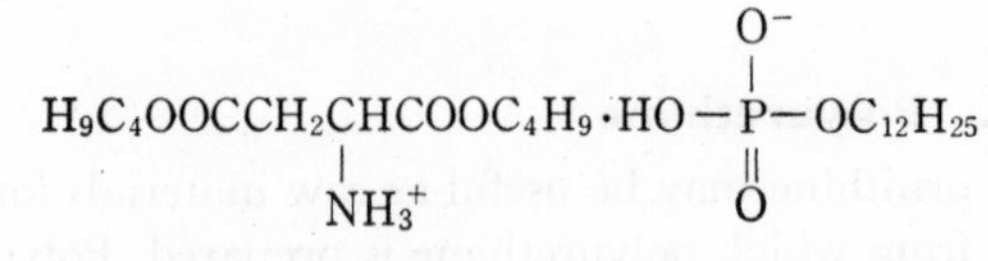

$$H_9C_4OOCCH_2\underset{\displaystyle NH_3^+}{\underset{|}{C}}HCOOC_4H_9\cdot HO-\overset{\displaystyle O^-}{\overset{|}{\underset{\displaystyle O}{\underset{\|}{P}}}}-OC_{12}H_{25}$$

XI

Antistatic agents which are derived from amino acids and which are soluble in polyolefin resins have been studied to improve the antistatic properties of these resins. The condensation product (XII), which is prepared by the treatment of leucine lauryl ester with ethylene oxide, is an excellent antistatic agent for polyethylene.[14]

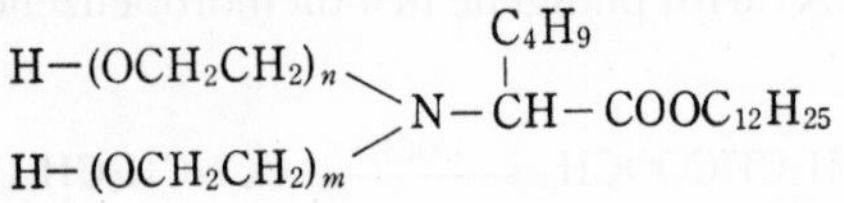

$$\begin{matrix} H-(OCH_2CH_2)_n \searrow \\ H-(OCH_2CH_2)_m \nearrow \end{matrix} N-\overset{\displaystyle C_4H_9}{\overset{|}{C}}H-COOC_{12}H_{25}$$

XII

There have been many patents sought for such agents, as shown in Table 6–5. The possibility of using amino acid derivatives as plasticizers in

TABLE 6–5

Antistatic agents derived from amino acids

Amino acid derivative	Example	Ref.
Amino acid ester	lysine ethyl ester	(15)
Aminosulfate of amino acid ester	*N*-butyl • butylamate acid • dibutyl glutamate	(16)
Higher alkylsulfate of amino acid or its ester	lysine monostearyl sulfate	(17)
	dibutyl aspartate cetylsulfate	(18)
N-Substituted acidic amino acid	*N,N*-dimethyl aspartic acid	
N-Acylglutamic acid ester		(19)
Amino acid amide	alanine dodecylamide	(20)
Pyrrolidonecarboxylic acid ester	2-pyrrolidone-5-carboxylic acid-2-ethylhexyl ester	(21)

TABLE 6–6

Amino acid derivatives used in the processing of synthetic polymers

Amino acid derivative	Use	Ref.
Nickel salt of amino acid	stabilizer (for polypropylene)	(22)
Metal salt of *N*-acylmethionine (*N*-lauroylmethionine–strontium)	stabilizer (for polyolefin)	(23)
	dye improver	
Glutamic acid ester (dioctyl glutamate)	plasticizer (for polyvinyl chloride)	(24)
N-Acylamino acid amide	plasticizer (for polyvinyl chloride)	(25)
N-Acylamino acid	butadiene polymer manufacture	(26)
N-Lauroylsarcosine	prevention of clouding by water drops (polyolefin films)	(27)

the processing of synthetic polymers has also been examined, as shown in Table 6–6.

6.4.3 Utilization of amino acids as agricultural chemicals

As environmental pollution by agricultural chemicals has become a serious problem, the development of chemicals not causing such pollution is desirable. These agents must not affect organisms other than the target plants, insects or microorganisms, i.e. they must exhibit selective attack, and must also be easily decomposed by microorganisms or sunlight so as not to cause environmental pollution. In other words, agricultural chemicals should not destroy the ecological equilibrium in nature. Chemicals satisfying such requirements may well be derived from amino acids, which are themselves natural compounds.

A. Amino acid derivatives effective against rice-plant disease caused by *Piricularia oryzae*

After examination of the activities of *N*-acyl derivatives of amino acids with higher fatty acids, or amino acid esters of higher alcohols, for suitability as agricultural chemicals, it was discovered that *N*-lauroylglycine, L-lauroyl-L-valine and lauryl esters of DL-alanine and valine are effective against rice-plant disease caused by *Piricularia* in the field,[28] though they do not exhibit any antimicrobial activity *in vitro*. As the effect of the derivatives is stronger when spread before the microorganism is inoculated than after, they are effective for preventive use. They are also said to be effective against fungal and spot diseases of tomato and downy mildew disease of cucumber. Since these amino acid derivatives are known to have low toxicity and to be easily decomposed by biological action, they may be the first of a new range of non-polluting agricultural chemicals.

Amino acid derivatives with special structures (XIII), such as dodecyl-di(aminoprolyl)glycine hydrochloride, have been reported to be effective in preventing rice-plant disease caused by *Piricularia*.[29]

$$R-[NHCH_2CH_2CH_2]_{2\sim3}-NH(CH_2)_{1\sim2}COOH\cdot HCl$$

$$R = C_8\sim C_{18}\text{ alkyl group} \qquad \text{XIII}$$

The effect of amino acids or their derivatives on the growth of *Piricularia oryzae* in culture media has also been investigated. Although many amino acids accelerate the growth of *P. oryzae*, inhibitory effects have been observed with L-methionine and L-arginine, and especially with benzoyl-L-valine and phenoxyacetyl-L-leucine.[30] It is not clear whether these compounds are really effective against the disease, but they represent another group of amino acid derivatives which may be potential agricultural chemicals not causing environmental pollution.

B. Effect of amino acid derivatives on the germination and growth of plants

The effects of amino acid derivatives on the germination of plants have been investigated. *N*-Benzoyl-L-aspartic acid and *N*-benzoyl-L-valine inhibit the germination of rice while N-α-phthaleneacetyl-L-methionine inhibits the germination of horse raddish. These facts suggest that the effect of amino acid derivatives depends on the type of plant, monocotyledonous or dicotyledonous.[31] Phenylacetyl-L-leucine inhibits an early stage of germination of dicotyledons but is ineffective in the case of monocotyledons. The mechanism of this inhibition is considered to involve lipase inhibition, which causes interruption of the energy supply for germination, which is derived from fat.[32] Amino acid derivatives suitable for use as weed killers may possibly be developed by continuing such studies.

The use of amino acids as growth-control agents for plants has been attempted. Tryptophan, a precursor of indoleacetic acid, which is a plant-growth hormone, has been reported to be effective in accelerating the growth of plants. It has also been reported that the growth of mushrooms is accelerated by spraying a solution of amino acids into the folds of the mushrooms.[33]

References

1. Colgate Palmolive Co., *U.S. Patent* No. 2689170 (1954).
2. Ajinomoto Co., Inc., *British Patent* No. 1261322 (1971).
3. Kawaken Fine Chemicals Co., Ltd., *Japanese Patent* No. 39–29444 (1964).
4. K. Volger, *Helv. Chim. Acta*, **47**, 526 (1964).
5. Ajinomoto Co., Inc., *Japanese Patent* No. 47–313 (1972).
6. Kyowa Hakko Kogyo Co., Ltd., *Japanese Patent* No. 43–6702 (1968).
7. Du Pont de Nemours and Co., *U.S. Patent* No. 2789973 (1957).
8. Courtaulds Ltd., *Japanese Patent* No. 40–5926 (1965); 41–6608 (1966).
9. Ajinomoto Co., Inc., *Japanese Patent* No. 42–7382 (1967).
10. Ajinomoto Co., Inc., *Japanese Patent* No. 41–12517 (1966).
11. S. Sakurai, *Kobunshi* (Japanese), **18**, no. 202, 9 (1969).
12. Melk and Co., *Japanese Patent* No. 39–1668 (1964).
13. Asahi Chemical Industry Co., Ltd. and Tanabe Seiyaku Co., Ltd., *Japanese Patent* No. 43–1240 (1968).
14. Matsumoto Yushi-Seiyaku Co., Ltd., *Japanese Patent* No. 41–12229 (1966).
15. Tanabe Seiyaku Co., Ltd., *Japanese Patent* No. 40–17130 (1965).
16. Kyowa Hakko Kogyo Co., Ltd., *Japanese Patent* No. 42–19571 (1967).
17. Tanabe Seiyaku Co., Ltd. and Asahi Chemical Industry Co., Ltd., *Japanese Patent* No. 42–14161 (1967).
18. Tanabe Seiyaku Co., Ltd., *Japanese Patent* No. 40–7367 (1965).
19. Tanabe Seiyaku Co., Ltd. and Asahi Chemical Industry Co., Ltd., *Netherlands Patent Appl.* No. 6406229 (1966).
20. Farbwerke Hoecht A. G., *Japanese Patent* No. 40–10707 (1965).
21. Tanabe Seiyaku Co., Ltd., *Japanese Patent* No. 42–22459 (1967).
22. Hercules Powder Co., *U.S. Patent* No. 3102107 (1963).

23. Toa Wool Spinning and Weaving Co., Ltd., *Japanese Patent* No. 42–26526 (1967).
24. Tanabe Seiyaku Co., Ltd., *Japanese Patent* No. 38–20969 (1963).
25. A. F. Bayer A. G., *Belgian Patent* No. 656263 (1965).
26. Ashland Oil and Refining Co., *U.S. Patent* No. 3162612 (1964).
27. Du Pont de Nemours and Co., *British Patent* No. 999809 (1965).
28. T. Misato, *Kagaku Keizai* (Japanese), 52 (1971–05); *Chemistry Today* (Japanese), no. 13, 18 (1972).
29. Mitsui Toatsu Chemicals Inc., *Japanese Patent* No. 41–11758 (1966).
30. S. Takano, T. Suzuki and K. Sahashi, *Nippon Nogeikagaku Zasshi* (Japanese), **46**, 309 (1972).
31. S. Ohmori, K. Yamada, T. Hasegawa, T. Suzuki and K. Sahashi, *ibid.*, **46**, 219 (1972).
32. S. Ohmori, T. Hasegawa, T. Suzuki and K. Sahashi, *ibid.*, **46**, 225 (1972).
33. Kyowa Hakko Kogyo Co., Ltd., *Japanese Patent* No. 38–25720 (1963).

CHAPTER 7

Future Prospects for the Amino Acid Industry

Y. Izumi*

This chapter will consider future prospects for the amino acid industry on the basis of economics, as well as of the chemical properties of amino acids. Amino acids may be characterized in terms of three features; firstly, amphoteric behavior due to the presence of both NH_2 and COOH groups in the molecule; secondly, enantiomerism due to the presence of the asymmetric α-carbon atom (except glycine); and thirdly, individual characteristics depending on the nature of the substituents on the alkyl group. For instance, although the L-forms are the only ones having biological activity, there is considerable scope for modification of the properties in various ways by modification of the α-alkyl substituents.

The use of amino acids for supplementing food and animal feed is likely to become one of the major outlets for amino acids. However since the supply of food and feed in the future will still depend largely on agricultural products, amino acids for supplementing food or feed will generally be those which are deficient in agricultural products such as grains, i.e. lysine, methionine and tryptophan. In the case of food, the L-isomer will be required, but for animal feed the relationship between nutritional efficiency and expense must be considered, and pure L-isomer may not be most economical. In general, since social needs depend mostly on economic conditions, the

* Yoshiharu Izumi, Institute for Protein Research, Osaka University, Suita-shi, Osaka 565, Japan

requirement for amino acids for food enrichment will be greatest in underdeveloped countries, whereas in developed countries the main demand wili be to supplement animal feed.

No amino acid derivatives have yet been produced on a commercial basis for use as food additives, but acylated amino acids or amides may become important in this respect in the future if they turn out to have superior nutritional value or taste, and can be produced at low cost. Amino acids are also important as additives for flavor, although glutamic acid will remain the most widely used amino acid for this purpose. Glycine, which has a simple structure with no asymmetric center, will continue to be used as a sweetener and palliative agent. Other amino acids or their derivatives may also be developed for this purpose as eating habits and flavor requirements change, but are unlikely to equal glutamic acid or glycine in importance.

Another possible outlet for amino acids is in surface-active agents and detergents. Although products with excellent properties have been developed, they are currently too expensive to gain wide acceptance. However, it is not necessary to use either naturally occurring amino acids or L-isomers alone, so the development of new synthetic derivatives of amino acids for use in surface-active agents or detergents at low cost may lead to the replacement of soaps based on fatty acids.

The injection of amino acids for medical purposes in some metabolic disorders is already practiced, and compounds such as Dopa are used as chemotherapentic agents in Parkinson's disease, for instance. Antagonistic amino acids could be potentially important drugs for cancer chemotherapy. The use of amino acids as starting materials for the synthesis of new hormonal peptides may also occur as a result of the development of efficient methods for the synthesis of high molecular weight peptides. In general, the medical use of amino acids will require the synthesis of many new derivatives, and although the scale of production will be very small, profitability is high on pharmaceutical products.

It is already known that acylamino acids are excellent agricultural chemicals. As amino acid derivatives tend to be non-toxic and readily biodegradable, they may be of value in this field. Their use as growth-regulating hormones for plants or as herbicides is also possible. The use of amino acids as an artificial manure for vegetables and fruits, in place of the fish manure used at present, is also possible if rising levels of national income provide sufficient stimulus for the emphasis of quality rather than quantity, even at increased cost.

Although nylon 6, an ω-amino acid polymer is commercially successful, synthetic polymers of α-amino acids have not had the same success. These polymers have excellent heat resistance, but are too brittle and too expensive for use as general-purpose fibers. However, some of these polymers have a texture similar to that of silk, and if the appearance and properties of the fibers can be improved by modification of the constituent amino acids, the fibers might well achieve acceptance for fashion goods where price is not the major consideration.

Various other potential uses exist for amino acid derivatives. Since polyamides are synthetic proteins, they have advantages as surgical materials and for artificial internal organs. Other applications include use as chelating agents or antioxidants, but these are based on newly synthesized derivatives rather than the common amino acids themselves.

At present, amino acids are generally thought of as expensive biological materials rather than as simple chemicals, and this inhibits the expansion of consumption. The future expansion of amino acid production will depend greatly on the development of non-biological outlets, particularly for DL-amino acids, and on the production of new types of amino acids and derivatives with special properties.

Subject Index

B

C

D

P

Q

R

S

T

U

V

W

X